Energy: Science and Technology

Energy: Science and Technology

Edited by Nora Ayling

SYRAWOOD
PUBLISHING HOUSE

New York

Published by Syrawood Publishing House,
750 Third Avenue, 9th Floor,
New York, NY 10017, USA
www.syrawoodpublishinghouse.com

Energy: Science and Technology
Edited by Nora Ayling

International Standard Book Number: 978-1-68286-470-8 (Hardback)

The publisher's policy is to use permanent paper from mills that operate a sustainable forestry policy. Furthermore, the publisher ensures that the text paper and cover boards used have met acceptable environmental accreditation standards.

Trademark Notice: Registered trademark of products or corporate names are used only for explanation and identification without intent to infringe.

Cataloging-in-Publication Data

Energy : science and technology / edited by Nora Ayling.
   p. cm.
Includes bibliographical references and index.
ISBN 978-1-68286-470-8
1. Power resources. 2. Energy facilities. 3. Power (Mechanics). 4. Renewable energy sources. I. Ayling, Nora.
TJ163.2 .E54 2017
621.042--dc23

Printed in the United States of America.

TABLE OF CONTENTS

PREFACE

This book on energy deals with the various branches of energy science ranging from energy storage to energy efficiency technologies. Energy engineering and management deal with all aspects of electrical energy generation and distribution. Most of the topics introduced in this book cover new techniques and the applications of energy technology. It is a compilation of topics which discuss technological innovation in this field and its future implications. This book aims to equip students and experts with the advanced topics and upcoming concepts in this area. In this book, using case studies and examples, constant effort has been made to make the understanding of the difficult concepts of energy engineering as easy and informative as possible for the readers.

Various studies have approached the subject by analyzing it with a single perspective, but the present book provides diverse methodologies and techniques to address this field. This book contains theories and applications needed for understanding the subject from different perspectives. The aim is to keep the readers informed about the progress in the field; therefore, the contributions were carefully examined to compile novel researches by specialists from across the globe.

Indeed, the job of the editor is the most crucial and challenging in compiling all chapters into a single book. In the end, I would extend my sincere thanks to the chapter authors for their profound work. I am also thankful for the support provided by my family and colleagues during the compilation of this book.

Editor

Professional ethics, engineering and energetic issues

Alice Ponchio[1], Alberto Mirandola[2, *]

[1]Department of Philosophy, University of Padua, Padua, Italy
[2]Department of Industrial Engineering, University of Padua, Padua, Italy

Email address:

aliponc@hotmail.com (A. Ponchio), alberto.mirandola@unipd.it (A. Mirandola)

Abstract: Engineering is an interdisciplinary area, where interaction of technical and socio-economic dimensions continuously occurs in the professional practice. The ethical aspects of the relationships between engineers and society have been particularly emphasized in the recent years. In fact, the main engineering societies set up their own codes of ethics and many engineering departments worldwide introduced engineering ethics in their curricula. After introducing the concept of professional ethics and its connection with the deontological codes of professional orders, this topic will be considered in close relation to the profession of engineering and with particular attention to the energetic and environmental issues. This not only in order to give an account of professional codes, but more over to highlight how professional ethics includes and at the same time goes beyond them. Professional codes derive from two features of the profession: autonomy and self-regulation. One of the main aims of the codes is to regulate the asymmetrical relationship between professional and client in order to maintain the fiduciary nature of this relationship. This is highlighted through the analysis of two professional engineering codes. The ethical dimension helps revitalize and give meaning to professional codes and law: if a good professional life is to be conducted, it is necessary to combine the use of legal and professional guidance with an ethical structure, which brings us back to the human being and his/her personal sense of responsibility.

Keywords: Professional Ethics, Ethical Codes, Engineering, Energy, Environment

1. Introduction

While each animal has been provided by nature with some of its own abilities, which are necessary to survive and are similar to those of the other animals of the same family, man1 has no intrinsic material ability. Nonetheless he is a *thinking being* and *has been equipped with hands*. He can thus conceive and build what is needed to survive and live in a given environment. In other words, he can use *technique* to compensate for his lack of specific abilities. In addition, man can not only think about the present, but also use his mind and experience to estimate what will be needed in the future and act accordingly. So, while animals adapt themselves to the environment, man can modify the environment to make it more suitable to his needs. These characteristics allow man to live in and face a great variety of different situations and to *live in a community to better exploit the available resources.*

The latter point expresses another feature of the human

being: he/she is not only a thinking being, but also a *social* one. Indeed he/she can and does live in a community. To do this, suitable organization is needed, with the proper diversification of abilities. Inside this organization, each individual assumes a defined role, which implies duties and rights. To live properly in a community, rules must then be developed and each person has to comply with them. As we know, these rules are of different kinds: behavioral, conventional, social, economic, legal, and ethical. Each kind of these rules regulates different aspects of human life, such as personal, public and contractual relationships, domestic and professional life.

This paper deals with the role of a specific kind of rules in a precise area of human life: the role of ethics in professional life. More specifically, the role of ethics in the profession of engineers and its importance in the education of engineering students will be analyzed.

The weight of these aspects is indeed increasing in the contemporary world. *Engineers* typically work in the field of science and technique. They are particularly suit to creating and building devices, instruments and systems useful to the

---

1 In this paper the word "man" is used to signify "human being" (man or woman), with no reference to sex.

community. The way of working of engineers has been characterized by a continuous and progressive evolution, starting from the activity of artisans (skilled manual workers) and inventors (having not only manual, but also mental skill and creativity) up to the interdisciplinary and integrated work, based on scientific knowledge, of modern engineers.

The complexity of the modern world makes the work of engineers highly diversified and ethically challenging. As will be analyzed in the first and second section, the profession of engineers as *profession* implies ethical regulations, which cannot be substituted by the legal ones. On the other hand, engineers' work brings about remarkable *responsibility* toward individuals as well as communities. The results of engineers' work often involve consequences, which raise ethical questions, such as, for instance, those related to the environmental impact of specific kinds of techniques and the sustainability of resources.

This paper is divided into eight sections. The first and the eighth ones have an introductory and conclusive function. The second one deals with the definition of the concept of profession and its implications; the third with the nature and the goals of professional codes as regard to the law. In the fourth section, two important ethical codes of engineering are analyzed (the NSPE and ASME codes). In the fifth section the energetic and environmental implications of the engineers' profession are examined, while the need for a wider ethical dimension in professional life (with specific attention to engineers' professional life), which goes beyond professional and legal codes, is highlighted in the sixth. The seventh section examines the structure of the course of ethics for engineers organized within a European University.

2. Professional Ethics

"Profession" is generally understood as an activity where a person provides a service on the base of his/her capacities, abilities and knowledge in order to obtain the goods to support him/herself. In these terms being a teacher, a baker, a plumber, a cashier, a secretary, a lawyer, etc., is having a profession. Besides this broad meaning of the term there is a more specific one. Narrowly *"profession" is a working activity of high qualified nature carried out by a human being who has acquired specialist competence through academic and practical training.* In this sense architecture, engineering, lawyer, medicine and the like are professions. The latter meaning of profession is relevant for ethical research and the following considerations will refer to it.

According to the most reputable studies on this matter, there are three key-criteria identifying a "profession" in its narrow sense: 1) intellectual education; 2) specialist competences and abilities; 3) institutional instruments that permit a responsible practice of these competences and abilities 2 . The expression "institutional instruments" indicates on the one hand obligations that constitute professional codes and that every professional as such

commits himself to honor. On the other hand, "institutional instruments" means the bench within the professional group that has the function of verifying the adherence to those codes.

Furthermore when we use the word "profession" what we mean is a practice. Quoting the philosopher Alasdair McIntyre, a practice is "any coherent and complex form of socially established cooperative human activity through which goods internal to that form of activity are realized in the course of trying to achieve those standards of excellence which are appropriate to, and partially definitive of, that form of activity"3.

A practice is therefore a *cooperative activity with internal goods and standards of excellence that define it.* Internal goods are not an individual's property and possessions (like fame or wealth, which are external goods) but shared and inclusive ones, i. e. goods for the whole community which participates in a practice. For example the development of a more sustainable and efficient energy system would be a good for all engineers and not only for the one who developed it.

A profession is then a common undertaking implying behavior to which practitioners conform. But this very comprehensive undertaking distinguishes itself from a mere technical activity. A technical activity is aimed at some further end and therefore has an instrumental value. On the contrary, *a profession has internal goods and thus an intrinsic value.* Indeed, while the value of a technical activity depends on its effectiveness in the achievement of the aimed end and on the value of the latter one (we praise the activity of a shoemaker, for example, calling it "good" if the shoes he/she has made are good, while we despise that activity, calling it "bad" if the shoes are bad), the value of a profession does not depend on a singular good it achieves, but on the practice of the profession itself. *A good professional activity produces the internal goods of the profession.* This means that the achievement of a good professional activity is inherent in the professional. Consequently, being a professional implies not only having technical knowledge and ability, but also knowing the internal goods of the profession and carrying them out.

Beyond these aspects of the concept at issue, its etymological derivation is also very important for indicating the characteristics of the professional. The word "profession" comes from the Latin *profiteri,* which means "to declare aloud", "to state openly". According to this meaning, a professional publicly accepts a special way of life: he/she declares aloud his/her competence and states openly he/she will use the necessary knowledge in the client's interest and not his/her own.

A profession is then a collective activity that is aimed at specific internal goods and implies a public declaration of competence and of specialist commitment in behalf of the client. This involves institutions of self-regulation and self-justice. These features are the main objects of reflection

---

2 Parsons [1] Vol. XII, pp. 536-541, 545-547.

3 McIntyre [2], p. 187.

about professional ethics.

3. Professional Codes

Professional codes are forms of self-regulation concerning professional groups. They arise from the intrinsic asymmetry characterizing every relationship between a professional and the person who addresses him. This asymmetry lays in the origin and in the essence of these relationships as such: when someone has a (medical, technical, juridical...) problem he/she appeals to a specialist in order to solve it. This specialist guarantees, "declares aloud", to have the necessary knowledge and ability to help him/her. There is always a disparity between the professional and the client: the latter is a person in need of help; he/she does not have the power to solve his/her problem by him/herself and so is dependent upon the former for the technical knowledge necessary for a choice and the professional's competence to carry out the choice once it is made. This asymmetrical relationship can be interpreted on the basis of four models:

1) According to a *commercial model*, the professional relationship is regulated by minimal market-place morality, which allows everything except for compulsion and deception. Therefore, someone who needs engineering consulting will address the engineer knowing that the only rule governing their relation will be that of the market.

2) The *corporative model* considers the relationship in a paternalistic way. This model emphasizes the role of the professional: he/she is the person who establishes what decision is right in that particular case. The client has no chance to decide on his/her own or to explain his/her values and priorities.

3) The *autonomist model* is essentially the opposite of the corporative one. According to it, the professional must totally represent the interest of the client with no possibility to give advice or opinions that differ from the client's point of view.

4) The *interaction-model* is based on an interaction between professional and client in order to reach the most equally shared decision: on one hand the professional provides the competence to satisfy the needs of the client, on the other hand the client makes his/her values and priorities known.

Since asymmetry derives from the specific nature of professional-client relationships, it cannot be eliminated. Nonetheless it can be limited and moderated; this moderation is the main goal of professional codes. This is not a question of mere regulation or balance, but a genuine *ethical task*. Indeed the special fiduciary character of professional relationships is at stake in the limitation of the asymmetry. In a state of vulnerability and inequality, the client is forced to trust the professional. As a matter of fact a professional relationship springs not only from someone's need for specialist competence and from the professional's competence-guarantee but also from needy person's belief in the professional. When we are in need of professional

competence we do not look for a professional in general, but for the one we can trust. The fact that a professional is successful and competent is not enough for us to choose him/her as a partner in a professional relationship. What we want is someone who is worthy of our confidence. Thus the professional relationship is not only an asymmetrical one, but also a fiduciary one.

The moderation of asymmetry aims to safeguard this fiduciary relation between the professional and the client within this non-eliminable asymmetry. Through this moderation, professional codes aspire to establish an interaction between professional and client according to the fourth model described above.

Since professional codes arise from a genuine ethical problem implicated in the nature of every professional relationship, they are ethical codes. These codes are also called "deontological codes". The adjective "deontological" derives from the ancient Greek *déon* which means "duty". A deontological code is therefore a set of ethical duties.

Emphasizing the ethical nature of professional codes is not trivial. Nowadays we are confronted with a progressive expansion of the role of law in the regulation of professional relationships. When the client thinks that a professional's performance has been not satisfactory, he/she often appeals to the jurisdictional system. In order to avoid legal worries, the professionals safeguard themselves in advance, taking out policies. This phenomenon is something problematic with regard to the very features of the relationship we are considering, namely its being asymmetrical and fiduciary.

Of course it is legitimate to use law to regulate the professional-client relationship, if the rights of the partners are at issue. Nonetheless, law is only able to regulate some aspects of a professional relationship, not the entire thing. As claimed above, this relationship is one between a person seeking help and a professional that possesses the expert knowledge that the client needs. Therefore there is an inherent inequality in this relationship that places the preponderance of power in the professional's hands. Law regulates relations between partners in a contract. Since in the contract the partners are equal and their relationship is a symmetrical one, the law can regulate it. On the contrary, the professional relationship cannot be considered a contract because one part (the client) is dependent upon the professional's services. This is why the appeal to law as the only regulatory instrument is not sufficient.

Moreover the recourse to legal remedies in every case where the partners disagree is dangerous for the relationship itself. This practice leaves no room for trust neither from the client, who is ready to use legal means to deal with every type of disagreement, nor from the professional, who takes measures in advance to safeguard himself. This leads to the failure of the fiduciary relationship between the professional and the client, as well as their professional relationship as such.

Since it cannot regulate the professional relationship as such, law cannot replace the professional codes. On the contrary, professional codes and law must collaborate. Indeed within the structure of these codes and legal regulation lie essential elements that guide and regulate professional life: examining the law and the professional codes, a professional can sufficiently and clearly identify obligations and responsibilities that must guide his/her professional activities. The awareness of the importance of this dual system of regulation and of the differences between strictly legal and professional obligations is very important.

4. Codes of Ethics of Engineering Societies

As emphasized in Section 1, Engineers have a great responsibility towards society and individuals. Through their work, they have a strong impact on the everyday and future life of men and animals; they can both greatly contribute to public welfare and cause damages; they can spread either good or bad information about technical issues, thus influencing the decisions of politicians; they can improve people's life style, but also negatively modify the environment.

The consciousness of this great amount of responsibility was the reason for many professional associations of engineers to establish ethical codes. One of the most advanced and detailed codes is that of the U.S. National Society of Professional Engineers (NSPE) [3]. Another, more concise, but extremely similar in the main concepts, is the Code stated by the American Society of Mechanical Engineers (ASME) [4].

The NSPE Code is divided into three parts:
• fundamental canons
• rules of practice
• professional obligations.

The ASME Code is divided into:
• fundamental principles
• fundamental canons
• the ASME Criteria for interpretation of the Canons

The main principles of both the NSPE and the ASME Code state that engineers must uphold and advance the *integrity*, *honor* and *dignity* of their profession.

In the following we will refer to the fundamental canons of these two codes, emphasizing the duties they impose on the profession of engineering. In general, these canons consider the relationships between engineers and society, engineers and clients, engineers and colleagues; they delineate how engineers should act and behave.

According to the NSPE-*Fundamental Canons*, engineers, in the fulfillment of their professional duties, shall:
1) Hold paramount the safety, health and welfare of the public;
2) Perform services only in areas of their competence;
3) Issue public statements only in an objective and truthful manner;
4) Act for each employer or client as faithful agents or trustees.
5) Avoid deceptive acts.

6) Conduct themselves honorably, responsibly, ethically and lawfully so as to enhance the honor, reputation and usefulness of the profession.

According to the ASME code the fundamental canons are that:
1) Engineers shall hold paramount the safety, health and welfare of the public in the performance of their professional duties.
2) Engineers shall perform services only in the areas of their competence.
3) Engineers shall continue their professional development throughout their careers and shall provide opportunities for the professional and ethical development of those engineers under their supervision.
4) Engineers shall act in professional matters for each employer or client as faithful agents or trustees, and shall avoid conflicts of interest or the appearance of conflicts of interest.
5) Engineers shall build their professional reputation on the merit of their services and shall not compete unfairly with others.
6) Engineers shall associate only with reputable persons or organizations.
7) Engineers shall issue public statements only in an objective and truthful manner.
8) Engineers shall consider environmental impact in the performance of their professional duties.

As we can see, the canons of the two codes are very similar. The first one is a milestone: engineers shall hold paramount the *safety, health* and *welfare* of the public in the performance of their professional duties. This sentence includes a variety of items, actions and rules of practice, because engineers work in widely diversified fields of activity: process industry; manufacture of machines, plants and products; design, construction and maintenance of buildings, roads, bridges; energy and the environment; public administration; self-employment; etc. Safety, health and welfare should be considered in all the steps of engineers' activity: design, construction or manufacture, operation, maintenance, monitoring and control, decommissioning. The main point of this canon is the requirement that engineers be aware of the dependence of the lives, health and welfare of the general public upon their judgment and act in accordance with this huge amount of responsibility. This implies working in conformity to safety standards, constantly reviewing the safety of their plans, designs or products, and informing the proper authorities when risks arise4 [ASME, first criterion of

---

4 In technical risk analysis, risk is commonly defined as *"the statistical expectation value of unwanted events, which may or may not occur"* (Hansson, 2004 [5]; see also Kermisch, 2010 [6]). A European Commission Report, (2000, p. 18 [7]) suggests that *"risk is widely recognized as a function of the probability and severity of an adverse effect/event occurring to man or the environment following exposure, under defined conditions, to a hazard"*. The notions of risk and responsibility are apparently distinct, because the previous definitions of risk are disconnected from the risk factors. But they relate to each other in the risk management process: risk assessment consists in the identification and quantification of risk and is disconnected from responsibility; but the acceptability and management of risk involves a question of "values" and is

interpretation].

This first canon includes particular attention to environmental health: this aspect of the codes will be discussed in the following section.

The second canon of the two codes is also the same: "Engineers should perform services only in areas of their competence". This duty is developed in the fifth canon of the ASME code: "Engineers shall build their professional reputation on the merit of their services and shall not compete unfairly with others". These statements require engineers to accept assignments only if qualified and to be honest about their competence and qualifications. Furthermore, they are required to continue their professional development through continuing education and also provide opportunities for the professional development of colleagues under their supervision.

The need for fairness and truthfulness in engineers professional life is also expressly stated in both codes: the NSPE code declares in its third canon and the ASME code in its seventh one that engineers shall "issue public statements only in an objective and truthful manner" This is a very important duty for engineers. It requires them to correctly inform people, decision makers and politicians about the issues of their competence and express technical opinions that are based upon their appropriate knowledge and competence, without being inspired by personal interest.

However, truthfulness and fairness are not only required when informing people and expressing professional judgements, but also in professional relationships: "Engineers shall act in professional matters for each employer or client as faithful agents or trustees, avoiding conflicts of interest and respecting the proprietary information and the intellectual property" [ASME 4th canon; see also NSPE 4th canon]. This canon commits engineers to avoiding every source of conflict of interests with employers and clients and to act with fairness and justice to all parties. This implies on the one hand that, "before undertaking work for others in which engineers may make improvements, plans, designs, inventions, or other records which may justify seeking copyrights, patents, or proprietary rights, engineers shall enter into positive agreements regarding the rights of respective parties" [ASME Criteria for Interpretation 2.4.l], and, on the other, that they admit their own errors [ASME Criteria for Interpretation 2.4.m].

5. Ethical Aspects of the Energetic and Environmental Issues

As seen in the previous section, the first canon of both codes implies particular attention to environmental health, which is closely related to the exploitation and use of energy resources. This concept is explicated in the eighth canon of the ASME Code:"Engineers should consider environmental

impact and sustainable development in the performance of their professional duties".

The fact that the ASME Code makes it a precise duty of engineers to consider environmental impact and the sustainable development in their work is worth noting. Indeed this is a very modern issue, since people have become conscious of environmental problems only in relatively recent years. In particular, the concept of sustainability is very important, despite its only recent relevance, because it has fundamental implications and consequences on our lives and on those of future generations. This concept is closely connected to science, technology and economy in a way which it is worth briefly explaining in the following.

Sustainability is not simply referred to the impact of certain practices on the environment, but is a more complex concept, because it deals with:

- the resources needed to sustain the human population (more than 7 billion in 2014 and still increasing);
- the social and economic impact;
- the impact on the environment;
- the technological constraints.

According to N. Lior [8], "sustainable development is of vital importance to humanity's survival" and is based on the "adequate satisfaction of quantitatively defined and interrelated economical, environmental and social criteria". However "it is very difficult to quantify environmental and social metrics". Thus, "sustainability is more and more extensively used erratically and often improperly and even fraudulently over the entire social spectrum by governments, institutions, business, industry, schools and individuals".

These features are particularly challenging for engineers: they should take these concepts into account in performing their professional activities, but without falling into demagogy. For example, when it comes to resources, energy and social organization, some opinion leaders of the so-called "green people" think that generally it would be desirable *to let nature simply run with its own rhythms*. This would be possible in a world less crowded than the present one, but today it is necessary to use the possibilities offered by technology, which is a valuable means that can prevent natural disasters or other dangerous events to occur. Engineers should distinguish utopian projects and their effects from realistic ones. Indeed, rather than utopian projects, long-term energy policies are thus fundamental. Engineers should realize that energy conservation is to be a goal in all fields of activity: so, seeking high efficiency and rational organization of energy systems is paramount. As for renewable energy sources, their *real* potential must be explained to the people, to avoid unrealistic reliance upon them. These issues are generally very complex: in this case, it is very important to be able to "reason by systems", which is typical of modern engineers. In any case, the so-called "3 E's" (Energy, Environment, Economy) must always be considered, because these three aspects are closely connected to each other.

Of course, there is a conflict between the need for resources of our crowed world and the environmental impact:

---

strongly related to responsibility. On the other hand, the assessment step also involves technical skills. So, engineers are interested in the entire risk management process.

finding the most acceptable compromise between these two features is a very challenging problem for politicians, scientists, economists and engineers.

6. Beyond Law and Professional Codes

No doubt the canons and the duties briefly exposed are very important for engineers' professional life. However, many questions and dilemmas about the correct concrete professional conduct cannot often be answered by professional codes (and even less by law). Consider this case: in designing a plant, which could be dangerous for the environment, an engineer has to do the interests of the client as well as the ones of the society, to be truthful about the impacts of its work and to find a way to develop the plant standards of safety and environmental sustainability without forgetting economic, environmental and risk factors. Or consider this one: in a period of economic crisis the engineer manager of a factory faces the problem of dismissing some workers to save the company. Like many others, these are very difficult tasks, which no deontological rule says how to solve. In many cases the solution of a problem or of a problematic situation in professional life is indeed up to the professional itself. *This requires the professional to look at law and professional codes and to judge beyond them about what is the right thing to do.*

Judging beyond law and professional codes does not imply stating their invalidity but interpreting them within a wider ethical context. This wider context consists in a sound ethical structure of the person and in his/her sense of responsibility. In order to answer the moral questions being raised in professional life, legal and professional guidance has to be combined with an individual's sound ethical structure and with his/her sense of responsibility.

This involvement of the person in the identification of the morally right action in a certain situation is a very important, often neglected aspect of professional ethics. As a matter of fact, we frequently speak about professional ethics identifying it with the professional codes, and neither the general inadequacy of the codes when faced with some ethical problems nor the need for a wider ethical context are regarded. That is a dangerous mistake, because it misunderstands the role of professional ethics. The latter does not only provide a set of duties that we must fulfill, but furthermore deals with formation and development of an ethical character in the agents. Using a term that perhaps is no longer seen as current, professional ethics aims for the development of certain virtues in the professional.

A virtue is not a natural but a developed trait of character that we achieve through exercise and personal commitment. Quoting Aristotle:"None of the moral virtues arises in us by nature; for nothing that exists by nature can form a habit contrary to its nature […].but the virtues we get by first exercising them, as also happens in the case of the arts as well. For the things we have to learn before we can do them, we learn by doing them, e.g. men become builders by building and lyreplayers by playing the lyre; so too we become just by doing just acts, temperate by doing temperate acts, brave by doing brave acts"5.

We can understand a *virtue* as *the motivation to act rightly and appropriately. The rules, even those of a deontological code, cannot guide a professional activity without being interpreted in the light of the moral structure and motivation of the agent.* The development of virtues is therefore indispensable for the rules to be effective.

Nevertheless the development of a virtuous character is not only relevant in cases when the guidance of deontological code or of law is lacking, but also indispensable for the practice of the profession itself. In the introduction the term "profession" has been defined as indicating a common activity that is aimed at specific internal goods. The achievement of these internal goods is closely related to the development of virtues. Quoting Alasdair McIntyre again,"A virtue is an acquired human quality the possession and the exercise of which tends to enable us to achieve those goods which are internal to practices and the lack of which effectively prevents us from achieving any such goods"6.

This means *that being a professional, and not a mere technician, implies the development of motivations to act in a morally correct and appropriate way. Only through these motivations or virtues the goods internal to the profession can be achieved. Virtues are then necessary conditions for the profession.*

Consider now the profession of engineering. By mutual consent it could be generally qualified as the discipline and profession that uses intelligence (ingenium), applying results in order to solve issues and to satisfy human needs. The internal good (or end) of this practice is then the solution and satisfaction of human problems and needs. What kind of virtues are required for the achievement of this internal good? In the following, four main virtues are listed, but others could be added to them.

The first virtue is *benevolence*. This is the virtue of being disposed to act for the benefit of others, quoting Immanuel Kant "an active practical benevolence"7 which makes the welfare of others its end, as well as its practical counterpart. Benevolence implies then the capacity to recognize others' needs or problems and the readiness to solve them. This capacity does not merely imply being able to find the best technical solution for the case in question, but above all the capacity to have a comprehensive insight on the best technical solution that has moral relevance for the situation. This involves taking into consideration the client's needs and values, the economic factors as well as the environmental impact of the prospected solution and the possibility to use alternative energy resources.

Responsibility is also a relevant virtue. We can see it as a future-oriented virtue. Being responsible means indeed, on one hand, being able to foresee the effects of our deeds and, on the other, being committed to justify our actions and to

---

5 Aristotle [9], pp.28 f. (1103).

6 McIntyre [2], p. 191.

7 Kant [10], pp. 199 (6:449-452).

pay for their consequences.

The third virtue which is required is *courage*. It consists in the capacity to not succumb to difficulties and to calmly face the risks and the uncertainties of some decisions and situations. This does not mean fearlessness, but, as Aristotle defined it, a "mean with regard to feelings of fear and confidence"8. Courageous is indeed not he who has no fear but he who finds the middle way between fear and fearlessness: "Though courage is concerned with feelings of confidence and of fear, it is not concerned with both alike, but more with the things that inspire fear; for he who is undisturbed in face of these and bears himself as he should towards these is more truly brave than the man who does so towards the things that inspire confidence. It is for facing what is painful, then, as has been said, that men are called brave. Hence also courage involves pain, and is justly praised; for it is harder to face what is painful than to abstain from what is pleasant"9.

The fourth virtue is *practical wisdom*, the virtue of deliberation and discernment10. It consists in the capacity to recognize the appropriate course of action that leads to the internal good of the profession in a very specific situation. This ability implies knowledge of the circumstances of the situation and of moral relevant features, the capacity to recognize human needs, the awareness of general legal and ethical duties and the skill to apply them to the very specific context as well as the awareness of the internal good of the profession. It also implies the skill of identifying the appropriate means for the intended end, the capacity to foresee the consequences of the action and to not succumb to difficulties. Practical wisdom is therefore the virtue that summarizes the others ("With the presence of the one quality, practical wisdom, will be given all the virtues"11 wrote Aristotle) and orients them to the practice.

These four virtues are significant for the profession of engineering. If being an engineer means using his/her intelligence to solve human problems and satisfying needs, such a professional cannot lack the capacity to see human needs and the readiness to solve them as well as the disposition to do good for someone. As a matter of fact, for example, we do not regard as a good professional the person who does the profession only in order to earn money or to have a great reputation, but not in order to satisfy human needs. He/she may be successful and do great deeds but we do not consider him/her as someone who expresses the profession's internal good. Furthermore we call a "good

professional" someone who is able to recognize the appropriate means to the good at issue, to foresee their consequences, to pay for them and who does not succumb to difficulties and uncertainties of everyday professional life. A good professional is then a person who can deliberate well, paying attention to the numerous factors of a situation.

Many professionals work in the area of self-employment, but others, as for example a lot of engineers, work within economic or industrial companies and may be leaders or have an important role within their organisation chart. Consequently, they are often involved in the so-called Corporate Social Responsibility (CSR) of the company, which is a form of self-regulation integrated into a business model, showing that the company monitors and ensures its active compliance with ethical standards, aiming at reaching some social good, beyond the interests of the firm and also beyond the spirit of the law. CSR covers both legal and moral responsibilities of the organization, which can be certified by accepting to be periodically controlled.

An international recognized voluntary certification is SA8000 (Social Accountability 8000), which is a management system standard modeled on ISO standards; it concerns some very important areas to be considered within the business activity: child labor, forced and compulsory labor, health and safety, freedom of association and right to collective bargaining, discrimination, disciplinary practices, working hours, remuneration. To gain and maintain this certification the company must go beyond a simple compliance to the standard, but also integrate it into its management systems and practices and demonstrate ongoing conformance with the standard, that is based on the principles of international human rights. As seen for a single professional, this management strategy of the company, which is voluntary, highlights an ethical attitude that goes beyond law, codes and rules.

Of course, a professional involved at high level in this kind of organization should have the characteristics that have been outlined in this paper; his/her behavior and way of working will be in agreement with practical wisdom.

7. Ethics in Engineering Education

The urgent need for a wider ethical structure also in engineering professional life has determined a lot of phenomena. The perhaps most striking are:

- The establishment of the scientific journal called "Science and Engineering Ethics", where discussions about the ethical issues involving science and technology are developed.
- The organization of curricula including engineering ethics throughout the world12.

The following will be focused on this second phenomenon. Generally speaking, in these curricula two teaching approaches might be used:

1) introducing ethical concepts and examples during some

---

8 Aristotle [9], p. 65 (1115a).

9 Aristotle [9], p. 71 (1117a).

10 "The man who is without qualification good at deliberating is the man who is capable of aiming in accordance with calculation at the best for man of things attainable by action. Nor is practical wisdom concerned with universals only-it must also recognize the particulars; for it is practical, and practice is concerned with particulars. This is why some who do not know, and especially those who have experience, are more practical than others who know; for if a man knew that light meats are digestible and wholesome, but did not know which sorts of meat are light, he would not produce health, but the man who knows that chicken is wholesome is more likely to produce health." Aristotle [9], p. 146 (1141a-b).

11 Aristotle [9], p. 156-158 (1145a).

---

12 See, for example, Berry et al. [11].

engineering modules;

2) introducing a specific module into the curriculum of engineering students.

Yet both approaches are problematical. The problem with the first one is that the ethical topics should be discussed by engineering professors, but most of them are generally not prepared to introduce ethics into their courses: they are most comfortable with quantitative concepts and often do not believe they are qualified to lead discussions on ethics. The problem with the second one consists in the fact that this module should be taught by an expert teacher, for example a teacher of philosophy, and this choice could provide the false message that ethics is something detached from engineering profession.

In order to not incur in these two problems, in a European University a course of "Ethics for Engineering Students" has been organized with a close collaboration between a professor of philosophy and a professor of engineering. The basic theoretical concepts are introduced and deepened by the professor of philosophy in the first part of the module, which has three main targets:

1) To introduce the engineering students to the specific language, problematic and particularities of ethics;

2) To pay attention to ethical questions often related to the profession of engineers. The main aspect which is analysed is the necessity to keep faithful relationships between engineers and clients in spite of the non-eliminable asymmetry characterizing them;

3) To show that the ethical thinking should not be confused with the legal implications of one's actions or with deontological codes: ethical thinking must go beyond legal and professional duties and lead engineers to use judgment in order to find the ethically right action-strategies. This requires the development of a virtuous character in a process, which considers the man not only as a professional, but as a whole.

The second part of the course, coordinated by a professor of engineering, consists of a series of speeches and interviews of professionals who are leaders in different fields, in order to give the students a set of experiences related to diverse professional situations. These lectures are developed as "open discussions" about typical ethical issues occurring in industry, public administration, self-employment, etc.

An important issue discussed during this second part of the course is *sustainability*, a topic that involves men, animals and the environment and the importance of which has already been emphasized in section 5. Different viewpoints can be expressed about it, so lively discussions can arise13. Another central item is the concept of *leadership*. The reason for choosing this special topic lays in the fact that many engineering students will probably become leaders of a company, an industry, a group of people in their future working life. In engineering education important issues are thus to explain what it means to be a leader and to consider how one can be a leader, and what characteristics a leader

should have.

The concept of leadership is not presented as indicating a merely hierarchical position. According to the content of the lectures, a leader should be a source of inspiration and confidence, willing to sustain and guide the subordinate individuals, to give sense to their work with the example of his/her own behaviour. Leadership is not a target to be reached, but a road to walk day by day; a road that requires self-sacrifice, patience and also humility; as a consequence, a mature system of relations between colleagues and co-workers will be established. Of course, the results cannot be obtained immediately, but in the long term, which requires perseverance and patience. In this way human and material resources will be better exploited, conflicts will be more easily reconciled, the general organisation of the team (or company, etc.) will be improved, and the atmosphere inside the team will be serene and relaxed.

The module "Ethics for Engineering Students" is offered to all Engineering students as a free-choice module: they can choose it or not. It is worth noting that the students who seem to be more interested in this module are generally those who attend the Energy Engineering Degree Course: it probably means that the students who are inclined to deepen energetic and environmental topics are also interested in ethical issues, which are closely connected to the profession of Energy Engineers.

8. Conclusions

In this paper professional ethics has been analyzed in close connection to the profession of engineering, with particular emphasis on Energy Engineering. The analysis has first stressed the twofold dimension of professional ethics. This dimension is rooted in two very ethical issues implied in the concept of profession as such: the need to preserve the fiduciary relation that characterizes every client-professional relationship and the achievement of the internal goods of the profession itself. Thus on the one hand professional ethics reminds us what we ought to do in order to encourage the fiduciary relation between client and professional through the professional codes. From this point of view two engineering codes and their main topics have been analyzed: the NSPE code and the ASME code. On the other hand, it has been stressed that the role of ethics in professional life goes beyond recommending what we ought to do and moreover suggests us how we have to act and what dispositions or motivations we are required to develop in order to be good professionals. This second aspect of professional ethics is unfortunately often overlooked or completely misunderstood by moralists who work on professional ethics as well as by professionals. But if a profession is a practice, if a practice is aimed to its internal goods and if the achievement of these goods needs the development of specific dispositions or virtues, virtues are then necessary conditions for the profession itself. If this argument is valid, it has two relevant implications. It firstly commits every professional to enhancing his/her own ethical character and to developing and cultivating virtues. Furthermore it

---

13 See, for instance, V. Miltojevic [12]

implies that the formation of new professionals should not be merely technical but also ethical. The validity of the argument and the acknowledgment of a relevant weight of ethical education seems to be confirmed by experience, that is by the fact that increasingly numerous curricula including ethics are organized in academic departments that traditionally are meant to be technical; an example has been examined in Section 7.

References

[1] Parsons T., Professions. In: Sillis D.L., editor. International Encyclopedia of the Social Sciences. New York USA: The Macmillan Press - The Free Press; 1968. Vol. Xii, p. 536-541, 545-547

[2] McIntyre A., After Virtue: A Study on Moral Theory. London UK: Duckworth 1985.

[3] National Society of Professional Engineers (NSPE): Code of Ethics for Engineers; Publication #1102, 2007.

[4] American Society of Mechanical Engineers (ASME): Code of Ethics of Engineers, June 10, 1998.

[5] Hansson, S., 2004. Philosophical perspectives of risk. Techné, 8 (1), 10-35.

[6] Kermisch, C., 2010. Risk and Responsibility: a Complex and Evolving Relationship. Science and Engineering Ethics.

[7] European Commission, 2000. First report on the harmonisation of risk assessment procedures, http://ec.europa.eu.

[8] Lior, N, 2011, Sustainability Ethics: a Call for Damage Control and Prevention; Proceedings of ECOS 2011, Novi Sad, Serbia, July 2011.

[9] Aristotle, Nicomachean Ethics. Trans. by D. Ross, Oxford University Press, 1980.

[10] Kant I., Metaphysics of Morals. Edit. by Gregor M. Cambridge UK: Cambridge University Press; 1998.

[11] Berry, B.E, White, G.K. and Arnas, A.O.: Engineering Ethics Education: a military Academy Point of View; Proceedings of ECOS 2011, Novi Sad, Serbia, July 2011.

[12] Miltojevic, V.: Education and Engineers' Environmental Ethics, Proceedings of ECOS 2011, Novi Sad, Serbia, July 2011.

Natural Antioxidant Changes in Fresh and Dried celery (*Apium graveolens*)

Manal A. Sorour[1], Naglaa H. M. Hassanen[2], Mona H. M. Ahmed[2]

[1]Food Engineering and Packaging Dept., Food Technology Research Institute, Agric. Research Center, Giza, Egypt
[2]Special Food and Nutrition Dept., Food Technology Research Institute, Agric. Research Center, Giza, Egypt

Email address:
manal.sorour@yahoo.com (M. A. Sorour), naglaahassaneen@yahoo.com (N. H. M. Hassanen),
monahanafiahmed@yahoo.com (M. H. M. Ahmed)

Abstract: The effect of temperature on natural antioxidant changes in fresh and dried celery was studied. Celery herbs were dried at 50 and 90°C using a laboratory scale hot air dryer. Fifteen phenolic components (gallic acid, protocatechuic acid, catechol , chlorogenic acid, syringic acid, caffeine , p-coumaric acid, ferulic acid, salycilic acid, cinnamic acid, chrysin, pyrogallol, ellagic acid , catechin and caffeic acid), five flavonoids components were identified in celery herbs (apignen, hesperitin, luteolin, quercetrin and rosmarinic) and three isoflavones components were identified in celery herbs (daidzein, genistein and isorhamnetin) were identified in celery herbs at 50 and 90°C. The chemical constituents of *apium graveolens* volatile oil were determined, the results observed that eleven components were isolated from *apium graveolens* essential oil and classified into five chemical categories namely, monocyclic terpenes (78.24%), bicyclic terpenes (14.88%), aliphatic hydrocarbons (1.79%), ketones (0.19) and sesquiterpene (2.89%). These identified compounds accounted for 97.99 % of the composition of *apium graveolens* essential oil. Organoleptic evaluation of *Apium graveolens* represented the mean scores and their statistical analysis indication for color, aroma, taste, texture and overall acceptability for biscuit treatments mixed with different concentrations of dried *Apium graveolens* at 50°C and 90°C.

Keywords: Drying of Celery, Phenolic Compounds, Flavor, Application of Dried Celery, Flavones

1. Introduction

Air-drying is a traditional, low cost technique that is used to lower the water content of herbs at low temperatures. The drying at low temperatures protects against the degradation of the active constituents, but it is slow and metabolic processes may continue longer, which may lead to quality loss of the aromatic plants and subsequently of the produced added value products [1].

Temperature is one of the most important factors affecting antioxidant activity. Generally heating causes an acceleration of the initiation reactions and hence a decrease in the activity of the present or added antioxidants [2].

Phenolics, in particular, are thought to act as antioxidant, anti-carcinogenic, anti-microbial, anti-allergic, anti-mutagenic and anti-inflammatory, as well as reduce cardiovascular diseases [3].

Flavonoids show a strong antioxidant and radical scavenging activity and appear to be associated with reduced risk for certain chronic diseases, the prevention of some cardiovascular disorders and certain kinds of cancerous processes. Flavonoids exhibit also antiviral, antimicrobial, and anti-inflammatory activities, beneficial effects on capillary fragility and an ability to inhibit human platelet aggregation, antiulcer and antiallergenic properties [4].

Apium graveolens Linn. (Apiaceae) has a long history of use in Ayurveda and Unani system of medicine. Apium graveolens L (Apiaceae) grows wild at the base of the north western himalyas and outlying hills in Punjab and in western India. A. graveolens has been used as a food, and at various times both the whole plant and the seeds have been consumed as a medicine. Celery seeds or celery seed extracts are used as flavoring agents and also in anti rheumatic formulations as the seeds have significance as arthritic pain relief, for treating rheumatic conditions and gout. Apart from the role in rheumatism, celery seeds proved its use in asthma, bronchitis and inflammatory conditions [5].

The aim of the work was to study the effect of drying air temperature (50 and 90°C) on the quality of celery herbs (phenolic, flavonoids, isoflavones and flavor of *apium graveolens)*, and the evaluation of *Apium graveolens* when added to biscuits with different concentrations.

2. Materials and Methods

2.1. Materials

Fresh plant celery (*Apium graveolens*) was obtained from Agriculture Research Centre, Giza, Egypt during September, 2013.

2.2. Methods

2.2.1. Drying of Celery Herbs

Drying of celery herbs were carried out using a laboratory scale hot-air dryer installed in Department of Food Engineering and Packaging, Agricultural Research Center at 50 and 90°C.

2.2.2. Determination and Identification of Phenolic and Flavonoids Fraction of Celery Herb Using HPLC

Phenolic and flavonoids fraction compounds were identified determined by the method described by Schieber *et al.,* 2001 [6]. A high performance liquid chromatography system equipped with a variable wave length detector (Agilant, Germany) 1100, autosampler, quaternary pump degasser and column compartment. Analyses were performed with a C18 reverse phase packed stainless-steel column (Zorbax ODS 5 µ m 4.6 x 250 mm) . HPLC method started with linear gradient at a flow rate of 1.0 ml / min with mobile phase of water / acetic acid (98:2 v/v, solvent A) and methanol / acetonitril (50 : 50,v/v, solvent B) , starting with 5% B and increasing B to levels of 30% at 25 min,40 % at 35min,52 % at 40 min,70 %at 50 min,100 % at 55 min. The initial conditions were re-established by 5 min wash in both solvents. All chromatograms were plotted at 280 nm to estimated phenolic acids and at 330 nm for flavonoids. All components were identified and quantified by comparison of peak areas with external standards.

2.2.3. Determination and Identification of Isoflavones Fraction of Celery Herb Using HPLC

Isoflavones fraction compounds were determined by the method described by Mantovani *et al.,* 2011 [7]. HPLC Samples of the purified extract had their bulk adjusted with methanol 80%, and filtered in polyethylene filters with PTFE membrane (Millipore Ltd. Bedford, E.U.A.) of 0.45 µm pore, before the injection. Isoflavones compounds identification and quantification were carried out through high performance liquid chromatography (Gilson 321), with a secondary pump, deaerator, automatic injector, detector (UV-Visible), and the software program Boriwn version 1.5. The chromatographic conditions described by Song et al. (1998) were used C18 coated column Lichrospher, of Merck (250 x 4,6 mm, 5 µm) was used, at 30°C; mobile phase was constituted of acetic

acid and methanol (19:1, v v-1) with 1 mL min.-1 initial flow; detection at UV-Visible 254 nm; and injection volume of 20 µL. The calibration curve was prepared using authentic standards in concentration of 0.25 to 0.1 mg mL-1 diluted in mobile phase.

2.2.4. Extraction of Essential Oils

The essential oil of *apium graveolens* was extracted by water distillation using a (Clevenger-type apparatus) for 4 hours. The obtained volatile oil was dried over anhydrous sodium sulphate and then holds in completely filled a glass bottle at -20°C until use [8].

2.2.5. Separation and Identification of Essential Oils Constituents

The GC/MS technique (HP. 5890A) was used to identify the *apium graveolens* essential oil constituents, under the following conditions: packed capillary column (50m×0.2mm×0.3 thickness film of carbowax 20M), Helium was used as a carrier gas at flow rate 20 cm/sec, HP 7673A automatic injector was used to inject 2.0 µL of diluted samples in ethyl alcohol (1:10, v/v) with split ratio 100:1, at 150°C, the oven temperature (programmed) was set at 60°C for l0min and increasing gradually by the rate of 2.8°C/min to the final temperature (200°C) during 60 min. Mass spectrum was used to identify the constituents by comparing the samples spectrum with the data stored in Chemstation library which containing over 43,000 compounds.

2.2.6. Biscuit Preparation

Biscuits were prepared according to the method reported by Naglaa and Gehad, 2011 [9].

2.3. Organoleptic Evaluation of Biscuit Matricaria Chamomilla

The sensory evaluation of the obtained biscuit were carried out by a panel consisted of 10 panelists who asked to evaluate taste, aroma, color, texture and overall acceptability attributes according to the method reported by Amerine *et al.,* 1965 [10]

2.4. Statistical Analysis

The obtained results were subjected to statistical analysis using the standard analysis of variance as outlined by Snedecor and Cochran, 1980 [11]

3. Results and Discussion

3.1. Chemical Characteristics of Dried Celery Herbs

The phenolic, flavonoids and isoflavones of *apium graveolens* fractionated, and identified by using HPLC.

3.2. Determination of Phenolic Components

The phenolic of celery herb are tabulated in Table (1). The results observed that fifteen phenolic components were identified in celery herbs (gallic acid, protocatechuic acid, catechol , chlorogenic acid, syringic acid, caffeine , p-coumaric acid, ferulic acid, salycilic acid, cinnamic acid, chrysin, pyrogallol , ellagic acid , catechin and caffeic acid). Meanwhile, celery was dried at 50°C and 90°C caused detectable decrease in the total contents of the eight phenolic in celery. The increment of protocatechuic acid was (99.84, 64.2%), catechol was (150.33, 34.74%), chlorogenic acid was (31.16, 34.96%), salycilic acid was (3.10, 8.21%), chrysin was (313.04, 53.68 %), pyrogallol was (478.97, 40.11%) and ellagic acid was (97.76, 60.72 %) at 50°C and 90°C, respectively.

On the other hand, The increment percentages were 427.73 and 45.56% of gallic acid, 19.70 and 38.76% of syringic acid, 63.15 and 53.87 % of caffeine, 60.86 and 93.52% of p-coumaric acid, 48.80 and 95.93% of ferulic acid, 47.91 and 64% of cinnamic acid, 68.80 and 61.21% of catechin and 68 and 50.37% of caffeic acid for dried celery at 50°C and 90°C, respectively. The obtained data were in harmony with finding of [12-14].

Table (1). *Phenolic acids of fresh and dried celery herbs using HPLC (ppm)*

Phenolic acids	Fresh	Dried at 50°C	Dried at 90°C
Gallic	20.85	110.45	160.78
Protocatechuic	90.50	35.76	12.77
Catechol	7.53	18.85	25.40
Chlorogenic	65.30	85.65	115.60
Syringic	25.22	20.25	12.40
Caffeine	6.65	2.45	1.13
P-Coumaric	17.76	6.95	0.45
Ferulic	16.80	8.60	0.35
Salycilic	3.54	3.65	3.95
Cinnamic	0.48	0.25	0.09
Chrysin	0.23	0.95	1.46
Pyrogallol	75.26	435.74	610.54
Ellagic	235.56	465.85	748.75
Catechin	18.43	5.75	2.23
Caffeic	4.22	1.35	0.67

3.3. Determination of Flavonoids Components

Flavonoids constitute the largest group of plant phenols and account for over half of the 8000 naturally occurring phenolic compounds [15].

Table (2). *Flavonoids of fresh and dried celery herbs using HPLC (ppm)*

Components	Fresh	Dried at 50°C	Dried at 90°C
Apignen	94.74	97.68	104.74
Hesperitin	92.76	136.54	194.65
Luteolin	43.45	76.48	110.26
Quercetrin	110.35	574.82	993.27
Rosmarinic	96.58	43.66	19.82

The flavonoids of dried celery herb are tabulated in Table (2). The results observed that five flavonoids components were identified in celery herbs (apignen, hesperitin, luteolin, quercetrin and rosmarinic). Celery was dried at 50°C and 90°C caused detectable increase in the total contents of the four flavonoids. The increment percentages were (3.1, 7.22%) for apignen, (47.19, 42.55%) for hesperitin, (76.01, 44.16%) for luteolin and (420.90, 72.79%) for quercetrin at 50°C and 90°C, respectively. Meanwhile, the decrement percentages

were (54.79, 54.60%) for rosmarinic at 50°C and 90°C, respectively. The obtained data were in harmony with finding of Wach et al, [16].

3.4. Determination of Isoflavones Components

The isoflavones of dried celery herb are tabulated in Table (3). The results observed that three isoflavones components were identified (daidzein, genistein and isorhamnetin). Celery was dried at 50°C and 90°C caused detectable increase in the total contents of the two isoflavones. The increment percentages were (136.47, 96.42%) for genistein and (96.94, 82.45%) for isorhamnetin at 50°C and 90°C, respectively. Moreover, the decrement percentages were (37.61, 39.95%) of daidzein at 50°C or 90°C, respectively. The obtained data were in harmony with finding of Manal and Sahar [13]

Table (3). *Isoflavones of fresh and dried celery herbs using HPLC (ppm)*

Components	Fresh	Dried at 50°C	Dried at 90°C
Daidzein	1820.45	1135.72	681.92
Genistein	30.84	72.93	143.25
Isorhamnetin	484.32	953.85	1671.36

3.5. Chemical Composition of Essential Oil

Essential oils of food products affect their nutritional availability due to their direct responsibility for consumer acceptance or rejection. Thus, the changes occurred in the chemical composition of the essential oil of two collective samples of celery dried at 50°C and 90°C.

The percent of essential oil for fresh celery was 0.72%, the obtained data are in harmony with the findings of Fazal and Singla [5]. The chemical constituents of *apium graveolens* volatile oil are tabulated in Table (4). From these results, it could be indicated that eleven components were isolated from *apium graveolens* essential oil. These components were identified and classified into five chemical categories namely monocyclic terpenes (78.24%), bicyclic terpenes (14.88%), aliphatic hydrocarbons (1.79%), ketones (0.19) and sesquiterpene (2.89%). The percent of these identified compounds were 97.99% of *apium graveolens* essential oil and the remainder portion was 2.01% representing seven unknown constituents. The first chemical group as shown in Table (4) was monocyclic terpenes which consists of two compounds (d-Limonene, 77.65% and γ-terpinine, 0.59%). Limonene was reported as the major constituent of *apium graveolens* essential oil by Misic et al., and Fazal & Singla [5 and 17]. The second recorded chemical group was bicyclic terpenes which consists of four components n (α-pinene, 1.36%, camphene, 0.29%, Sabinen, 1.72% and β-pinene, 11.55%. The third identified chemical group was aliphatic hydrocarbons which consists of one compound (β– myercene, 1.79%). This compound was reported as constituent of *apium graveolens* essential oil by Misic et al., 2008 [17]. The fourth identified chemical group was ketones which consist of one compound (L-carvone, 0.19%). The five chemical group was sesquiterpene which consists of three compounds (β-caryophyllene, 1.11%, β-selinene, 0.93% and α- selinene

0.85%).

The results observed that dried celery at 50°C and 90°C caused detectable decrease in the total contents of five chemical groups. The percent of monocyclic terpenes was decreased to (5.77, 11.37%), (27.75, 41.66%) for bi-cyclic terpenes, (20.67, 50.27%) for aliphatic hydrocarbons, (42.10, 63.15%) for ketones and (19.03, 41.17%) for sesquiterpene at 50°C and 90°C, respectively. The obtained data were in harmony with finding of Okoh *et al.,* 2008 [18]

Table (4). Chemical components of celery (apium graveolens) essential oils fractionated and identified by GC/Mass technique.

Chemical compounds	Area%		
	fresh	*50°C*	*90°C*
1-Monocyclic terpenes:			
d-Limonene	77.65	73.34	69.13
γ-terpinene	0.59	0.38	0.21
Total :	78.24	73.72	69.34
2-Bi cyclic terpenes:			
α-pinene	1.36	0.97	0.75
Camphene	0.29	0.15	0.08
Sabinen	1.72	1.45	1.22
β-pinene	11.51	8.18	6.63
Total :	14.88	10.75	8.68
3-Aliphati hydrocarbons:			
β-Myercene	1.79	1.42	0.89
Total :	1.79	1.42	0.89
4- Ketones:			
L-Carvone	0.19	0.11	0.07
Total:	0.19	0.11	0.07
5-Sesquiterpene:			
β-Caryophyllen	1.11	0.90	0.77
β-selinene	0.93	0.78	0.54
α- selinene	0.85	0.66	0.39
Total:	2.89	2.34	1.70
6-Unknown:	2.01	11.66	19.32

3.6. Organoleptic Evaluation of Celery (apium graveolens)

Table (5) represents the mean scores and their statistical analysis indication for color, aroma, taste, texture and overall acceptability for biscuit treatments mixed with different concentrations of dried celery *(apium graveolens)* herbs at 50°C and 90°C. The data in Table (5) indicated that there are no significant differences in the quality attributes i.e., texture of biscuit which manufactured with or without (control) addition of different levels of *apium graveolens.* According to the results, the mean score of aroma, taste and overall acceptability of biscuit samples increased as the concentration of *apium graveolens* increased from 0.1 to 0.3%. Therefore, high significant differences were found among these samples and their control sample. However, no significant differences were found in the acceptability among all biscuit samples incorporated different concentrations of *apium graveolens* dried at 90°C (0.1, 0.2 and 0.3%). On the other hand, the biscuit samples formulated with 0.1% , 0.2% and 0.3% of *apium graveolens* dried at 50°C had higher aroma and taste scores in comparison with biscuit samples of the control and celery dried at 90°C.The obtained data are in harmony with the findings of Lawiess and Naglaa & Gehad, [13 and 19]

Table (5). The mean scores of organoleptic evaluation of biscuits manufactured by addition of dried celery apium graveolens herbs at 50°C and 90°C.

Treatments	Color	Aroma	Taste	Texture	Overall quality	Accept ability
Control	10.0[a]	7.0[d]	7.0[d]	9.0[a]	8.0[e]	+
Dried celery at 50°C						
Apium graveolens (0.1%)	9.0[b]	9.0[b]	9.0[b]	8.0[b]	9.00[b]	+++
Apium graveolens (0.2%)	9.0[b]	10.0[a]	10.0[a]	8.0[b]	9.50[a]	++++
Apium graveolens (0.3%)	8.0[c]	10.0[a]	10.0[a]	8.0[b]	9.50[a]	++++
Dried celery at 90°C						
Apium graveolens (0.1%)	8.0[c]	8.0[c]	8.0[c]	8.0[b]	8.0[e]	++
Apium graveolens (0.2%)	7.0[d]	9.0[b]	9.0[b]	8.0[b]	8.5[cd]	++
Apium graveolens (0.3%)	7.0[d]	9.0[b]	9.0[b]	8.0[b]	8.5[cd]	++
LSD	0.01779	0.02516	0.0251	0.00562	0.03081	

(+) The sample is acceptable.
(+++) The sample is very good acceptable.
(++) The sample is good acceptable.
(++++) The sample is excellent acceptable.
*Any two means, at the same column , have the same letter did not significantly different at 5% or 1% level of probability.

5. Conclusion

Celery herbs were dried at 50 and 90°C using a laboratory scale hot air dryer and the phenolic, flavonoids and isoflavones of *apium graveolens* fractionated, and identified by using HPLC. Drying of celery herbs at 50°C and 90°C caused detectable decrease in the total contents of the eight phenolic acids (protocatechuic, catechol, chlorogenic, salicylic, chrysin, pyrogallol and ellagic acid), while it caused an increment in the total content of gallic, syringic, caffeine, p-coumaric, ferulic, cinnamic, catechin and caffeic. Five flavonoids components were identified in celery herbs (apignen, hesperitin, luteolin, quercetrin and rosmarinic) and three isoflavones components were identified (daidzein, genistein and isorhamnetin) in dried celery herbs. The chemical constituents of *apium graveolens* volatile oil were determined, the results observed that eleven components were isolated from *apium graveolens* essential oil and classified into five chemical categories namely, monocyclic terpenes, bicyclic terpenes, aliphatic hydrocarbons, ketones and sesquiterpene. Organoleptic evaluation of *Apium graveolens* represented the mean scores and their statistical analysis indication for color, aroma, taste, texture and overall acceptability for biscuit treatments mixed with different concentrations of dried *Apium graveolens* at 50°C and 90°C.

References

[1] M. Keinänen and R. Julkunen-Tiitto, Effect of sample preparation method on birch (Betula pendula Roth leaf phenolics, Journal of Agricultural and Food Chemistry, 44: 2724–2727, 1996.

[2] J. Pokorny, Addition of antioxidants for food stabilization to control oxidative rancidity, Czech Journal of Food Sciences, 4: 299–307, 1986.

[3] D.O. Kim, S.W. Jeong and C. Y. Lee, Food Chem. 81:321–326, doi:10.1016/S0308-8146(02)00423-5, 2003

[4] G. Giuseppe, B. Davide, G. Claudia, Ugo Leuzzi and C. Corrado: Flavonoid Composition of Citrus Juices. Molecules, 12, 1641-1673, 2007

[5] S. S. Fazal and R. K. Singla, Review on the Pharmacognostical & Pharmacological Characterization of *Apium Graveolens* Linn. Indo Global, Journal of Pharmaceutical Sciences, 2(1): 36-42, 2012.

[6] A. Schieber, P. Keller, R. Carle, Determination of phenolic acids and flavonoids of apple and pear by high performance liquid chromatography system, J. of Chromatography A, 910:265-273, 2001.

[7] D. Mantovani, L. C. Filho, L. C. Santos, V.L.F. de Souza and C.S. Watanabe, The use of HPLC identification and quantification of isoflavones content in samples obtained in pharmacies. Acta Scientiarum. Biological Sciences. Maringá, v. 33, n. 1, pp. 7-10, 2011.

[8] M. Guenther, The essential oils, Vol. III, IV, 4th Ed. D. Van Nostrand Company, Inc. Princeton, New Jersy. Tornto, New York, London, 1961.

[9] Naglaa H.M. Hassanen and Gehad, F.A.Fath El-bab. Anti-microbial activities of *Matricaria chamomilla* L. flower. Egypt. J. Biomed. Sci., 37: 189-211, 2011.

[10] M. A. Amerine, R. M. Panglorn, and Roessler, E. B. (): Principals of sensory evaluation of food"Academic Press, New York, 1965.

[11] G.W. Snedecor and W.G. Cochran, Statistical methods.7th Ed., p. 420. Iowa Stat. Univ. Press, Ames, Iowa, USA, 1980.

[12] A. Crozier, B. J. Indu and N. Michael, Dietary phenolics: chemistry, bioavailability and effects on health. Nat. Prod. Rep., 26, 1001–1043, 2009.

[13] M. S. M. Manal and S. A. Sahar, The Effects of Purslane and Celery on Hypercholesterolemic Mice, World Journal of Dairy & Food Sciences, 7 (2): 212-221, 2012.

[14] Liga Priecina and Daina Karklina, Natural Antioxidant Changes in Fresh and Dried Spices and Vegetables, International Journal of Biological, Veterinary, Agricultural and Food Engineering Vol. 8, No. 5, 480-484, 2014.

[15] N. Balasundram, K. Sundram and S. Samman, Phenolic compounds in plants and agri-industrial by-products: Antioxidant activity, occurrence, and potential uses, Food Chemistry, 99, 191- 203, 2006.

[16] A. Wach, K. Pyrzyn´ska and Magdalena Biesaga, Quercetin content in some food and herbal samples. Food Chemistry 100 :699–704, 2007.

[17] Misic Dusan, Irena Zizovic, Marko Stameni, Ruzica Asanin, Mihailo Risti, Slobodan Petrovi and Dejan Skala, Antimicrobial activity of celery fruit isolates and SFE process modeling Biochemical Engineering Journal, 42, 148–152, 2008.

[18] O. O. Okoh, A. P. Sadimenko, O. T. Asekun and A. J. Afolayan, The effects of drying on the chemical components of essential oils of *Calendula officinalis* L. African Journal of Biotechnology Vol. 7 (10), pp. 1500-1502, 2008.

[19] Lawiess and Julia, The Encyclopaedia of essential oils.Published in Great Britain in 1992 by Element Books Limited Longmead, Shaftesbury, Dorset, pp.81-84 and 103-105, 1992.

[20] Van Wassenhove, F., Dirinck, P., Vulsteke, G. and Schamp, N., Aromatic volatile composition of celery and celeriac cultivars. Hort. Sci., 25, 556-559, 1990

[21] Fazal, S. S., Ansari, M. M., Singla, R. K., Khan, S., Isolation of 3-n-Butyl Phthalide & Sedanenolide from *Apium graveolens* Linn. Indo Global Journal of Pharmaceutical Sciences, 2(3): 258-261, 2012.

3

A Comparison of Solar Power Systems (CSP): Solar Tower (ST) Systems versus Parabolic Trough (PT) Systems

Huseyin Murat Cekirge[1], Ammar Elhassan[2]

[1]Department of Mechanical Engineering, Prince Mohammad Bin Fahd University, Al Khobar, KSA
[2]Department of Information Technology, Prince Mohammad Bin Fahd University, Al Khobar, KSA

Email address:
hmcekirge@usa.net (H. M. Cekirge), aelhassan@pmu.edu.sa (A. Elhassan)

Abstract: Comparison of Comparison of Solar Power System (CSP) power plants will be introduced and discussed; Solar Tower (ST) plants and Parabolic Trough (PT) plants are subjects of this comparison. The comparison will be made possibly analytical or quantitatively instead of qualitatively. Examples will be presented and explained in detail. The main issues such efficiency, area of the plant, environmental issues, molten salt storage, the cost of the plants, dust and humidity, maintenance and operation cost and total investment are discussed.

Keywords: Solar Tower Power, Parabolic Troughs, Comparison of CSP Power Plants, Environmental Impact of Solar Power, Environmental Impact of Parabolic Troughs, Raspberry Pi, Arduino, Arm Architecture

1. Introduction

Several Concentrated Solar Plants (CSP) are at the design, research and development phases, these are mainly, Parabolic Trough (PT), Solar Tower and Linear Fresnel systems. Design limitations combined with economic reasons mean that the building of Linear Fresnel Power systems was adjourned globally; therefore PT and Solar Tower systems will be compared in this study. Qualitative comparisons of these systems are presented in Table 1.

Table 1. Qualitative comparison of Solar Tower and PT systems.

Technology type	Description	Pros	Cons
SOLAR TOWER	Fixed centralized receiver tower surrounded by field of surrounding heliostats	Higher efficiency	Few commercial applications ; mostly due to high sophistication of tracking and heliostat allocating software; and heliostat designs which are not spread worldwide in great extent
		Minimal piping and fitting	
		Fixed receiver unit	
		Flat mirrors	
	Dual-axis control tracking		
	Receiver contains water or molten salt	WI-FI control and single power cable	
		Proof of concept at first generation sites	
	Storage potential is applicable	Ideal for hybrid plants	
		Field set-up flexibility	
PARABOLIC TROUGHS	Field or long rows of parabolic mirrors	Proven development and operational track record	Relatively low efficiency thermal to electric conversion
	Mirrors concentrate sunlight onto movable receiver system	Employs single-axis trackers "Off-Use-shelf" systems available	Utilizes more expensive curved mirrors
			Durability of piping and ball joints
	Receiver contains oil or water		Environmental issues of oil-based platforms

The reason of minimal commercial utilization of Solar Tower is the sophisticated tracking and control software and WIFI system cannot be offered by many commercial companies. PT's have implementation history in the renewable energy industry. PT's generate energy through heating of the transfer fluid. ST system is based on a number of heliostats and tower boiler, more efficient than the PT systems with the added bonus of not having toxic heat transfer fluid, and efficient use of storage using molten salt.

2. Comparison of Solar Tower (ST) and Parabolic Troughs (PT)

2.1. Energy Calculations, ST and PT

Considering an Integrated Power Plant, IPP, and Solar Tower can contribute all energy that is needed by the steam turbine, Figure 1, if enough land is provided for hosting the power plant structures.

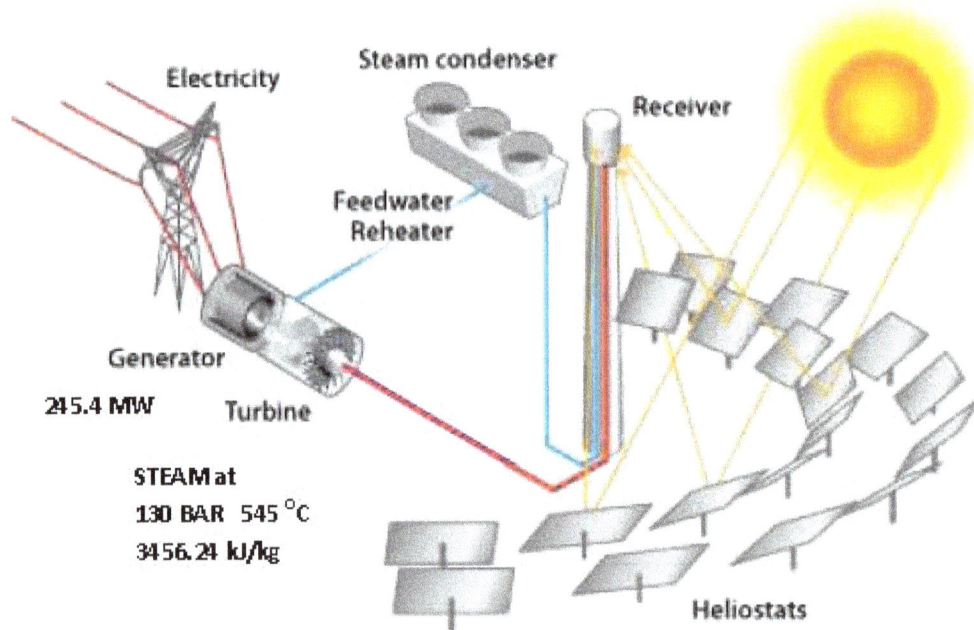

Figure 1. Steam goes to turbine directly from the Solar Tower, [1].

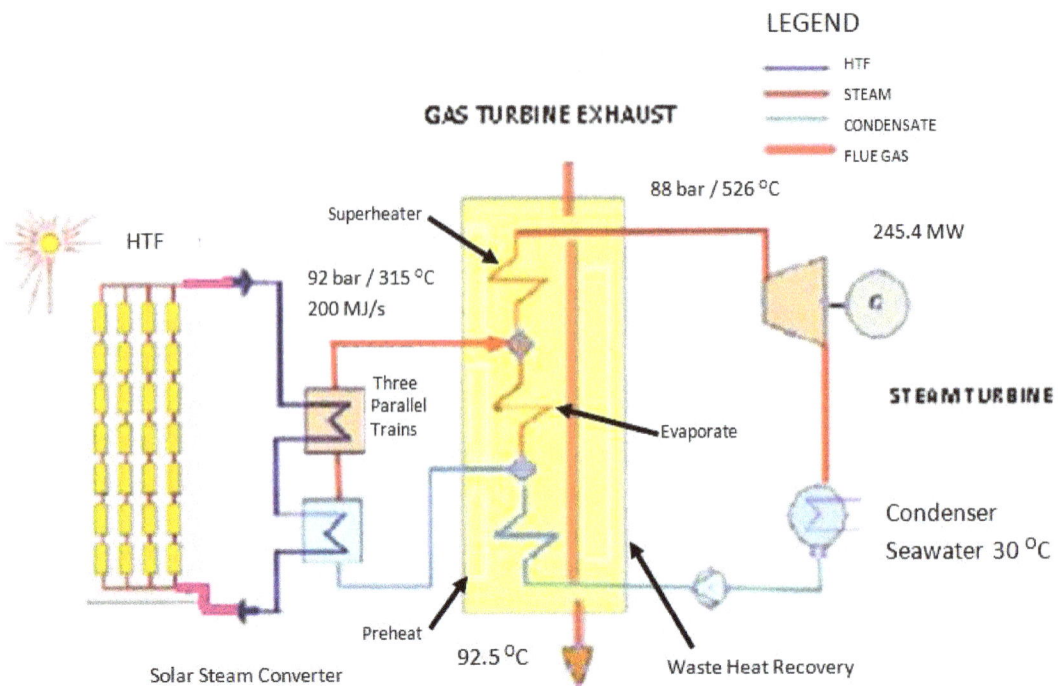

Figure 2. PT with Waste Heat Recovery Boiler, WHRB [1].

In the case of the PT system as in Figure 2, the contribution of solar power is 40 MW and the thermodynamic calculations of enthalpy, entering to and exiting from Waste Heat Recovery Boiler WHRB, were made through [2]. The steam from the PT cannot be used without the utilization of a Waste Heat Recovery Boiler WHRB. As such, gross efficiency is reduced by approximately twenty percent because the heat transferring fluid, HTF, cannot exceed 400 °C.

PT systems with thermal fluid are not as efficient as CSP Solar Tower systems (ST) that inject steam directly into the turbine; ST systems are approximately 30-40 percent more efficient. In Parabolic Trough type PT plant, the heat transfer is based on thermal oil which can be heated to no more than to 400°C; and degrades and loses efficiency at higher temperatures.

Table 2. Thermodynamic calculations of Solar Tower, [2].

IPP_Solar_PT				
Enthalpy_kJ/kg_(exit_pt)_at_bar_	92	°C	315	2791.11
Enthalpy_kJ/kg_(entrance_ST_turbine)_at_bar_	88	°C	526	3453.50
PT_power_mJ/s	200			
Steam_turbine_output_MWe	245.4			
Total_energy_PT_MWe	40.0			

In the case study of a PT system, solar field is operated with a maximum temperature of 393 °C; it delivers heat to the power block where the working fluid is water/steam using a couple of heat exchangers (HX's). At the exit of HX on the power block, the steam temperature measures 315 °C and 92 bar pressure. The saturation temperature of dry steam at 92 bars is 305 °C, i.e., steam is slightly superheated. In order to utilize steam in the

Table 3. Summary of results.

Type	Group	Entry power MW	Steam	Useable MWe
SOLAR ENERGY_PT	IPP	200	92 BAR 315 °C	40
SOLAR ENERGY_ST	IPP	200	130 BAR 545 °C	66

Turbine more efficiently, the temperature needs to be higher; therefore, the exhaust gas from the turbine, at 550 °C, is utilized. The inlet steam of the steam turbine has a temperature of 526 °C at 88 bars. If we consider enthalpies of the steam both at superheated conditions,

$$h(315°C, 92 \text{ bar}) = 2796.65 \text{ kJ/kg},$$

$$h(526°C, 88\text{bar}) = 3453.81 \text{ kJ/kg}.$$

In other words, a difference of 657.16 kJ energy per kg of steam is produced by the exhaust gas in the waste heat recovery boiler. This kind of upgrade of solar heat is necessary in PT type installations in order to utilize the steam produced at 315 °C in more efficient high temperature steam turbines. On the other hand, the exhaust gas of gas turbine with a temperature of 550 °C; is sufficient to produce steam in the waste heat recovery boiler (WHRB) with the desired quality to be delivered to the steam turbine. This will be utilized power production without the need of solar generated steam. In addition to lower energy conversion efficiency, limited hybridization capability, long rows of pipes in the field necessary to transport heat transfer oil, and durability of movable heavy structures are taken into account for PT systems.

In the alternative, solar tower (ST) system, CSP plant is already available to produce steam directly deliverable to the high efficiency steam turbine with steam inlet conditions of 88 bars and 526 °C temperature. Therefore, in this scenario,

gas turbine exhaust gas in the Waste Heat Recovery Boiler (WHRB) will produce the same quality steam with the ST CSP plant. This will improve the system performance by about 657 kJ per kg of steam circulating in the intermediate loop. As this is indicated in the block associated with PT plant, 200 MJ/s power requires about 71.5 kg steam circulation per second, i.e.,

$$657 \text{ kJ/kg} \times 71.5 \text{ kg/s} = 46975.5 \text{ kJ/s} = 46.98 \text{ MW}$$

of thermal power is lost due to the inclusion of low quality steam produced by PT system, Table 3.

For implementation purposes, different schemes may be developed for better coupling of WHRB and Solar Tower with steam turbines. In these terms, various scenarios might be developed for comparisons.

2.2. Comparison of Land Use ST Versus PT

Land use refers to the land area directly occupied by a power plant structure, ST and PT, and is expressed in units of m2/(MWh/y), that is square meter over megawatt hour per year. The visual impact indicating the area over which a power plant disturbs the landscape is measured in units of m2/ (MWh/y)). Both ST and PT systems are normally setup in remote areas with negligible visual impact. There is always misunderstanding that ST power plants require larger land areas, a typical land use of a ST is presented in Figure 3.

ST system has an algorithm to locate the heliostats in the

field in an optimum way. Besides these algorithms are being tested and perfected in the field for efficient use of heliostat area, [3].

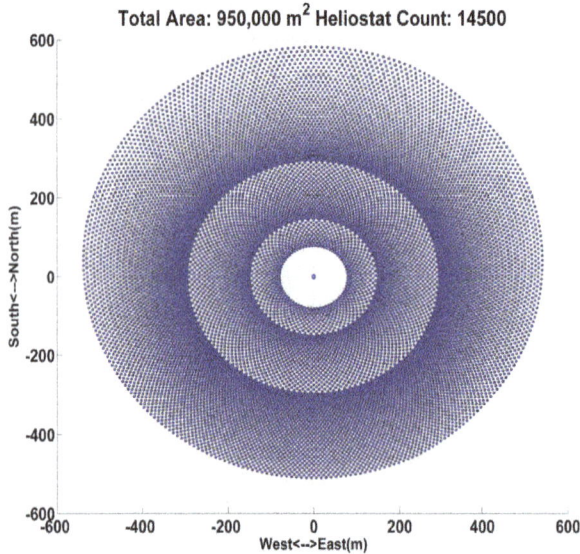

Figure 3. *A Solar Tower, ST, "Heliostat" area. [4].*

ST system can produce more power than PT system installed equivalent area. ST power system has higher efficiency since the control software has dominant factor to provide maximum solar energy input, [3], into solar power plant.

2.3. Environmental Impact of ST Versus PT

The potential environmental pollution risks relating to leaks or emissions of HTF (heat transfer fluids) impact soil, ground and surface water and air quality thus affecting human life. In the case of synthetic oil, as HTF, is used and compared to other possible HTFs, it is friendlier to the environment. PT systems consist of a widespread distribution of the receivers, i.e. tubes or fittings in the solar fields, thus increasing the risk of HTF leakages with the obvious environmental risks associated with it. This aspect is more problematic if the land/site area is not exclusive utilized by to the TES (thermal energy storage) subsystem and the process equipment. In this case, a higher pollution risk threatens the area if leakages of HTF occur in the storage system. There is also an unavoidable HTF odor in installations from leakages since the system has many pipes, fittings and ball joints. The synthetic oil, which it is highly toxic, can pollute the soil and could pass very rapidly to the water systems; MSDN of two HTFs can be found in [4] and [5]. This suggests that oil should be avoided in case of existing vulnerable aquifers. In ST based systems, this kind of pollution is not a factor.

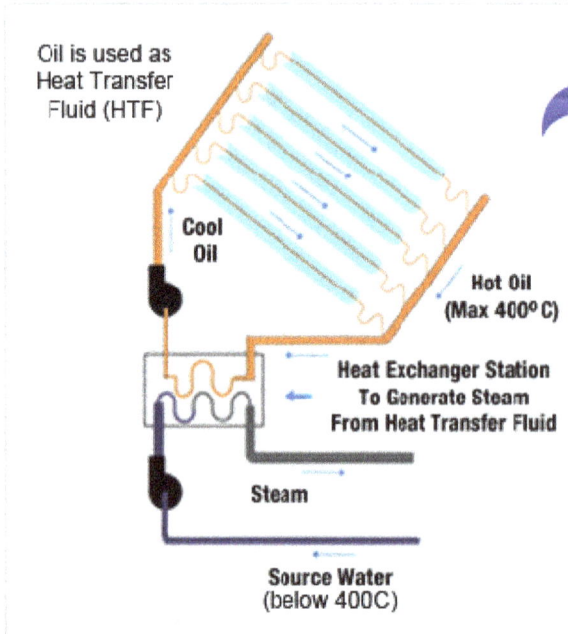

For a solution of this problem in PT systems, steam should be obtained directly [6], which is not common practice, figure 4 illustrates. In these PT's, heat transfer fluid is replaced by water, Hittite Parabolic Toughs, Direct Steam Generation (DSG). The heat storage and efficiency solutions of this system are in development.

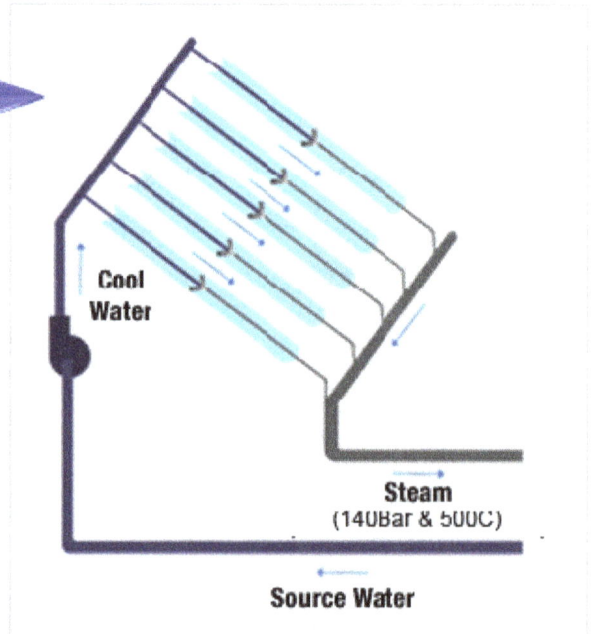

Figure 4. *Direct steam generation (DGS), the Hittite PT's, [6].*

Molten Salt for ST is less hazardous to the environment since again, HTF is not a factor in the heat exchange process. Water will be used for both systems for cleaning and there must be a source of water close to the plant. In the case of desalination, the water problem can be solved without any difficulty. In desert regions, dust is a high impact factor and hence, the CSP requires regular cleaning of the panel surfaces; this can be done by wipers and water spraying. The daily consumption of water for washing mirrors of ST is quite minimal. Humidity does not play an important role on the efficiency of ST power systems respect to PTs, since STs have flat surfaces.

Other environmental issues such as impact on wildlife have to be analyzed. For example, desert tortoises became a major factor and they were relocated during construction of the Ivanpah plant [7], risk to birds is very minimal and no more than the risk from windows and domestic cats [8]. Many bird mortalities apparently were caused by flight through the hot solar flux, burning feathers and beaks, and others are caused by hitting the heliostats, or anomalies in bird navigation habits.

2.4. Maintenance of ST Versus PT

PT systems consist of a number of pipes, fittings and connections, in other words the system is quite fragile. As an example: Daggett, 103 MW, PT Solar Power Station, located at Daggett, California, USA, the maintenance cost including parts, and excluding Thermal Exchange System, TES and Power Block as at May 2009 is : $1,289,786.00. This number is quite high, even if loss of production during repairs is notwithstanding [9].

PT systems are made from a disparate set of materials including metal, glass and plastic which are highly susceptible to thermal load thus rendering these systems very fragile. The major problem for all CSP plants is the frequent breakage of expensive parabolic glass mirror panels which are manufactured using highly specialized fabrication processes. These issues must be eliminated, in order to reduce operation and maintenance (O&M) costs. Also, heat receivers must be redesigned to decrease (O&M) costs of the whole system by abandoning traditional "metal to glass attachments".

Maintenance costs are estimated at around fifteen percent in the favor of ST systems over PT systems, Figure 5 illustrates. It can be seen that breakdown risks are higher in the case of PT systems thus causing loss of income from total production, [10].

2.5. Investment of ST Versus PT

Tables 4 and 5 are show investments costs and returns from ST and PT systems. It can be seen that the initial investment for ST is 16 percent less than PT for a typical 100 MW plant. If the storage facilities are considered, the PT system is 42 percent more costly, [11], for the same size plant.

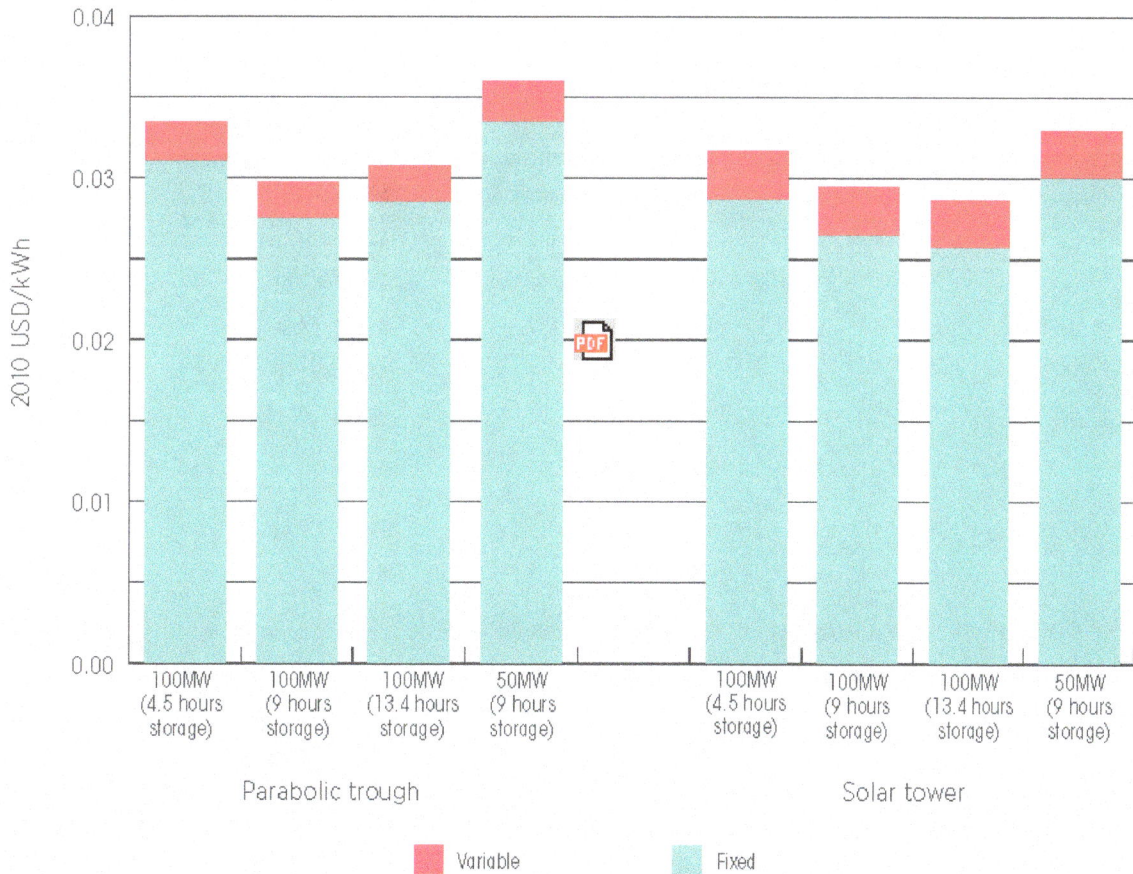

Figure 4. Operations and maintenance costs for PT and ST, [10].

Table 4. Current and projected plant cost for 100 MWe power tower with 6 hours storage at Longreach, Queensland, (unit costs based on power tower road map, [12]).

All costs in AUD 2010	Current Cost	Future Cost (2010)	Generic plant sizing	Current Capital Cost	Future Capital Cost (2020)
Site Improvements ($/m^2)	27	27	1,010,046 m	$27,422,752	$27,422,752
Solar Field ($/m^2)	217	130	1,010,046 m^2	$219,382,013	$131,629,208
Receiver & Tower ($/kWh$_{th}$)	217	185	507353 kWh$_{th}$	$110,197,072	$93,667,511
Storage ($/kWh$_{th}$)	33	22	1,691,180 kWh$_{th}$	$55,098,644	$36,732,430
Power Block ($/kW$_e$ gross)	1086	869	115,000 kW$_e$	$124,890,000	$99,912,000
BOP ($/kW$_e$ gross)	380	272	115,000 kW$_e$	$43,711,500	$31,222,500
Indirect Cost			35%	$203,245,693	$147,205,240
Indicative Cost ($/kW$_e$ net)				$7,836	$5,675
O&M ($/kW-yr)	71	54	100.05 MW$_{net}$	$7,062,530	$5,432,715
Solar Multiple			1.8		
Storage Hours			6		
Capacity Factor (%)			40.9		
LCOE ($/MWh) (20 yr life, WACC 7%)				$226	$164

Table 5. Current and projected plant cost for 100 MW parabolic trough plant at Longreach, Queensland, current and future unit costs based on, [13], and parabolic trough road map respectively, [14].

All costs in AUD 2010	Current Cost	Future Cost(2017)	Reference plant sizing	Current Capital Cost	Future Capital Cost (2017)
Site Improvements ($/m2)	27	27	918,026 m	$24,786,702	$24,786,702
Site Improvements ($/m2)	320	217	918,026 m2	$293,768,320	$199,211,642
Solar Field ($/m2)	98	46	918,026 m2	$89,966,548	$42,229,196
Storage ($/kWhth)	87	29	1,877,110 kWhth	$163,308,570	$54,436,190
Power Block, BOP ($/kWe gross)	1021	884	111,000 kWe	$113,331,000	$98,124,000
Indirect Cost	18.50%	16%		$126,754,811	$70,334,032
Indicative Cost ($/kWe net)				$8,119	$4,891
O&M ($/kW-yr)	80	51	100 MWnet	$7,980,000	$5,130,000
Solar Multiple			2		
Storage Hours			6		
Capacity Factor (%)			43.20%		
LCOE ($/MWh) (20 yr. life, WACC 7%)				$223	$135

3. Solar Tracking Sub-System

To maintain maximum efficiency of the system, it is imperative to keep the solar panels pointing towards the sun for as long as possible. There are several methods for achieving this task, varied by cost and complexity. It is proposed that the use of the new breed of ARM-Architecture [16, 17] microcomputers such Arduino [15] and Raspberry Pi [18] coupled with one of three methods of calculating sun location and ray direction. In both cases below, we propose that solar panels be mounted on motors/actuators that are controlled via relays and microcomputers in addition to a power source (photovoltaic panel or power grid) to power the motors.

3.1. GPS Based Calibration

In this method, the microcomputer with its GPS and Compass sensors will be programmed with the necessary calibration code to position each solar panel at the correct angle for maximum exposure to sunlight. The cost of this approach is probably the least but only after the algorithm to track in relation to the GPS sensors has been written correctly.

3.2. Voltage Differential Based Calibration

By using twin solar panels mounted at right angle to each other, it is possible to read the two harvested voltage values from them and the microcomputer controlled motor/actuator is powered to rotate the platform such that the 2 voltage values from the two solar cells are kept with minimum differential thus guaranteeing that they are pointed at the sun in an optimum way. The cost of this method is higher because it uses two mirrors and requires voltage differential sensors.

3.3. Image Based Calibration

By utilizing a Raspberry Pi with a webcam, it is possible to track the location of the sun in the sky using an algorithm to take snapshots of the sky or of a bar shadow [19] and contrast them and rotate the platform accordingly [20] and [21]. The cost of this approach is in-line with the method in 3.1, although this technique has the advantage of being well documented and tested.

4. Conclusions

Solar thermal power plant technologies are important sources for providing a significant part of the clean and renewable energy needed in the future. Among these technologies, the qualitative comparison was made between ST systems and PT systems. It has been shown that the ST systems are superior to PT systems in various aspects including:

- ST systems are more efficient, at least 30 percent
- Land area per energy output is 20 to 30 percent in favor of ST systems
- No pollutants or environmentally hazardous materials are utilized in ST systems, hence the energy produced to pollution ratio is much higher
- Operating and Maintenance expenses are around 15 to 20 percent less in ST systems
- Without heat storage sub-systems, ST systems require 15 to 20 percent less upfront investment when considering output based calculations of ST and PT plants. With storage sub-system factored in, this figure is around 30 to 40 percent in favor of ST systems.

At the moment; ST systems are not being commercialized in wide extent, since technology of software and hardware of ST systems are in the hands of few commercial establishments. The software is the major factor; specifically solar tracking and heliostat allocating algorithms are needed to run ST system efficiently; however it is being possessed by limited number of organizations. In addition to all these factors, the ST systems can fit many climatic and geographic conditions.

References

[1] S. Erturan, Presentation of Greenway Solar, Istanbul, (http://greenwaycsp.com), 2014.

[2] Michael J. Moran and Howard N. Shapiro, Fundamentals of Engineering Thermodynamics, 6th Edition, John Wiley & Sons, New Jersey, 2008.

[3] S. Erturan, Greenway, private communication, 2015.

[4] Paratherm, Heat Transfer Fluid, MSDS No. 14439240-5531120-5149421-102103, 30 October 2014.

[5] MultiTherm IG-1, Heat Transfer Fluid, MultiTherm LLC, January 1, 2013.

[6] O. Capan, Hittite, private communication, (http://www.hittitesolarenergy.com), 2015.

[7] Ivanpah Brightsource Plant, (http://www.brightsourceenergy.com/ivanpah-solar-project#.VS4AvPmUd8E), 2015.

[8] J. Desmond, BrightSource, private communication, 2015.

[9] C. Turchi, Parabolic Trough Reference Plant for Cost Modeling with the Solar Advisor Model, NREL, Boulder, CO, 2010.

[10] Concentrating Solar Power, Renewable Energy Technologies: Cost Analysis Series, Volume 1: Power Sector Issue 2/5, IRENA, International Renewable Energy Agency June 2012.

[11] Jim Hinkley, Bryan Curtin, Jenny Hayward, Alex Wonhas (CSIRO), Rod Boyd, Charles Grima, Amir Tadros, Ross Hall, Kevin Naicker, Adeeb Mikhail (Aurecon Australia Pty Ltd) , Concentrating solar power – drivers and opportunities for cost-competitive electricity, CSIRO, March 2011.

[12] C. Turchi, M. Mehos, C. K. Ho, and G. J. Kolb, Current and future costs for parabolic trough and power tower systems in the US market. SolarPACES 2010. Perpignan, France, 2010.

[13] C. Kutscher, M. Mehos, C. Turchi, And G. Glatzmaier, Line-Focus Solar Power Plant Cost Reduction Plan, NREL and Sandia, 2010.

[14] G. J. Kolb, C. K. Ho, T. R. Mancini and J. A. Gary, Power tower technology roadmap and cost reduction plan. Sandia National Laboratories (Draft, Version 18, December 2010), 2010.

[15] Arduino, (www.arduino.cc), 2015.

[16] ARM Processors Guide, (http://www.arm.com/products/processors), 2015.

[17] ARM Architectur, (https://www.scss.tcd.ie/~waldroj/3d1/arm_arm.pdf), 2015.

[18] Raspberry Pi, (https://www.raspberrypi.org), 2015.

[19] H. Arbab, B. Jazi, M. Rezagholizadeh, A computer tracking system of solar dish with two-axis degree freedoms based on picture processing of bar shadow, Renewable Energy, 34:1114–1118, 2009.

[20] R. Abd Rahim, M. N. S. Zainudin, M. M. Ismail, M. A. Othman, Image-based Solar Tracker Using Raspberry Pi Journal of Multidisciplinary Engineering Science and Technology (JMEST) ISSN: 3159-0040, Vol. 1 Issue 5, December–2014.

[21] P. Omar Badran and Ismail Arafat, The Enhancement of Solar Distillation using Image Processing and Neural Network Sun Tracking System International Journal of Mining, Metallurgy & Mechanical Engineering (IJMMME) Volume 1, Issue 3 (2013) ISSN 2320-4052; EISSN 2320-4060, 2013.

4

Combustion Characteristics and Energy Potential of Municipal Solid Waste in Arusha City, Tanzania

Halidini Sarakikya[1, *], Jeremiah Kiplagat[2]

[1]Department of Electrical Engineering, Arusha Technical College, Arusha, Tanzania
[2]Department of Energy Engineering, Kenyatta University, Nairobi, Kenya

Email address:

sarakikyazablon@yahoo.com (H. Sarakikya), jeremykiplagat@gmail.com (J. Kiplagat)

Abstract: Municipal Solid Waste (MSW) generation has been increasing due to population growth, changing life style, technology development and increased consumption of goods. The increase of waste generation combined with the use of waste dumps may lead to environmental and social problems such as water contamination, land and atmospheric pollutions, resulting to breeding grounds for vermin, cause risk of fire, bad smell and potentially are the cause of illness. Energy recovery from municipal solid waste can alleviate these problems while providing a source of energy. The objective of this study is to evaluate the combustion properties and energy potential from municipal solid waste of Arusha, Tanzania. Incineration is among the methods for MSW treatments, and therefore, the data and information provided shows that energy can be recovered from Arusha MSW during incineration process. Energy flow (exothermic and endothermic) and thermal degradation analysis were carried out using Differential Scanning Calorimetry (DSC) and thermo – gravimetric analysis (TGA) respectively. The sample of composition of municipal solid waste examined included paper, cardboard, wood, textile, rubber, polyethylene Teraphthalate (PETE), low density polyethylene (LDPE) and food waste. These materials were heated in a combined DSC and TGA analyser and experiments were performed at heating rate of 10°C/min, in a pure nitrogen atmosphere at temperatures between room temperature and 1100 °C. The results observed from TGA and DTG show that the highest reactivity was the samples from Central Market, followed by those from Sakina and Ngarenaro market. It was observed that municipal solid waste is less reactive to combustion compared to dry biomass, thus its reactivity can be improved by removing non- combustible materials such as metals and food scraps or by pre-treating the MSW so as to reduce the amount of oxygen present in it. The final analysis of the municipal solid waste showed that, the average percentage of nitrogen, sulfur, chlorine and phosphorus in the waste were 2.36%, 0.37%, 0.04% and 0.11% respectively, which is low and therefore, emissions released by this MSW during combustion are also low. The energy content of the solid waste tested was about 12MJ/kg on dry basis. The elemental composition shows that municipal solid waste contains 50% and 5% of carbon and hydrogen respectively.

Keywords: Municipal Solid Waste, Thermal Behavior, Thermo Gravimetric Analysis

1. Introduction

The generation of municipal solid wastes (MSW) in Tanzania has grown steadily in recent years. It has been estimated by Dar es Salaam Local Authority (DLA), that approximately 4200 tons per day of solid waste were generated in Dar es salaam, Tanzania in 2011, which represent a generating rate of 0.93kg/cap/day basing on a population of 4.5 million [1].The increasing waste generation among others is due to the increase of population in urban areas which is caused by migration of people from rural areas seeking employment for poverty alleviation. Mundi [2], points out that

over past 51 years, urban population in Tanzania has increased from 528,508 in 1960 to 12,359,930 in 2011. Globally, 1.2 billion people live on less than $1.25 a day, where a large number of them live in southern Asia and sub Sahara Africa, Tanzania inclusive [3].The lack of both awareness and technologies for the utilization of MSW into valuable assets such as energy conversion, recycling, incineration and composting is still a problem to society [4]. Due to that reason, Arthur et al [5] reports that the disposed solid waste at Murieti dumpsite in Arusha, Tanzania amounted to 120 tons per day in 2012, and according to Breeze [1], about 63% of the solid waste was found disposed in unplanned areas in the city of Dar es Salaam in 2011.

The utilization of fossil fuels has been in practice for many years now as a source of energy, however this utilization produces greenhouse gases such as CO_2, and generate other pollutants such as SO_2, and NO_x which increase generation of acid rain [5]. Globally, fossil fuel resources are depleting while the consumption is increasing. Terra Symbiosis [6], explain that, discoveries of oil peaked around 1960 and have since slowed significantly. In fact, the time will come when the costs of extraction will increase whilst production will start to decline. World energy consumption doubled between 1970 and 2000, and it is expected to double again between now and 2050. Global consumption of coal increased by 5.4 percent in 2011, to 3.72 billion tons of oil equivalent, while natural gas use grew by 2.2 percent, to 2.91 billion tons of oil equivalent [7].

Satisfying the energy demands through the use of renewable energy sources is on the main agenda now days because of the fossil fuel depletion and environmental issues. Municipal solid waste is the result of human activities which if an appropriate management system is not used, it may lead to environmental pollution and endanger mankind's health. However, the MSW generations increases can be taken as an opportunity for the source of energy for power generation in domestic or industrial use. The European Commission's Environmental Data Center [8], clearly states that the principal definition of municipal waste is "Municipal waste include households and similar wastes". The bulk of the waste stream originating from households and other similar wastes from various sources such as commerce, private, and public service are also included. Imed [9], argues that in low income countries Tanzania inclusive, the solid wastes originated from home, commerce and other areas such as public sectors are more than 50% of the collected solid waste. It is opposite in high income countries where life style favors fewer home cooking by generating only about 30% of solid waste from these areas. Generally, these MSW can be grouped into organic and inorganic materials [10, 11].

Combustible materials from municipal solid waste can be used for energy recovery as they can reduce the utilization of fossil fuels, thus assisting in minimizing global warming [12].The emission of CO_2 coming from biogenic combustion of municipal solid waste is renewable, hence reducing the global warming as it completes the carbon cycle as it does in biomass [13, 14]. Many studies also show that, energy recovery from municipal solid waste can be a better way of managing environment from pollution [15, 16, 17].

There has been much research conducted on the utilization of MSW for the energy recovery. Sushmita [18], for example conducted a research on Technological Options for treatment of MSW of Delhi which finally found out that the landfills used have a lot of potential for energy recovery that can be used for commercial applications such as power generation, whereby the methane gas from this landfills could be transported as pipeline gas. Amin and Yang [19], did a research and concluded that MSW is a domestic energy resource with a potential to provide a significant amount of energy and suggested to have a plan of incineration in Tanjung

Langsat landfill area which could use the high average amount of heating value about 23000KJ/Kg of collected MSW. Surroop and Juggurnath [20], investigated the energy potential from co- firing coal with municipal solid waste by comparing the electricity units generated and the amount of Greenhouse Gas emission when using MSW alone as a fuel and when it is co-fired with coal. Sveta Angelova et al [21], carried out a research on municipal solid waste utilization and disposal through gasification. The syngas from this process has a calorific value of 7.5 – 17.5 MJ/m^3 which can be used to generate electricity and heat or for chemical conversion into various products. The study concluded that, converting MSW to energy has the environmental advantages of reducing the number of landfills, preventing water/air contamination, and lessening the dependence on oil and other fossil fuels for power generation. JD Nixon et al [22], conducted a study and compared five technologies to evaluate options of energy recovery from MSW in India. In the study, Anaerobic Digestion (AD) was identified as the preferred technology for generating electricity from MSW. Both Analytical hierarchy process (AHP) and hierarchical analytical network process (HANP) models gave priority to Anaerobic Digestion and Gasification processes for the energy recovery from MSW.

The scope of this paper is thus to evaluate and analyze the combustion characteristics and energy potential of Arusha City. The use of WTE technologies in Tanzania for converting MSW into energy are still not yet employed. The information provided in this paper is useful for energy recovery during MSW disposal through incineration process. In Tanzania, solid wastes are thrown away in landfills and in various damps instead of utilizing them and convert what would otherwise be a waste product into a high value product. Energy in the form of heat can be recovered from these solid wastes and used in various applications such as warming or steam production for electricity generation. In order to achieve this, the study focused on determining the characteristics of municipal solid waste from three different areas in Arusha; Ngarenaro, Sakina and the Central market.

The paper is structured as follows. Section 2 provides methodology and materials applied in the study, section 3 describes results obtained and discussion of the proximate analysis, ultimate analysis and the calorific value of the solid waste materials. Section 4 provides detailed information on the Thermo Gravimetric Analysis (TGA) curves and Derivative Thermal Gravimetric (DTG) analysis curves of the solid waste materials. A short conclusion is provided in section 5.

2. Material and Methods

2.1. Methodology

This section presents the methodology applied to undertake a study. The methodology consisted of sampling and selection, sorting and laboratory analysis to determine the chemical and physical properties of municipal solid waste of Arusha city. The method of sampling based on ASTM D5231 namely random truck sampling and quartering [10]. Figure 1 shows one of the trucks used to carry the waste to the dumping site, where sorting and sampling was conducted. Wastes were randomly collected from different collecting sources of Central market, Sakina and Ngarenaro markets within Arusha as shown in Figure 2. The random track sampling is shown in the flow chart of Figure 3.

Figure 1. One of the trucks carrying MSW from the city to the site.

Figure 2. Two different collecting points at Ngarenaro market.

The wastes were sorted and weighted by using weighing balance and then separated according to defined classification such as food waste, plastics, mixed papers, frozen fruits, diapers, local news papers, cardboards, yard wastes, leathers, glass and metal as shown in Figure 4.The non-combustible wastes were removed from the rest of the wastes. The combustible waste was availed for analysis in accordance to the method developed by [10]. In order to accurately

determine the waste composition, an average weight of about 200kg of municipal solid waste was taken. This was assumed to be a good representative of the total municipal solid waste composition at each collecting point under this study. The samples were subjected to standards test methods of proximate and ultimate analysis in accordance to ASTM D3172 and ASTM D3176 respectively.

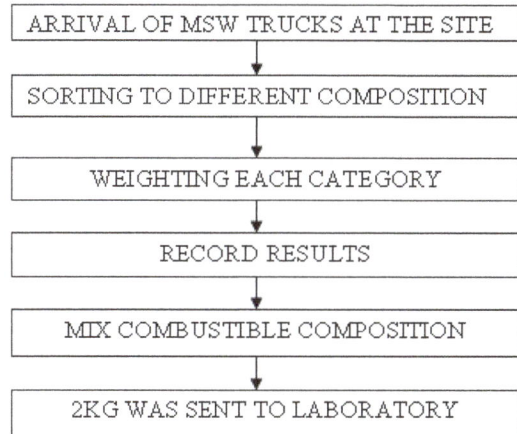

Figure 3. Flow of random truck sampling.

The thermal degradation analysis was studied under Nitrogen condition using a thermo gravimetric analyzer type NETZSCH STA 409 PC Luxx connected to power unit 230V, 16A. High purity nitrogen of 99.95% used as carrier gas controlled by gas flow meter was fed into the thermo gravimetric analyzer with flow rate of 60ml/min and a pressure of 0.5 bars. In the STA 409 PC Luxx, proteus software was used to acquire, store and analyze the data.

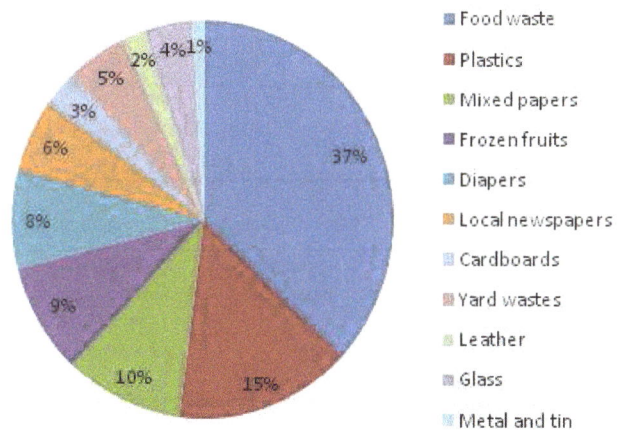

Figure 4. MSW composition.

2.2. Sample Preparation

The samples were shredded into smaller pieces of approximately 30mm size, mixed up and grounded in a grinding machine of less than 1mm size, for the purpose of increasing surface area of the sample that will allow easy penetration of heat [23, 24].

Then a sample of 30 ± 0.1 mg with average particle size less than 1mm was loaded to crucible, dried in the oven at $100^{\circ}C$

for 1hr in a VECSTAR 174799 FURNACE, Model-F/L in accordance to standard (ASTM D3176), and then analyzed in the TG equipment in the temperature ranging from room temperature to 1100°C, at heating rate of 10°C/min.

The heating rate variations change the peak temperature of the decomposition. The more the heat rate, the more increase of temperature [24]. The calculated thermo-gravimetric output from proteus software was obtained as thermal decomposition profile, thermo-gravimetric (TG), Derivative Thermo-gravimetric (DTG) and Differential Scanning Calorimetry (DSC) curves.

The heat released and absorbed by the municipal solid waste degradation was determined from the differential scanning calorimetry curves. The DSC monitors heat effect associated with phase changes transitions and chemical reactions as a function of temperature [24]. The heat was determined by calculating the area between the baseline and the curve. The heat can be positive or negative. When the heat is positive the process is endothermic and when the heat is negative the process is exothermic [24].

3. Results and Discussion

3.1. Proximate and Ultimate Analysis

This section presents the results obtained from the analysis. The results of proximate and ultimate analysis are shown in Table 1 & 2. The moisture content of the municipal solid waste as received ranged between 55.70 and 63.99 wt. %, which is more than 50 wt. % of the total weight of the sample. This high moisture content hinders the combustion process as it rises the ignition temperature [5]. The volatile matter released from

MSW samples for Ngarenaro, Sakina and Central market were 74.43, 84.00 and 78.31wt %, respectively.

This can be compared with the volatile matter contained in pure biomass such as forest residue, oak wood, and pine which are 79.9%, 78.1% and 83.0 wt. % respectively [20]. Generally, solid wastes which contain high volatile, have low fixed carbon, the same case for the Arusha municipal solid waste from Sakina which has volatile matter of 84.00% and fixed carbon of about 6.00 wt. %, compared to that of Ngarenaro and Central market. The advantage of high volatile and low fixed carbon is rapid burning of fuel, while fuel with low volatile and high fixed carbon like coal need to be burned on a grate as it takes long time to burn out, unless it is pulverized to a very small size [25].

Basing on proximate analysis of Table 1 and values of volatile matter and fixed carbon obtained, the study shows that, the Arusha municipal solid waste is combustible. The ash content ranged between 3.29 to 5.97 wt. %, which is relatively low, and advantageous for waste management and environment preservation due to possibility of having small quantity of heavy metals, salts, chlorine and organic pollutant [24]. The ultimate analysis of the municipal solid waste in Table 2 shows that, the average percentage amount of nitrogen, sulfur, chlorine and phosphorus in the waste were 2.36, 0.37, 0.04 and 0.11 respectively which is low, therefore, emissions released by these MSW during combustion are also low. The carbon and hydrogen content are above 50% and 5% respectively which may contribute to high calorific value of Arusha municipal solid waste. The oxygen content is more than 34%. Sulfur content is about 0.29%, which is low compared to those of bituminous coal which is 1.1 wt. % [5].

Table 1. Proximate analysis of municipal solid waste from different areas of Arusha City.

Location	Moisture of received MSW (wt %)	Volatile (wt %) Dry basis	Ash (wt %) Dry basis	Fixed Carbon (wt %)	HHV (MJ/kg)
Ngarenaro	59.67	74.43	8.16	17.41	11.00
Sakina	63.99	84.00	10.00	6.00	11.37
Central Market	55.70	78.30	13.48	8.22	12.7

Table 2. Ultimate analysis of municipal solid waste from different areas of Arusha City.

Location	C (wt %) Dry basis	H (wt %) Dry basis	O (wt %) Dry basis	N (wt %) Dry basis	S (wt %) Dry basis	Cl (wt %) Dry basis	P (wt %) Dry basis
Ngarenaro	55.57	5.38	34.88	2.09	0.31	0.04	0.10
Sakina	55.70	5.29	34.27	2.13	0.22	0.07	0.13
Central Market	53.20	5.24	34.71	2.86	0.37	0.04	0.11

3.2. Calorific Value

The average energy content of municipal solid waste from Arusha city as measured by using a bomb calorimeter was 12MJ/kg on dry basis. This is about 40% of energy contained in coal which is 27MJ/kg and 30% less than energy contained in other biomasses which is about 17MJ/kg [5, 20].This means energy release during combustion of MSW is lower than that of coal and biomass combustion, and therefore, one needs to burn larger amount of MSW to get the same amount of energy. The energy content of MSW can be improved by pre-treating

the MSW so as to reduce the amount of oxygen, since oxygen reduces the energy content of fuel [24]. The MSW can also be co-fired with coal for improving energy content of the feedstock to the combustion plant [20]. Other efficiency processes which convert MSW to energy are pyrolysis, gasification and torrefaction, which are used to produce bio-oil, syngas and char.

4. TGA Curves

This section provides detailed information of the Arusha

MSW obtained by using Thermo-gravimetric Analysis (TGA) and Derivative Thermo-gravimetric analysis (DTG). Thermo-gravimetric Analysis (TGA) is a reliable and widely used laboratory technique employed to study the extent of mass changes due to volatilization and combustion of fuel components. In addition to that, it allows great flexibility in controlling the composition of the combustion gases. The municipal solid waste tested degraded from 79 to about 88 wt. % in the thermo gravimetric analyzer as shown in Figure 5. In this case, the MSW has been burned up in their mixed state. Therefore it is important to study the combustion characteristics of the mixed waste. Figure 5 shows the curves that, gives the burning temperature of the mixed MSW, and from the figure it seems that the burning temperature of MSW

of Arusha city is around 600 °C.

The MSW from the Central market degraded by 88 wt. %, while those from Ngarenaro degraded by 79 wt. %. The mass change of solid wastes was observed at a range between 31 and 34 wt% and from this, the residues were formed. These residues contain fixed carbon and ash. The high amounts of residues were observed in MSW from Ngarenaro which is 18.54 wt. % whereas those from the Central market and Sakina had lower amount of residues of about 13.95 wt % and 12.89 wt % respectively. The char available in the residues can be used as fuel. However, MSW which has high ash content hinder the combustion of char due to the layer of ash that is formed on the surface and which inhibits the diffusion of oxygen into the char [26].

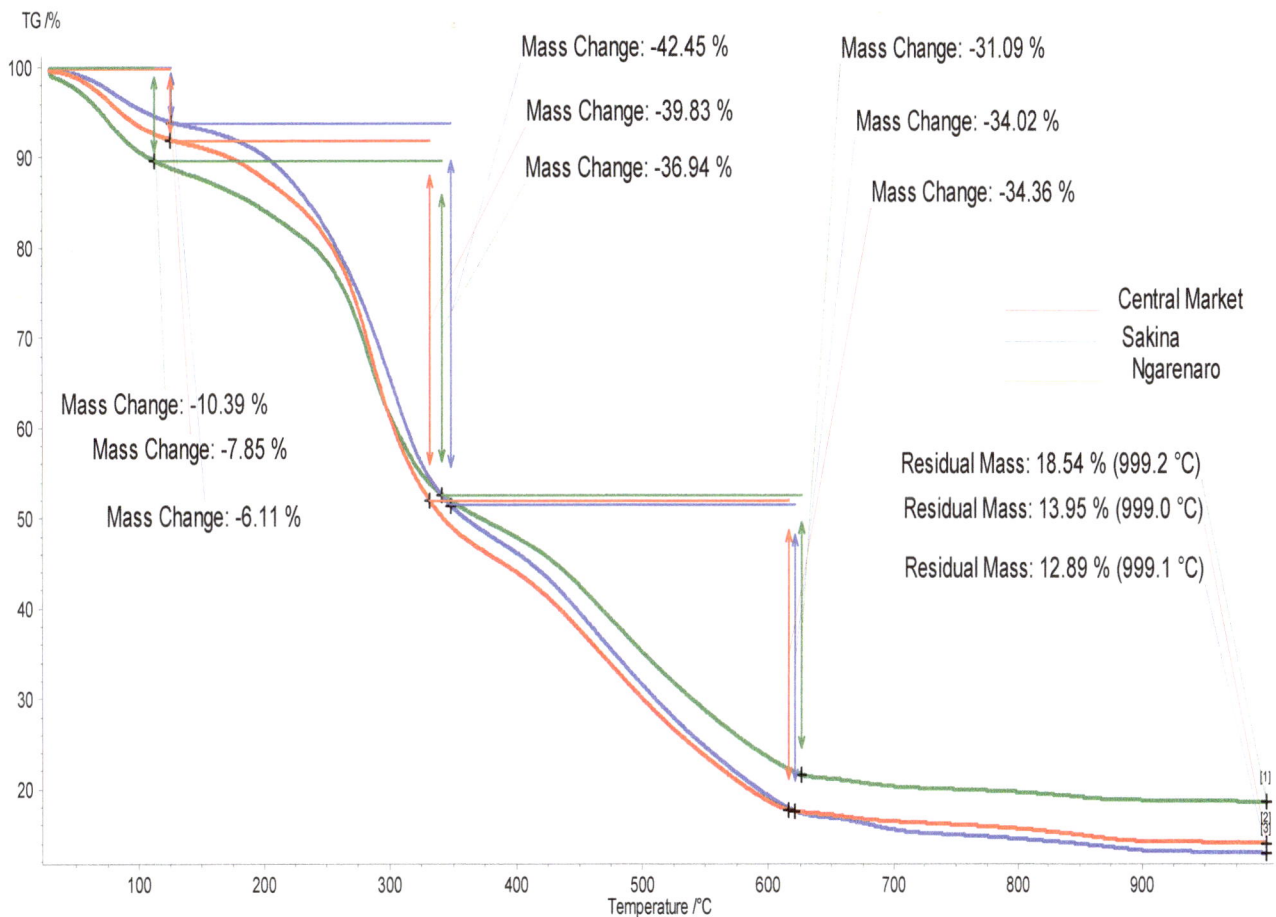

Figure 5. TGA graphs of Municipal solid waste from three locations in Arusha city.

4.1. DTG Curves

Derivative Thermo-gravimetric analysis (DTG) is the analysis which gives the trend of reactivity of particles with the increase of temperature in the furnace. The point of the burning profile in which the maximum weight loss comes about to the combustion is called peak temperature. This point is considered as an indicator of the reactivity of the sample. Another purpose of DTG is to assist in understanding of chemical reactions occurring in the furnace, combustor design, temperature profile of the combustion chamber, retention time and giving information on the change of mass and volume of

the MSW as it travels down the grate.

4.2. Burning Phases of the MSW

Figure 6 shows the derivative of thermo-gravimetric analysis (DTG), which has four visible zones; these are moisture release zone, lignocellulosic degradation zone, plastic degradation zone and char pyrolysis zone [27]. Lai et al [28], also observed and identified these four zones. In the Figure therefore, thermal destruction of solid waste is accomplished in four phases. The first phase is the drying phase which occurs in the initial heating of the solid materials.

Here moisture is driven off as the materials are heated past the evaporation temperature of water and in this case the temperature is around 150°C. The second phase is the volatilization of vapors and gases which occur as the temperature of the waste continues to rise. Vapors and gases diffuse out as their respective volatilization temperature are attained. Here exothemicity property is shown by the waste materials as they are releasing out energy, and temperature is around 380°C to 640°C. This is the zone where the temperature of the particles is greater than the temperature of the environment and is where the particles release energy. The municipal solid waste shows exothemicity property at the devolatilization zone. The devolatilization is that zone where

the temperature of the particles is greater than the temperature of the environment and in this case, it shows that the municipal solid waste can be easily ignited at temperature above 380°C.

The third phase in the burn down of solids is the oxidation of the burnable solids left after the vapors and gases are volatilized where the temperature is between 640°C to 900°C. The fourth phase is the process which involves the final burn down of char and the consolidation and cooling of the inert residues known as ash, the temperature being higher than 900°C. Generally in this case, Figure 6 gives the trend of reactivity of the Arusha municipal solid wastes with temperature.

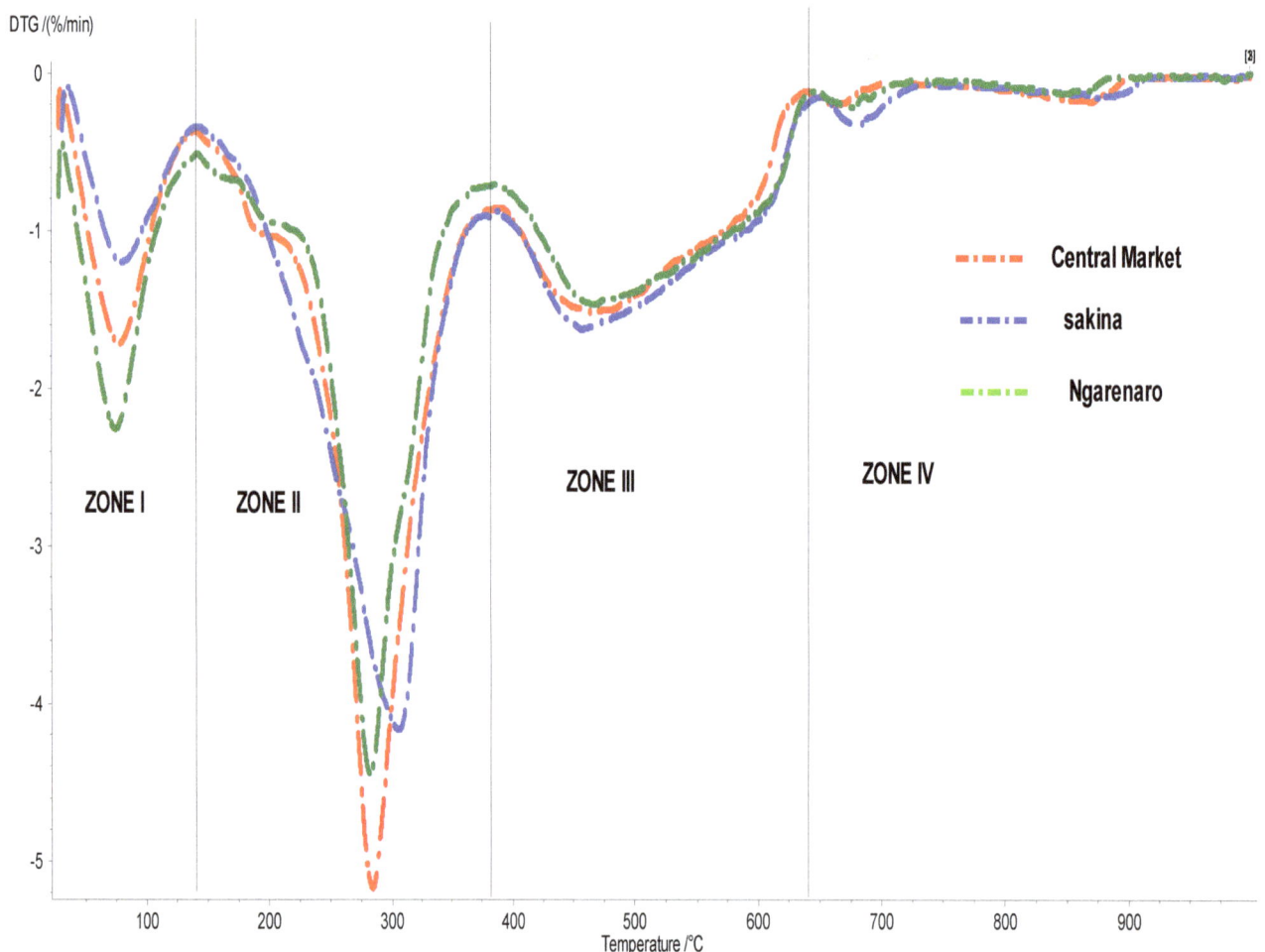

Figure 6. DTG of municipal solid waste from three locations in Arusha City.

5. Conclusion

This paper presents the findings of municipal solid waste characterization of Arusha city. The solid waste from three areas within the city was analyzed to determine the potentiality for energy recovery. The ultimate analysis of the waste showed that the waste contains more than 50% and 5% of carbon and hydrogen respectively which may contribute to high calorific value of Arusha municipal solid waste. The average energy content of waste was 12MJ/kg on dry basis, which is equivalent to 40% of energy contained in coal and 30%

less than energy contained in other biomasses which exhibit that municipal solid waste can be used for energy recovery. The average percentage amount of nitrogen, sulfur, chlorine and phosphorus in the waste were 2.36%, 0.3%, 0.05% and 0.11% respectively which is relatively low. Therefore emissions released by these MSW during combustion are also low. The results obtained from TGA and DTG showed that the MSW has the burning temperature of around 600°C, which means they are suitable for combustion processes. The most reactivity of solid waste materials was obtained from Central market compared to that of Ngarenaro and Sakina. This indicates that MSW from Sakina has higher energy contents compared to others. Thus based on the results in this paper, this municipal solid waste are suitable for energy conversion.

Acknowledgements

The authors wish to thank Arusha City Council for allowing use of their facilities during waste characterization and Arusha Technical College for financial assistance during the study.

References

[1] Breeze R. Municipal Waste Management in Dar es Salaam. Draft Baseline analysis prepared for the World Bank, Washington DC. (2012).

[2] Mundi. Tanzania Urban Population (2011). Accessed at: http://www.indexmundi.com/facts/tanzania/urban-population

[3] World Bank. Poverty Headcount Ratio (2011). Accessed at: http://data.worldbank.org/indicator/SI.POV.DDAY.

[4] Ntakamulenga R. The status of solid waste management in Tanzania. A paper presented during the coastal East Africa on solid waste workshop (2012), Mauritius.

[5] Arthur M et al. Potential of Municipal Solid Waste as renewable energy source: A case study of Arusha- Tanzania. International journal of renewable energy technology research (2014). Vol 6, page 8.

[6] Terra Symbiosis. Exhaustion of fossil fuels. Nature at the heart of human development, (2014)

[7] World Watch Institute. Coal and natural gas consumption and production, (2013)

[8] European Commission, Environmental Data Center. Guidance on municipal Waste data collection (2012) Euro stat – unit E₃.

[9] Imed A Khatib. Municipal Solid Waste Management in Developing Countries. Future Challenges and Possible opportunities

[10] Amin K and Go Su Yang. Identification of the Municipal Solid Waste characteristics and potential of plastic Recovery at Bakri landfill, Muar, Malasya. Journal of Sustainable Development (2012) Vol. 5 No_7.

[11] Ministry of Finance (Government of Tanzania). National Audit Office: A performance audit on the management of solid waste in big cities and region(s) in Tanzania. Mbeya, Dar es Salaam, Mwanza and Arusha (2012).

[12] Ryu C. Potential of Municipal Solid Waste for Renewable Energy Production and Reduction of Greenhouse gas emissions in South Korea, Air and Waste management Association vol. 60 (2010) pages 176-183.

[13] Cheng H. and Hu Y. Municipal solid waste (MSW) as a renewable source of energy. Current and future practices in China, Bioresource Technology, vol. 101 (2010) pages 3816-3824.

[14] Sharholy M.Ahmad K. Mahmood G and Trivedi R. Municipal solid waste management in Indian cities –A review, Waste management vol. 28 (2008).

[15] Alexander K. and Nickolas J M. Energy recovery from Municipal Solid Wastes by gasification. North American Waste to Energy Conference (NAWTEC 11), 11 proceedings, ASME International, Tampa FL (2003). pages 241- 252.

[16] American Society for Mechanical Engineers (ASME).Waste to Energy and materials recovery. An executive summary for a white paper submitted to congress by ASME – SWPD (2007), Washington DC.

[17] Yang N. Zhang H. Chen M. Shao LM. and He PJ. Greenhouse gas emissions from municipal solid waste incineration in China. Impacts of waste characteristics and energy recovery, Waste Management, (2012).

[18] Sushmita Mohapatra. Technological Options for Treatment of Municipal Solid Waste of Delhi. International Journal of Renewable Energy Research (2013). Vol. no_3.

[19] Amin K and Go Su Yang. Energy potential from municipal solid waste in Tanjung Langsat landfill, Johor, Malasya. International Journal of Engineering Science and Technology (2011) vol. 3 no_12.

[20] Surroop D. and Juggurnath A. Investigating the energy potential from co firing coal with municipal Solid Waste. University of Mauritius research journal (2011) vol. 17-2011.

[21] Sveta Angelova, Dilyana Yordanova, Vanya Kyose and Ivan Dombalov. Municipal Waste utilization and disposal through gasification. Journal of Chemical Technology and Metallurgy (2013). vol.2 no_49.

[22] JD Nixon, PK Dey, SK Ghosh and PA Davies. Evaluation of options for energy recovery from municipal solid waste in India using the hierarchical analytical network process, Energy, 2013 - Elsevier (2013).

[23] Eleftheriou P. Energy from waste: A possible alternative energy source for Cyprus municipalities. Journal of Energy Conversion and management (2002) 43, Page 4.

[24] Mohd H and Ridzman Z. Combustion of Municipal Solid Waste in Fixed Bed combustor for energy recovery. Journal of Applied science (2012) 12(11)page 1177.

[25] Inesa B. Agnese L. Maija Z. Alexandr A. Valentin S and Galina T. Effect of main characteristics of pelletized renewable energy resources on combustion characteristics and heat energy production.Chemical Engineering transactions, 29 (2012), pages 901-906.

[26] Masaharu K. Tooru D. Shinya T. Masao T and Takehiro K. Development of new stroker incinerator for Municipal Solid Waste using oxygen enrichment. Mitsubishi Heavy Industries, Technical review (2011) vol. 2.

[27] Yong – hua LI et al. Challenges of power engineering and environment. International conference on power engineering (2007), Hangzhou, China.

[28] Lai Z. Ma X. Tang Y and Lin H. A study on municipal solid waste (MSW) combustion in NO/O_2 and CO_2/O_2 atmosphere from the perspective of TGA, energy, 36 (2): 819 – 824.

5

Optimization of Process Parameters for Biodiesel Production Using Response Surface Methodology

Emmanuel I. Bello[1], Tunde I. Ogedengbe[1], Labunmi Lajide[2], Ilesanmi. A. Daniyan[3, *]

[1]Department of Mechanical Engineering, Federal University of Technology, Akure, Nigeria
[2]Department of Chemistry, Federal University of Technology, Akure, Nigeria
[3]Department of Mechanical & Mechatronics Engineering, Afe Babalola University, Ado-Ekiti, Nigeria

Email address:
afolabiilesanmi@yahoo.com (I. A. Daniyan)
[*]Corresponding author

Abstract: The effect of five process parameters namely: reaction time, reaction temperature, stir speed, catalyst concentration and methanol-oil ratio on the transesterification process of waste frying oil to biodiesel were investigated. Optimization of the five process parameters and their quadratic cross effect was carried out using a four level-five factor central composite experimental design model and response surface methodology with each factor varied over four levels. Taking the biodiesel yield as the response of the designed experiment, the data obtained were statistically analysed to get a suitable model for optimization of biodiesel yield as a function of the five independent process parameters. The optimization produced 30 feasible solutions whose desirability equals to 1 and the selected (most desirable) condition was found to be: reaction time (3 hrs), reaction temperature (58°C), stir speed (305.5 rpm), catalyst concentration (1.4 wt%) and methanol to oil ratio (6:1), while the optimum yield of biodiesel for this condition was found to be 91.6%. The developed model was tested and validated for adequacy by substituting random experimental values as input parameters and the output parameters from the developed model were close to the experimental values. The biodiesel properties were characterized and the results obtained were found to satisfy the standard for both the ASTM D 6751 and EN 14214.

Keywords: Biodiesel, Central Composite Design, Optimization, Transesterification, Yield

1. Introduction

Biodiesel is one of such renewable alternative fuel derived from triglycerides by transesterification of vegetable oils and animal fats (Nie, *et al.*, 2006; Shibasaki-Kitakawa *et al.*, 2007, Aworanti *et al.*, 2013)

Biodiesel is sustainable, renewable, biodegradable, safe to handle and simple to use, environmental friendly, non-toxic, and essentially free of sulphur and aromatics. (Monyem *et al.*, 2001, Highina *et al.*, 2012). Chemically, biodiesel is a fuel composed of mono-alkyl ester of long chain fatty acid derived from vegetable oil or animal fat, designated as B100 and meeting the requirements of ASTM (American Society for Testing and Materials) D 6751 or EN (European Norm) 14214 (Hai, 2002, European Biodiesel Board, 2006). It can

be used either in the pure form (B100) or as blends with fossil diesel in diesel engines (Canacki and Van Gerpen, 2005; Basiron and May, 2005). According to Bello (2008), biodiesel can be produced from used frying oil, coconut oil, palm oil etc. however high demand of diesel fuel and the availability of waste cooking oil indicate that biodiesel from used oil cannot completely replace fossil diesel fuel but can contribute to reduce the dependency on petrol based diesel (Martin and Grossman, 2011).

2. Physical Experiment

Waste cooking oil was obtained from Afe Babalola University, Ado Ekiti restaurants in Ado Ekiti, Nigeria and was filtered and pre-heated to remove impurities. Reagents namely methanol (99.8% purity), hydrogen chloride (HCl), sodium hydroxide (NaOH) and anhydrous sodium sulphate (Na$_2$SO$_4$) of analytical grade were purchased from Brilliant Chemical store in Lagos, Nigeria. Methanol was the choice of alcohol because it is cheap and it is a short chain alcohol that reacts faster. The catalyst employed for both acid and alkali transesterification were hydrogen chloride (HCl) and sodium hydroxide (NaOH) prepared at different concentration ranging from 1.0wt% - 2.0wt%. Acid transesterification was used for converting used cooking oil to biodiesel using 99.8% pure methanol, HCl and NaOH as catalyst. Waste cooking oil was filtered and pre-heated at a temperature of 100°C to remove water and other volatile impurities. HCl at a concentration of 1.4 wt% was added to methanol with methanol in excess in a molar ratio of 6:1 to oil. The mixture was thoroughly stirred in the mixing tank for 1 hour to form methanolic HCl. The pre-heated oil was thereafter transferred to the reactor alongside with methanolic HCl and the mixture was stirred at 305 rpm and a temperature of 60°C for 2 hours. The product was discharged into the separating funnel and allowed to settle for 24 hours. After the settling, the product is observed to have separated into two distinct layers. The top layer being a mixture of water and methanol while the bottom layer the transesterified oil. This is the first process aimed at breaking the long chain of fatty acid from used oil. This process is followed by alkali transesterification. The transesterified oil was transferred back into the reactor where prepared sodium hydroxide (NaOH) at a concentration of 1.4 wt% and methanol at a molar ratio of 6:1 to oil (Methanol in excess to oil) forming sodium methoxide was added. Continuous stirring of the mixture was done in the reactor at 305 rpm at 60°C for 3 hours. The product was thereafter discharged into the separating funnel and allowed to settle for 24 hours after which the products separated into two distinct layers. The lighter biodiesel at the top and the heavier glycerol at the bottom. The methyl ester (biodiesel) was then washed with distilled water at a volume ratio of 3:1 by stirring gently. The methyl ester was dried by passing it through anhydrous sodium sulphate (Na$_2$SO$_4$). The dried biodiesel was stored in a refrigerator to prevent oxidiation.

Waste cooking oil was obtained from Afe Babalola University, Ado Ekiti restaurants in Ado Ekiti, Nigeria and was filtered and pre-heated to remove impurities. Reagents namely methanol (99.8% purity), HCl, Sodium hydroxide, anhydrous sodium sulphate of analytical grade were purchased from Brilliant Chemical store in Lagos, Nigeria. Methanol was the choice of alcohol because it is cheap and it is a short chain alcohol that reacts faster. The catalyst employed for both acid and alkali transesterification, HCl and sodium hydroxide was prepared at different concentration ranging from 1.0 wt% - 2.0wt%. Acid transesterification was used for converting used cooking oil to biodiesel using 99.8% pure methanol, HCl and NaOH as catalyst. Waste cooking oil was filtered and pre-heated at a temperature of 100°C to remove water and other volatile impurities. HCl at a concentration of 1.4 wt% was added to methanol with methanol in excess in a molar ratio of 6:1 to oil. The mixture was gently stirred in the mixing tank for 10

minutes to form methanolic HCl. The pre-heated oil was thereafter transferred to the reactor alongside with methanolic HCl and the mixture was stirred at 305 rpm and a temperature of 60°C for 3 hours. The product was discharged into the separating funnel and allowed to settle for 24 hours. After the settling, the product separated into two distinct layers. The top layer being a mixture of water and methanol while the bottom layer the transesterified oil. This is the first process aimed at breaking the long chain of fatty acid from used oil. This process is followed by alkali transesterification. The transesterified oil was transferred back into the reactor where prepared sodium methoxide was added. NaOH is at a concentration of 1.4 wt% and methanol at a molar ratio of 6:1 (Methanol in excess to oil). Continuous stirring of the mixture was done in the reactor at 305 rpm at 60°C for 3 hours. The product was thereafter discharged into the separating funnel and allowed to settle for 24 hours after which the products separated into two distinct layers, the lighter biodiesel at the top and the heavier glycerol at the bottom. The methyl ester (biodiesel) was then washed with distilled water at a volume ratio of 3:1 by stirring gently. The methyl ester was dried by passing it through anhydrous sodium sulphate (Na$_2$SO$_4$). The dried biodiesel was stored in a refrigerator to prevent oxidiation.

3. Numerical Experiment

Modelling and Optimization of the production process variables were carried out with Design-Expert® (version 7) software for experiment design using a four-level-five factor central composite design model and response surface methodology to study the effect of independent variables such as reaction time (hours), temperature (°C), stir speed (rpm), catalyst concentration (wt%) and methanol-oil ratio on the biodiesel yield with the response in terms of percentage yield. The Response Surface Methodology (RSM) is a viable statistical tool for the optimization of process variables as it simplifies complex nature of many experimental runs. Besides, it studies the interactive effect of two or more variables and the effect on the response (target). The following input process parameters were varied.

i. Reaction time: 1-5 hours.
ii. Reaction temperature: 40-90°C.
iii. Stir Speed: 200-400 rev/min.
iv. Catalyst concentration: 1-2 wt%.
v. Methanol to Oil ratio: 4:1-9:1.

4. Central Composite Design

The Central Composite Design is the most widely used RSM model. Response surface methodology (RSM) is a mathematical/statistical based technique which is useful for analyzing the effects of several independent variables on the response (Box and Drapper, 1987; Enweremadu and Rutto, 2015). It is used to investigate the quadratic cross effect of five input process parameters namely: time, temperature, stir speed, catalyst concentration and methanol-oil ratio on biodiesel yield. Each numeric factor is varied over 4 levels: plus and minus alpha (axial point). Plus and minus 1 (factorial point). Uosukaimen *et al.,* (1999), Ghadge and Raheman *et al.,* (2006), Kansedo *et al.,* (2009), Salamatina *et al.,* (2013 a & b), Jeong and Part (2009), Jeong and Park *et al.,* (2009), Silva *et al.,* (2006), Aworanti *et al.,* (2013) and Enweremadu and Rutto (2015),

were some of the authors who have investigated and reported the optimization of biodiesel production using the response surface methodology. However, the quadratic cross effect of five process parameters has not been investigated.

Table 1. *Numeric Factors and Levels.*

s/n	Factor	Name	Unit	-1 Level	+1 Level	-alpha	+alpha
1.	A	Reaction time	hours	1	5	0.00930244	5.9907
2.	B	Reaction temperature	°C	40	90	27.6163	102.384
3.	C	Stir speed	rpm	200	400	150.465	449.535
4.	D	Catalyst concentration	wt%	1	2	0.752326	2.24767
5.	E	Methanol to oil ratio		4	9	4.74302	0.256977

Table 1 shows the input values for process parameters denoted as numeric factors over 4 levels. This generated a runs of 15 experiments and the data obtained was statistically analysed with Design-Expert® (version 7) software to get a suitable model for biodiesel yield (%) as a function of the independent variables. The biodiesel yield is taken as the response of the designed experiment for the

transesterification process.

Table 2 summarizes the experiment designed in terms of study type which is response surface using central composite as initial design and a quadratic design model. The numbers of experimental runs was 50. From Table 2, the mean and standard deviation for each process variable as well as biodiesel yield was calculated.

Table 2. *Summary of Design for Numerical Experimentation Analysis.*

Design Summary			
Study type	Response surface	Runs	50
Initial design	Central composite	Blocks	No blocks
Design model	Quadratic		

Factor	Name	Units	Type	Low Actual	High Actual	Low Coded	High Coded	Mean	Std. Dev.
A	Time	hours	Numeric	1.00	5.00	-1.000	1.000	3.000	1.708
B	Temperature	°C	Numeric	4.00	90.00	-1.000	1.000	65.000	21.352
C	Stir speed	rpm	Numeric	200.00	400.00	-1.000	1.000	300.000	85.407
D	Catalyst concentration	w%	Numeric	1.00	2.00	-1.000	1.000	1.500	0.427
E	Methanol-oil ratio		Numeric	4.00	9.00	-1.000	1.000	6.500	2.135

Response	Name	Units	Obs.	Analysis	Minimum	Maximum	Mean	Std. Dev.	Ratio	Trans	Model
Y1	Yield	%	15	Polynomial	70.000	91.600	84.140	5.916	1.300	None	None

Mathematical model was formulated from each response (Biodiesel yield) which correlates the response (Yield) to the process variables through first and second order as well as interactive terms according to equation 1

$$Y = \beta_0 + \sum_{i=1}^{k} \beta_i x_i + \sum_{i>j}^{k} \sum_{j}^{k} \beta_{ij} x_i x_j \qquad (1)$$

where

Y=Response (Biodiesel Yield)

$\beta_{i=}$ Linear regression coefficient

β_{ij}=Quadratic coefficient when i=j and interactive effect coefficient when $i \neq j$

β_0 =Regression coefficient

x_i, x_j = Independent process variables (uncoded)

k=Number of factors optimized in the experiment

Biodiesel was also produced from the developed pilot plant at optimum transesterification process condition. Yield of biodiesel from used and ununsed oil from transesterification process is calculated as shown in Table 3. Table 3 shows the feasible combinations of the five process variables with their yield determined experimentally for the first 15 runs according to equation 2

$$Yield = \frac{Weight\ of\ biodiesel}{oil\ weight} \times 100\% \qquad (2)$$

Table 3. Process Variables and Yield.

s/n	Standard deviation	Block	A: Reaction time (hrs)	B: Reaction temperature (°C)	C: Stir speed (rpm)	D: Catalyst concentration (wt. %)	E: Methanol to Oil	Yield (%)
1	29	1	1.00	40.00	400.00	2.00	9.00	88.8
2	13	1	1.00	40.00	400.00	2.00	4.00	89.9
3	18	1	5.00	40.00	200.00	1.00	9.00	90
4	8	1	5.00	90.00	400.00	1.00	4.00	81.5
5	36	1	3.00	102.38	300.00	1.50	6.50	80
6	31	1	1.00	90.00	400.00	2.00	9.00	84.2
7	25	1	1.00	40.00	200.00	2.00	9.00	91
8	10	1	5.00	40.00	200.00	2.00	4.00	70
9	26	1	5.00	90.00	200.00	2.00	9.00	78
10	14	1	5.00	40.00	400.00	2.00	4.00	86.7
11	9	1	1.00	40.00	200.00	2.00	4.00	90.2
12	45	1	3.00	65.00	300.00	1.50	6.50	78
13	16	1	5.00	90.00	400.00	2.00	4.00	90
14	4	1	5.00	90.00	200.00	1.00	4.00	82.8
15	32	1	5.00	90.00	400.00	2.00	9.00	81

5. Statistical Analysis

Using response surface quadratic model, the values of standard error, Variance Inflation Factor (VIF), and standard deviation were gotten for possible combination of process parameters as shown in Table 4.

Table 4. Design Matrix Evaluation for Response Surface Quadratic Model.

s/n	Coded Factor	Actual Factor	Standard Error	VIF	0.5% Standard deviation	1% Standard Deviation	2% Standard deviation
1	A	Time	0.17	1.00	30.9	83.1	99.9
2	B	Temperature	0.17	1.00	30.9	83.1	99.9
3	C	Stir speed	0.17	1.00	30.9	83.1	99.9
4	D	Catalyst concentration	0.17	1.00	30.9	83.1	99.9
5	E	Methanol-oil ratio	0.17	1.00	30.9	83.1	99.9
6	AB	Time × temperature	0.18	1.00	27.7	78.0	99.9
7	AC	Time × stir speed	0.18	1.00	27.7	78.0	99.9
8	AD	Time × catalyst concentration	0.18	1.00	27.7	78.0	99.9
9	AE	Time × methanol-oil ratio	0.18	1.00	27.7	78.0	99.9
10	BC	Temperature × stir speed	0.18	1.00	27.7	78.0	99.9
11	BD	Temperature × catalyst concentration	0.18	1.00	27.7	78.0	99.9
12	BE	Temperature × methanol-oil ratio	0.18	1.00	27.7	78.0	99.9
13	CD	Stir speed × catalyst concentration	0.18	1.00	27.7	78.0	99.9
14	CE	Stir speed × methanol-oil ratio	0.18	1.00	27.7	78.0	99.9
15	DE	Catalyst concentration × methanol-oil ratio	0.18	1.00	27.7	78.0	99.9
16	A^2	Time^2	0.29	1.31	38.0	91.1	99.9
17	B^2	Temperature^2	0.29	1.31	38.0	91.1	99.9
18	C^2	Stir speed^2	0.29	1.31	38.0	91.1	99.9
19	D^2	Catalyst concentration^2	0.29	1.31	38.0	91.1	99.9
20	E^2	Methanol-oil ratio^2	0.29	1.31	38.0	91.1	99.9

Table 4 shows, the five process parameters namely: reaction time, temperature, stirs speed, catalyst concentration and methanol-oil ratio are represented by variables A, B, C, D and E. The polynomial analysis and errors were also calculated. The standard errors are similar within type of coefficient. The Smaller values are most significant. The Ideal Variance Inflation Factor (VIF) is 1.0, the result obtained is significant. VIF's above 10 depicts poor model indicating coefficients are poorly estimated due to multicollinearity. Also, Ideal R-squared is 0.0, high R-squared means terms are correlated with each other, possibly leading to poor models. The predictive capability and significance of the model is further validated with the value of R-Squared (0.9895) and the Adjusted R-Squared (0.9263) close to 1. R-squared is a multiple correlation coefficient indicating the degree of relationship of the response variable to the combined linear predictor variable. The correlation coefficient values being a measure of goodness of fit of the model indicates high degree of correlation between the observed and predicted values (Aworanti, *et al.,* 2013)

The final equation in terms of coded factors is given as:

$$Yield = 78.00 + 6.59*A + 1.34*B - 0.10*C + 14.36*D - 2.56*E - 5.10*A*B + 4.49*A*C - 12.83*A*D$$
$$+ 4.71*A*E - 3.96*B*C + 2.18*B*D - 7.19*B*E$$

(3)

The desirability equals to 1 indicating the significance of the process variables to amount of yield and viability of the developed model for prediction.

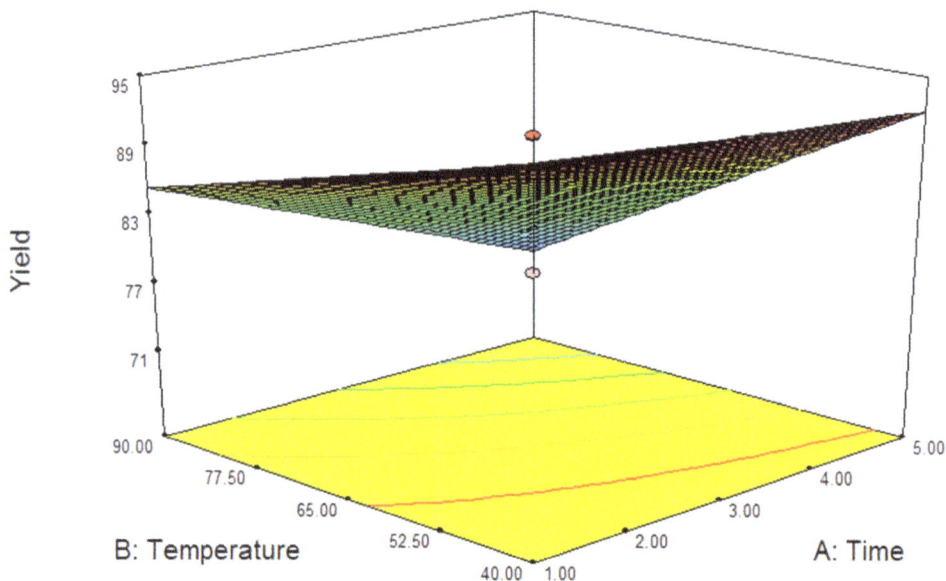

Figure 1. Effect of Interaction of Time and Temperature on Biodiesel Yield.

Figure 1 is a 3D Response Surface plot of the interaction effect of time and temperature on biodiesel yield when catalyst concentration is 1.50, methanol-oil ratio: 6.50 and stir speed: 300 rpm. From Figure 1, biodiesel yield increased significantly with increase in temperature and reaction time until a sharp decrease in yield above 58°C and 3 hours. This may be due to evaporation of methanol or reverse glycerolysis.

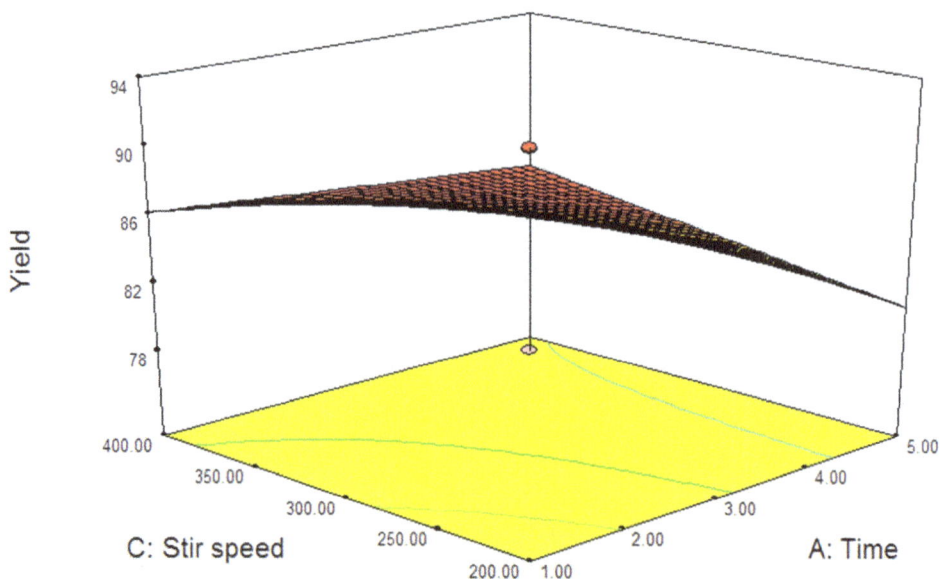

Figure 2. Effect of Interaction of Time and Stir Speed on Biodiesel Yield.

Figure 2 is a 3D response surface plot of the interaction effect of time and stir speed on biodiesel yield when catalyst concentration is 1.50, methanol-oil ratio: 6.50 at 65°C. From Figure 2, biodiesel yield increased significantly with increase in reaction time and stir speed until a decrease in yield above 3 hours and 305.5 rpm. This may be due to excessive agitation.

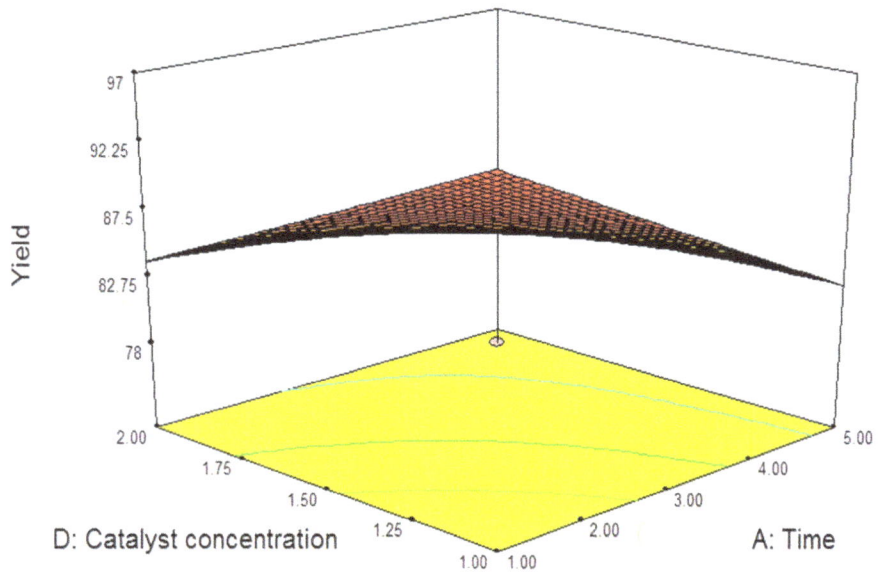

Figure 3. *Effect of Interaction of Time and Catalyst Concentration on Biodiesel Yield.*

Figure 3 is a 3D response surface plot of the interaction effect of time and catalyst concentration on biodiesel yield when methanol-oil ratio is 6.50, stir speed: 300 rpm at 65°C. Biodiesel yield increases with increase in catalyst concentration. Addition of excessive catalyst favours saponification reaction and reduces biodiesel yield (Goyal *et al.,* 2012).

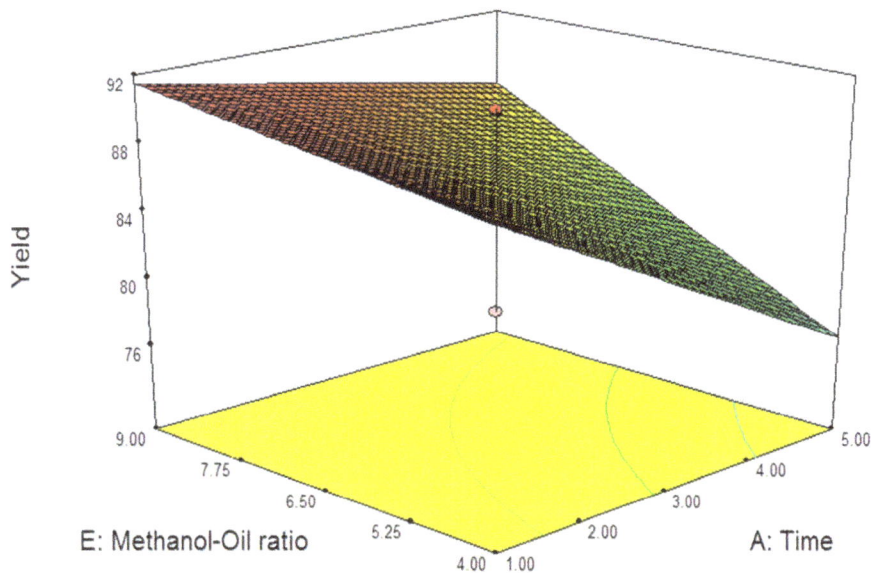

Figure 4. *Effect of Interaction of Time and methanol-oil ratio on Biodiesel Yield.*

Figure 4 is a 3D response surface plot of the interaction effect of time and methanol-oil ratio on biodiesel yield when catalyst concentration is 1.50, stir speed: 300 rpm at 65°C. Methanol-oil ratio of 6:1 (methanol in excess) for 3 hours brought about optimum yield of biodiesel while ratio greater than 6:1 brought about significant decrease in yield. Too much methanol reduces the flash point thus eroding an important advantage of biodiesel.

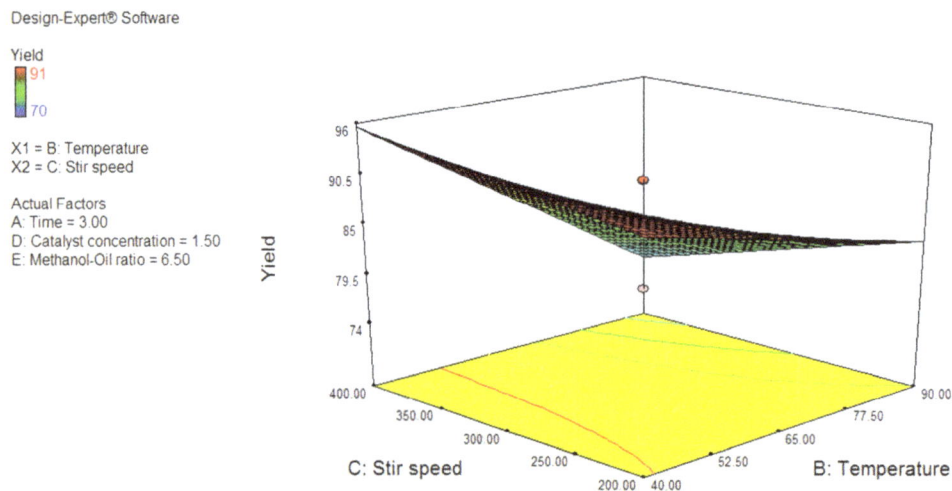

Figure 5. Effect of Interaction of Temperature and Stir speed on Biodiesel Yield.

Figure 5 is a 3D response surface plot of the interaction effect of temperature and stir speed on biodiesel yield when catalyst concentration is 1.50, methanol-oil ratio: 6.50 for 3 hours. Excessive agitation causes splashing and the mixture tend to foam which may result in cavitation corrosion.

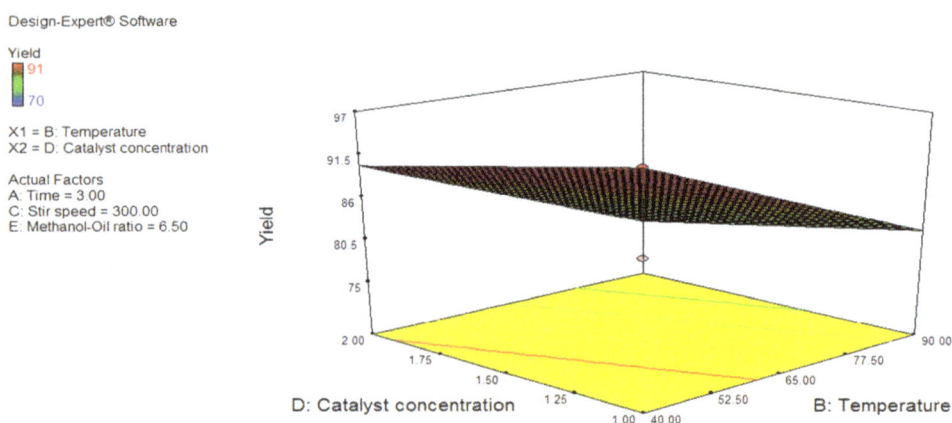

Figure 6. Effect of Interaction of Temperature and Catalyst Concentration on Biodiesel Yield.

Figure 6 is a 3D response surface plot of the interaction effect of temperature and catalyst concentration on biodiesel yield when methanol-oil ratio is 6.50 for 3 hours at a stir speed of 300 rpm. Excessive catalyst beyond 1.4 wt % decreases biodiesel yield as soap may be formed which prevents ester layer formation.

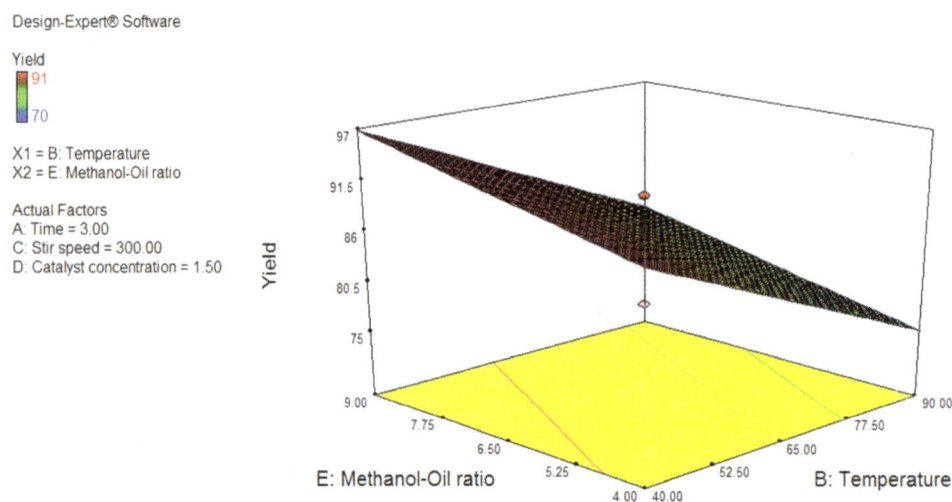

Figure 7. Effect of Interaction of Temperature and Methanol-oil ratio on Biodiesel Yield.

Figure 7 is a 3D response surface plot of the interaction effect of temperature and methanol-oil ratio on biodiesel yield when catalyst concentration of 1.50 for 3 hours at a stir speed 300 rpm. Biodiesel yield increases with increase in methanol oil ratio up to 6:1 after which the yield decreases with excessive temperature.

6. Result and Discussion

Taking the biodiesel yield as the response of the designed experiment, the data obtained were statistically analysed to get a suitable model for optimization of biodiesel yield as a function of the five independent process parameters. The optimization produced 30 feasible solutions whose desirability equals to 1 and the selected condition (most desirable) was found to be: reaction time (3 hrs), reaction temperature (58°C), stir speed (305.5 rpm), catalyst concentration (1.4 wt%) and methanol to oil ratio (6:1), the optimum yield of biodiesel for this condition was found to be 91.6%.

The developed model was tested and validated for adequacy by substituting random values (from Table 3) as input parameters and the output parameters from the developed model were close to the observed values from Table 3. The regression model was found to be highly significant at 95% confidence level as correlation coefficients for R-Squared (0.9895), adjusted R-Squared (0.9263) and predicted R-Squared (0.9653) was very close to 1 for variables CE, B, C and CD. This is an indication that very small deviation exists between the actual and predicted values. From the coefficient of correlation R (0.9947), the model is highly significant as it explains about 99.47% variance resulting from the interaction of process variables with negligible 0.53% which cannot be correlated. Since there is good agreement among values of correlation coefficient R, R-squared and predicted R-Squared, the selected model is adequate to describe the actual value as a more complicated model is not needed.

7. Conclusion

The following conclusions were drawn after the successful completion of this work:
i. The results obtained showed that used frying oil can be transesterified to biodiesel which solves oil disposal problems in restaurants and converts waste to wealth.
ii. A reaction time of 3 hours is sufficient to drive the transesterification reaction to completion with an optimum yield of 91.6%
iii. Increase in temperature speeds up the rate of transesterification reaction and reduces the reaction time for the conversion of oil to biodiesel. Excessive increase in temperature above the evaporating temperature of methanol will subsequently reduce yield due to gradual evaporation of methanol. Hence, the higher the temperature up to 58°C, the higher the

biodiesel yield, the lower the temperature the lower the biodiesel yield. When the temperature is above 60°C, lower yield of biodiesel is obtained, because of evaporation of methanol.
iv. The higher the stir speed up to 305.5 rpm, the higher the biodiesel yield, the lower the stir speed, the lower the biodiesel yield. This is because reaction occurs at the interface between the oil and methanol. Excessive increase in stir speed may cause splashing of raw materials due to excessive agitation hence decreasing the biodiesel yield.
v. The optimum catalyst concentration was 1.4% wt. Soap and gel may be formed when catalyst amount increases beyond the optimum, this prevents ester layer separation.
vi. The regression model was found to be highly significant at 95% confidence level as correlation coefficients for R-Squared (0.9895), adjusted R-Squared (0.9263) and predicted R-Squared (0.9653) was very close to 1 hence an indication of good correlation and predictive capabilities.

References

[1] Aworanti, O. A., S. E. Agarry and A. O. Ajani (2013). Statistical Optimization of Process Variables for Biodiesel Production from Waste Cooking Oil Using Heterogeneous Base Catalyst. British Biotechnology Journal. 3(2)116-132.

[2] Basiron, Y. and May, C. Y. (2005). Crude Palm Oil as a Source of Biofuel, Malaysian Palm Oil Board, Malaysia, Technical Report.

[3] Bello, E. I. (2008). Evaluation of Coconut Oil Methyl Esters as an Alternative Fuel for Diesel Engine. Ph.D thesis. Federal University of Technology, Akure.

[4] Box G. E. P. and N. R. Draper. (1987). Empirical Model-building and Response Surfaces, John Wiley and Sons, New York, p.663.

[5] Canacki, M. and Van Gerpen, J. (2005). "A Pilot Plant to Produce Biodiesel from High Fatty Acid Feedstock".

[6] Enweremadu, C. C. and Rutto, H. L. (2015). Optimization and Modelling of Process Variables of Biodiesel Production from Manula Oil Using Response Surface Methodology. (2015). Journal of Chemical Society, Pakistan, 37(2)256-265.

[7] European Biodiesel Board (2006). Biodiesel Chains: Promotng Faourable Conditions to Establish Biodiesel Market Actions WP 2 "Biodiesel Market Status" Deliverable 7: EU-27 Biodiesel Report.

[8] Ghadge S. V. and H. Raheman (2006), Optimization of Biodiesel Production by Sunflower Oil Transesterification, *Bioresour. Technol.*, 97, 379.

[9] Goyal, P., M. P. Sharma and S. Jain (2012). Optimization of Esterification and Transesterification of High Free Fatty Acid Jatropha Curcas Oil Using Response Surface Methodology. Journal of Petroleum Science Research, 1(3)36-43.

[10] Hai, T. C. (2002). The Palm Oil Industry in Malaysia, WWF, Malaysia.

[11] Highina, B. K., I. M. Bugaje and B. Umar (2012). Liquid Biofuel as Alternative Transport Fuel in Nigeria. Int. Journal of Petroleum Technology Development Vol. 1, pp 1-15.

[12] Jeong G. T. and D. H. Park. (2009). "Optimization of Biodiesel Production from Castor Oil Using Response Surface Methodology," Appl. Biochem Biotechnol., vol. 156, pp. 431–441.

[13] Jeong, G. T. H., S. Yang and D. H. Park. (2009). "Optimization of Transesterification of Animal Fat Ester Using Response Surface Methodology," Bioresour Technol., vol. 100, pp. 25–30.

[14] Kansedo, J. K. T. Lee and S. Bhatia. (2009). Biodiesel Production from Palm Oil via Heterogeneous Transesterification, Biomass Bioenergy, 33, 271.

[15] Martin M. and I. E. Grossman (2011). Optimization of Heat and Water Integration for Biodiesel Production from Cooking Oil and Algae. pp. 1-38

[16] Monyem A., J. H. Van Gerpen and Canacki (2001). The Effect of Timing of Oxidation on Emission from Biodiesel Fuelled Engines. Transactions of the ASAE 44(1) pp. 35- 42.

[17] Nie K, Xie F, Wang F, Tan T. (2006). Lipase catalyzed methanolysis to produce biodiesel optimization of the biodiesel production. J. Mol. Catalysis B: Enzymatic. 43:142-17.

[18] Salamatinia, B., I. Hashemizadeh and A. Z. Abdullah. (2013). Alkaline Earth Metal Oxide Catalysts for Biodiesel Production from Palm Oil: Elucidation of Process Behaviors and Modeling using Response Surface Methodology, Iranian J. Chemistry Chem. Eng. , 32.

[19] Salamatinia, B., I. Hashemizadeh and A. Z. Abdullah. (2013). Intensification of Biodiesel Production from Vegetable Oils using Ultrasonic-assisted Process: Optimization and Kinetics, Chem. Eng. Process.: Process Intensification, 73.

[20] Shibasaki-Kitakawa N, Honda H, Kuribayashi H, Toda T, Fukumura T, Yonemoto T. (2007). Biodiesel production using anionic ion-exchange resin as heterogeneous catalyst. Bioresour. Technol. 98:416-421.

[21] Silva, N. D. L. D., M. R. W. M Maciel, C. B. Batistella and R. M. Filho. (2006). "Optimization of Biodiesel Production from Castor oil," Appl Biochem Biotech., vol. 130, pp. 405–414. Uosukainen, E., M. Lamsa, Y. Y. Linko, P. Linko and M. Leisola. (1999). Optimization of Enzymatic Transesterification of Rapeseed Oil Ester using Response Surface and Principal Component Methodology, *Enzyme Microb. Technol.*, 25, 236.

6

Optimisation of acid hydrolysis in ethanol production from prosopis juliflora

Temesgen Atnafu Yemata[1,2]

[1]Institute of technology, Department of Chemical Engineering, Addis Ababa, Ethiopia
[2]Addis Ababa University, Chemical Engineering Department, Addis Ababa, Ethiopia

Email address:
atnafutemesgen16@gmail.com

Abstract: Lignocellulosic materials (eg.Prosopis juliflora) can be utilized to produce ethanol, a promising alternative energy source for the limited crude oil. This study involved optimization of acid hydrolysis in ethanol production from prosopis juliflora. The conversion of prosopis juliflora to ethanol can be achieved mainly by three process steps: pretreatment of prosopis juliflora wood to remove lignin and hemicellulose, acid hydrolysis of pretreated prosopis juliflora to convert cellulose into reducing sugar (glucose) and fermentation of the sugars to ethanol using Saccharomyces cerevisiae in anaerobic condition. A two level full factorial design with four factors, two levels and two replicas ($2^4*2=32$ experimental runs) was applied to optimize acid hydrolysis and study the interaction effects of acid hydrolysis factors, namely, acid concentration, solid fraction, temperature, and time. An optimization was carried out to optimize acid hydrolysis process variables so as to determine the best acid concentration, solid fraction, temperature, and contact time that resulted maximum ethanol yield. The screening of significant acid hydrolysis factors were done by using the two-level full factorial design using design expert® 7 software. The statistical analysis showed that the ethanol yield of (40.91% (g/g)) was obtained at optimised acid hydrolysis variables of 0.5%v/v acid concentration, 5%w/w solid fraction, 105.01°C temperature, and 10 minutes hydrolysis time.

Keywords: Prosopis Juliflora, Pretreatment, Hydrolysis, Fermentation, 2 Level Factorial, Optimization

1. Introduction

Oil prices are at all times high and there is growing zest to reduce our dependence on oil. It is finite resource, gas supplies and oil reserves are shrinking, will definitely run out in the future. World energy demand is expected to double by 2050 as it is shown in figure 1.1 below. The demand of energy is currently exponentially exceeding the rate of local supply sources, a look beyond the fossils is crucial for long term economic growth and energy security purpose. The volatile situations in the Middle East, where vast reservoirs are, are also creating uncertainties about the availability of the supply. There is also the greater environmental risks associated with exploitation of crude oil (IEA world energy outlook, 2004).

With the diminishing supply of petroleum oil and the political instability in countries where much of the world's oil reserves are found, the prices of petroleum-based fuels are irreversibly going up. As a result of concerns of sustainability, environmental protection, and national energy security, more

and more countries have prioritized the importance of renewable energy sources. Ethanol has once again become attractive in the energy marketplace and, in fact, the demand for ethanol has been increasing in recent years (Lin and Tanaka, 2006; Ford, 2004).

Ethanol as well as other bio-fuels produced from plant biomass is alternative to fossil fuels. Ethanol does not add to a net carbon dioxide atmospheric increase thus there is in theory no contribution to global warming. Combustion of ethanol results in relatively low emissions of volatile organic compounds, carbon monoxide and nitrogen oxides. Ethanol was used as transportation fuel at the beginning of 20th century in the U.S., but it was abandoned for fuels processed from petroleum (oil) after World War II because these were cheaper and had higher energy values (Lin and Tanaka, 2006; Ford, 2004). During the last two decades, technology for ethanol production from non-food-plant sources has been developed to the point at which large-scale production could be a reality in the next few years (Mosier, N., et al., 2005). Moreover, agronomic residues such as corn stover (corn cobs and stalks), sugar cane waste, wheat or rice straw, forestry and paper mill discards, the paper portion of municipal waste, and mainly dedicated energy crops collectively termed 'biomass' can be converted to fuel ethanol (Divya Paruchuri ,December 2008).

1.1. Statement of the Problem

Ethiopia is currently looking at growing high-yielding crops for the production of bio-fuels as alternatives to traditional fuels (petrol and diesel) to address imminent shortages and reduce impacts of climate change. Owing to such phenomenon, and indeed in view of the recent trends in the escalating price of the traditional petro-fuel, biofuel has been gaining greater attention by the Ethiopian government. But due to the increased cost of food crops, producing ethanol using Prosopis juliflora wood is an alternative feed stock: for one thing, Prosopis juliflora is a fast growing tree species and grows in Ethiopia mainly in arid and semi-arid areas of the Rift Valley. And the other reason is it is a highly invasive exotic tree that is spreading in the pastoralist areas of Ethiopia making vast areas of land unavailable for grazing and it is becoming difficult to remove it.Thirdly, when the plant is cut, new off springs is grown from the root in a short period (Hailu Shiferaw et al., 2004). Invasion of rangelands by Prosopis juliflora also caused shortage of grazing land for livestock, which resulted in drastic reduction of livestock number as well as product; thorns damage eyes and hooves of camels, donkeys, and cattle then by poisoning eventually lead to death. Prosopis juliflora is invading potential croplands forcing local farmers with less capital and machinery to abandon their farmland and settlement. In general, this is a matter of serious concern for the life of the local people as pastoralists depending on livestock for their livelihood (Senayit et al., 2004). Due to the above reasons and as Prosopis juliflora is widely available in Ethiopia; we can use Prosopis juliflora as Ethanol feed stock.

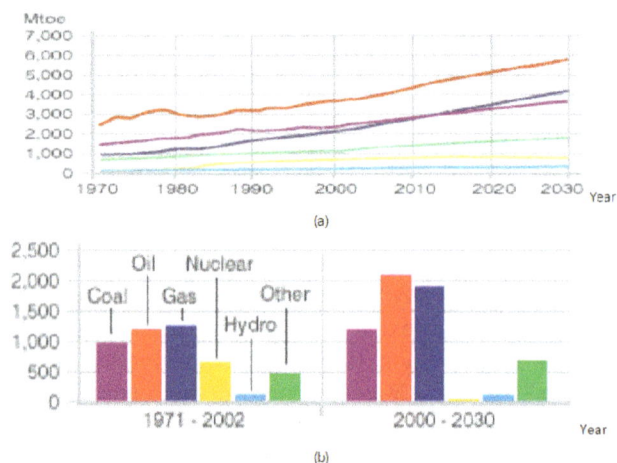

Fig. 1.1. Illustration of Projected World Energy Demand (a) projected world energy demand and (b) Increase in world primary energy demand by fuel (IEA world energy outlook, 2004).

MTOE= Million Tones oil equivalent. Method of assessing calorific value of different sources of energy in terms of one tone of oil

2. Lignocellulosic Biomass as Ethanol Feedstock

Lignocellulosic biomass refers to plant biomass that is composed of cellulose, hemicellulose, and lignin. The carbohydrate polymers (cellulose and hemicelluloses) are tightly bound to the lignin. Lignocellulosic biomass can be grouped into four main categories: agricultural residues (including corn stover and sugarcane bagasse), dedicated energy crops, wood residues (including sawmill and paper mill discards), and municipal paper waste(Wikipedia, the free encyclopedia).

2.1. Composition of Lignocellulosic Materials

Cellulose: is a linear polymer of D-glucose units linked by β-1, 4-linked glucose. Cellulose molecules are completely linear and have a strong tendency to form intra and intermolecular hydrogen bond.

Hemicellulose: Hemicelluloses are heterogeneous polymers of pentoses (xylose, arabinose), hexoses (mannose, glucose, galactose), and sugar acid (Saha et al., 1997).

Lignin: is a long-chain, heterogeneous polymer composed largely of phenyl propane units most commonly linked by ether bonds (Saha et al., 1997).

Extractives: are woody compounds that are soluble in neutral organic solvents or water. The extractives usually represent a minor fraction (between 1-5%) of lignocellulosic materials. They contain a large number of both lipophilic and hydrophilic constituents. The extractives can be classified in four groups: (a) terpenoids and steroids, (b) fats and waxes, (c) phenolics constituents and, (d) inorganic components (Taherzadeh, 1999).

2.2. Production Methods of Cellulosic Ethanol

There are two primary routes for the production of cellulosic ethanol - biochemical and thermochemical routes. The biochemical route relies primarily on the use of enzymes and other microorganisms and the thermochemical route relies on the application of heat and chemical synthesis. The below process flow diagram (fig. 2.1) shows the basic steps in production of ethanol from cellulosic biomass (Zhu JY et al, 2009).

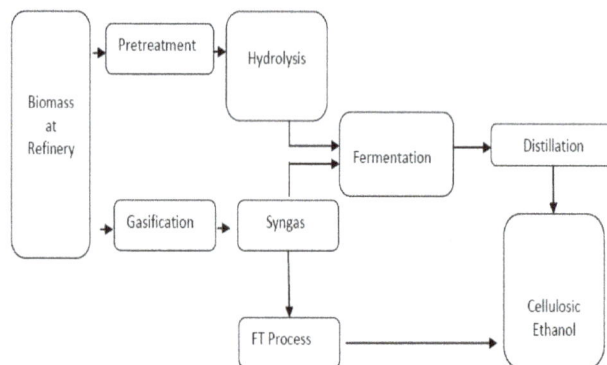

Figure 2.1. Schematic Diagram of Ethanol productions from lignocellulosic feedstocks (Zhu JY et al., 2009).

2.2.1. Biochemical Conversion (Sugar Platform)

The biochemical conversion process is similar to the process currently used to produce ethanol from corn starch. Enzymes or acids are used to break down a plant's cellulose into sugars, which are then fermented into liquid fuel. Four key steps are involved. First, feedstock is pretreated by changing its chemical makeup to separate the cellulose and hemicellulose from the lignin in order to maximize the amount of available sugar. Second, hydrolysis uses enzymes or acids to break down the complex chains of sugar molecules into simple sugars for fermentation. Third, fermentation is used to convert the sugar into liquid fuel. Fourth, the liquid fuel is distilled to achieve a 95% pure form (fig. 2.7) (Zhu JY et al., 2009).

3. Materials and Methods

3.1. Materials

The materials used to run all experiments are listed below:

Chemicals: Phenol,Sodium Hydroxide (NaOH, min. assay 98% BDH Chemicals Ltd Poole England cellulose),Sulphuric Acid (H_2SO_4, (98%, England)), Dextrose sugar, Yeast extract, Urea, $MgSO_4.7\ H_2O$, yeast (Saccharomyces cerevisiae) (manufactured in France by S.I. Lesaffre with the strain 'saf-instant').

Equipments: Pycnometer, pH-Meter ,Shaking incubator, Vertical Autoclave, Cutting mill, Autoclavable bio Reactor, Shaker, Ovens- Loading model 100 -800, Beschikung, Funnel, Sieves (mesh size of 2.0 mm, Sortmks-3332, PFEUFFR, Germany), Digital balances (model = Sartorius with 0.01 mg sensitivity, and model EP214C), Vacuum Filter (model = BN 3 STAATLICH, Berlin),Rotary Evaporator (model = D79219, Staufen, Germany).

3.2. Methods

3.2.1. Sample Preparation

Sample preparation process include: manual size reduction (Knife cutting), drying, grinding and sieving. Grinding of Prosopis juliflora into powder form gives the surface area of the sample increased which enhance the contact between hemicellulose and cellulose with dilute acid to reduce cellulose crystallinity.

3.2.2. Pretreatment of Prosopis Juliflora

Acid pretreatment involves the use of concentrated and diluted acids to break the rigid structure of the lignocellulosic material. The most commonly used acid is dilute sulphuric acid (H_2SO_4), which has been commercially used to pretreat a wide variety of biomass types switch grass , corn stover , spruce (softwood) , and poplar (B. Du et al., 2010). In this study dilute sulfuric acid pretreatment method with 1.2% concentration was used. The powder Prosopis juliflora was pretreated inside autoclave and heated at temperature of 135°C for 30 minutes. Prosopis juliflora powder was fed as batches and every batch contains 300 g of screened Prosopis juliflora powder with a ratio of 10:1(v/w) water to the sample.

In sample pretreatment for all batches acid concentration of 1.2%, temperatures of 135°C and retention time of 30 minutes were used.

3.2.3. Hydrolysis

The cellulose molecules composed of long chains are broken down to "free" the sugar, before it is fermented for alcohol production. Though hydrolysis is of many types, dilute acid hydrolysis is an easy and productive process. Also the amount of alcohol produced in case of acid hydrolysis is more than that of alkaline hydrolysis. Concentrated acid hydrolysis is not used as it is a hazardous and corrosive process and also acid has to be separated out after hydrolysis for the experiment has to be feasible.

The 2 level full factorial experimental design method using Design expert® 7 software was chosen to optimize acid hydrolysis in ethanol production from Prosopis juliflora and to determine the effect of four operating variables of the acid hydrolysis, including acid concentration, solid fraction, temperature (T), and time, and a level of two, with two replica ($2^4 *2 = 32$ experiment) and one response variables which were yield of ethanol.

Table 3.1. *Maximum and minimum values of variables of acid hydrolysis in ethanol production from Prosopis juliflora*

	Variables	Units	Low level (-)	High level (+)
1	Acid concentration	% v/v	0.5	2.5
2	Solid fraction	% w/w	5	10
3	Temperature	(°C)	105	125
4	Time	Minutes	10	20

3.2.4. Fermentation

Microorganism: All fermentations were carried out using yeast (Saccharomyces cerevisiae) manufactured in France by S.I. Lesaffre with the strain 'saf-instant') in an anaerobic condition.

Fermentation Medium: One liter of production medium was prepared according to the requirements of Saccharomyces cerevisiae, containing 100 gm dextrose, 2gm dry yeast extract, 10 gm Urea, 1gm $MgSO_4.7\ H_2O$ and 1000 ml make up distilled water.

3.2.5. Distillation

Distillation is the method used to separate two liquids based on their different boiling points. However, to achieve high purification, several distillations are required. In this study separation are made by rotary evaporator at a temperature of 85 °C.

4. Results and Discussions

To see how well the cubic polynomial model satisfies the assumptions of the analysis of variance (ANOVA), the plots of residuals and residual versus predicted values were analyzed.

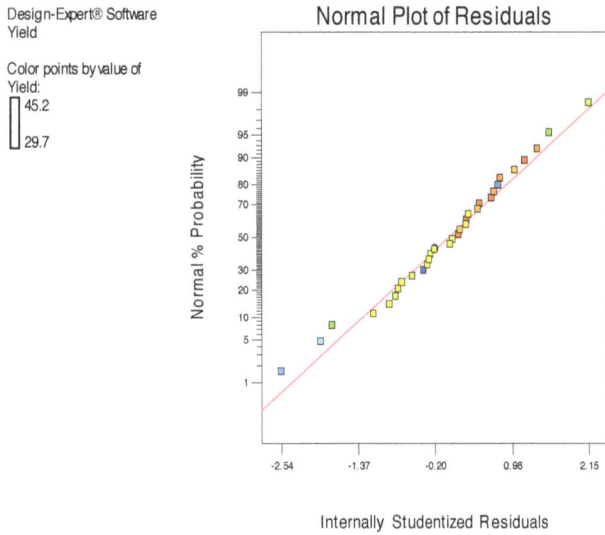

Figure 4.1. *Normal plot of residuals*

The normal probability plot, (Fig. 4.1), indicates the residuals following a normal distribution, in which case the points follow a straight line. This indicates the model satisfies the assumption of ANOVA.

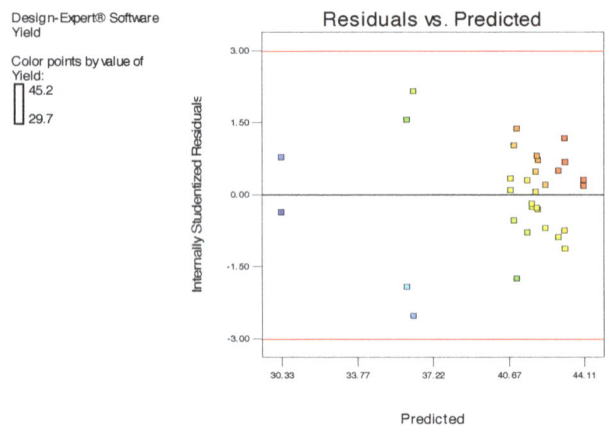

Figure 4.2. *Plot of residuals versus model predicted values*

The plot of the residuals versus the predicted response values (Fig. 4.2), tests the assumption of constant variance. The plot shows constant range of residuals across the graph which is welcome deserving no need for a transformation to minimize personal error.

4.1. Interaction Effects

Acid hydrolysis is influenced by different factors and the ethanol yield has a complex relationship with independent variables that contain first, second and third-order polynomials and may have more than one maximum point.

The best way of expressing the effect of any parameter on the yield within the experimental space under investigation was to generate response surface plots of the equation. The three dimensional response surfaces, contours and interactions were plotted in figures (4.3 a, b, c, and d), as a function of the interactions of any two of the factors by holding the other two at average value. In the interaction

plots the black line represents low level of variables and the red line represents high level of variables.

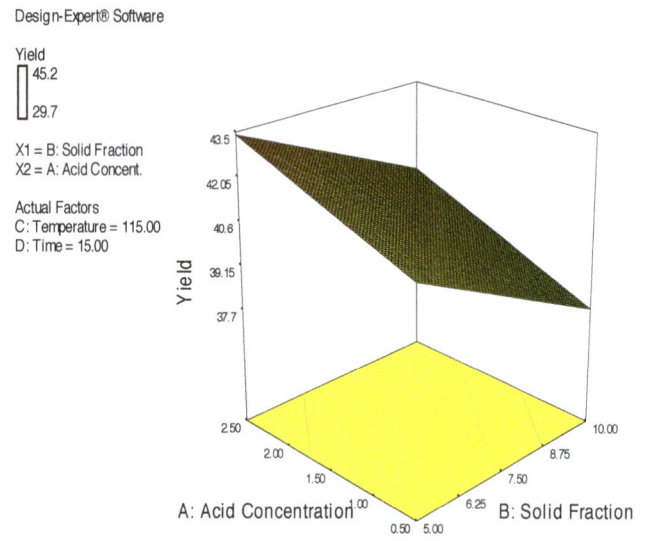

(a)

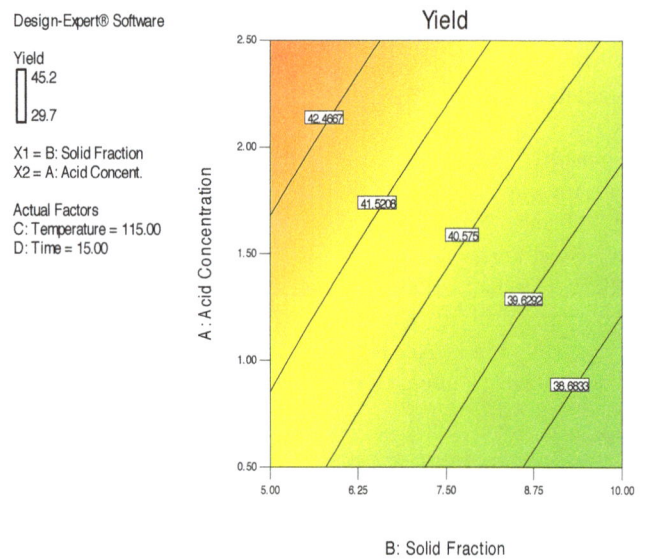

(b)

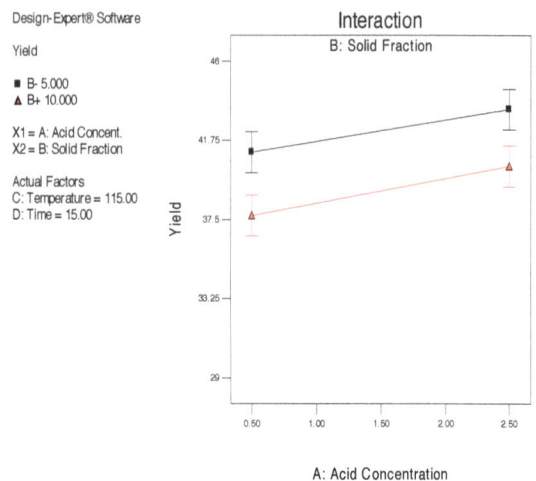

(c)

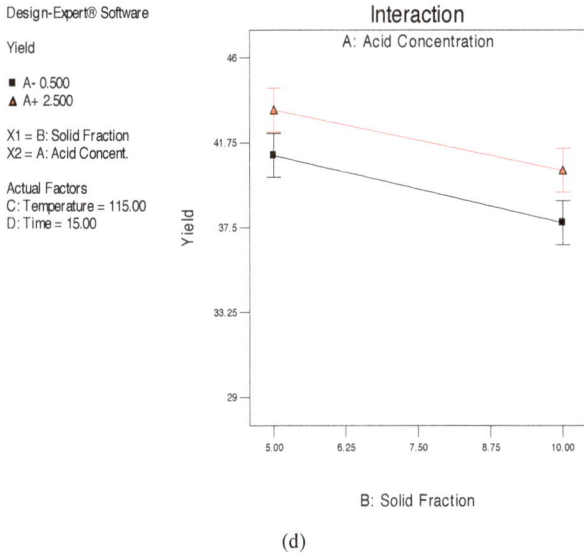

Design-Expert® Software

Yield

■ A- 0.500
▲ A+ 2.500

X1 = B: Solid Fraction
X2 = A: Acid Concent.

Actual Factors
C: Temperature = 115.00
D: Time = 15.00

(d)

Figure 4.3. Response surface plot(a), contour plot (b) and interaction plot (c)and (d) of ethanol yield as a function of acid concentration and solid fraction

4.2. Optimization

The optimum acid concentration, solid fraction, temperature and time for maximum ethanol yield are 0.50 %v/v, 5.00 %w/w, 105.01°C and 10.00 minutes respectively with 40.9 % ethanol yield.

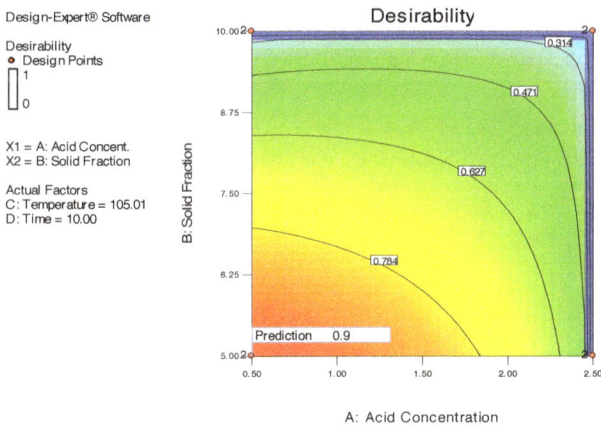

Design-Expert® Software

Desirability
● Design Points

X1 = A: Acid Concent.
X2 = B: Solid Fraction

Actual Factors
C: Temperature = 105.01
D: Time = 10.00

Prediction 0.9

Figure 4.4. Optimization contours on ethanol yield

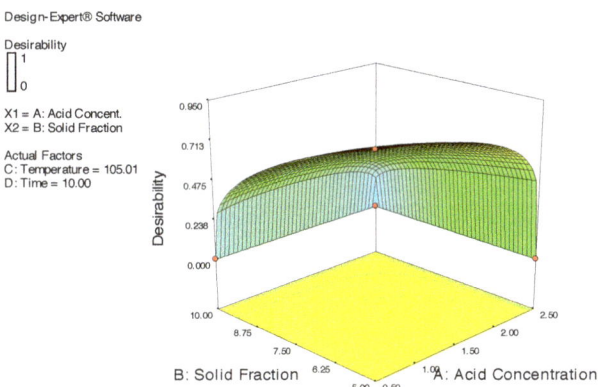

Design-Expert® Software

Desirability

X1 = A: Acid Concent.
X2 = B: Solid Fraction

Actual Factors
C: Temperature = 105.01
D: Time = 10.00

Figure 4.5. Surfaces of possible optimum solutions

4.3. Model Validation

As determined by the 2-level factorial design result using Design-Expert® v.7 software, an experiment with acid concentration ,solid fraction, temperature and time was conducted to carry out the effect of the design used. The optimal values test factors were 0.5 % v/v, 5 % w/w, 105.01°C and 10 minutes. The experiment was carried out at the optimized conditions. Ethanol yield of 40.91 (average) obtained and was in good agreement with the predicted one. Therefore the model is considered to be accurate and reliable for predicting the yield of ethanol.

5. Conclusion and Recommendation

5.1. Conclusion

Due to the diminishing of fossil fuel resources, production of ethanol from lignocellulosic material has acquired significance as a fuel for the future. This study examines the possibility of prosopis juliflora wood for ethanol production. The conversion of prosopis juliflora wood to ethanol was carried out with dilute acid pretreatment, dilute acid hydrolysis, fermentation and distillation process steps.

In this study, 2 level full factorial experimental design was used for the optimization of acid hydrolysis process conditions as well as to investigate interaction between acid hydrolysis process factors using Design Expert® 7 software. The effects of acid hydrolysis variables, namely acid concentration, solid fraction, temperature, and time on the ethanol yield were investigated. A cubic polynomial regression model was assumed for predicting response and the probability p- values of 0.0001 indicate the model was highly significance. The choice of the mathematical model was confirmed by variance analysis. It is concluded that the assumed cubic polynomial models satisfactorily explained the effects of the above-mentioned variables on the ethanol yield. Ethanol yield of 40.91% was obtained when optimum conditions were acid concentration 0.5%, solid fraction 5%, temperature of 105.01°C, time of 10 minute, which indicates that at this condition no inhibitors (furfural and HMF) are produced that inhibit the fermentation process steps. Validation experiments verified the availability and the accuracy of the model with desirability 90 %. The predicted value was in agreement with the experimental value (40.91 wt. %). Based on this study, it is evident that the chosen method of optimization was efficient, and reliable.

5.2. Recommendations

Producing ethanol from renewable resources is becoming an important issue for the whole world. Therefore, the work needs to be continued for further development of ethanol production from prosopis juliflora.

It is also, recommend that in this study acid hydrolysis variables are optimised; future studies should include optimisation of pretreatment process, optimisation of fermentation process and optimisation of distillation process variables to obtain maximum yield of ethanol from prosopis juliflora wood.

Additionally, it is recommend that preliminary design of pilot plant, process development and scale up has to be performed.

References

[1]   B.Du et al., Bioconversion of forest products industry waste cellulosic to fuel ethanol: A review, Bioresource Technology.

[2]   Divya Paruchuri (December, 2008). Conversion of Hardwoods to Ethanol: Georgia.

[3]   Du et al., 2010, Comparative growth, biomass production, nutrient use and soil amelioration by nitrogenfixing.

[4]   Lin, Y., and Tanaka, S., 2006, Ethanol fermentation from biomass resources: current state and prospects, Applied Microbiology and Biotechnology 69: 627-642.

[5]   Mosier, N., et al., 2005, Features of promising technologies for pretreatment of lignocellulosic biomass, Bioresource Technology 96: 673-686.

[6]   Saha, B.C. and Bothast, R.J., (1997) Enzymes in lignocellulosic biomass conversion. *ACS Symp.Ser.* 666 46-56.

[7]   Senayit, R., Agajie, T., Taye, T., Adefires, W. and Getu, E (2004), Invasive Alien Plant control and Prevention in Ethiopia, Pilot Surveys and Control Baseline Conditions. Report submitted to EARO, Ethiopia and CABI under the PDF-B phase of the UNEP/GEF Project, removing Barriers to Invasive Plant Management in Africa. EARO, Addis Ababa.

[8]   Shiferaw, H., Teketay, D., Nemomissa, S. and Assefa, F. (2004). Some biological characteristics that foster the invasion of *Prosopis juliflora* (Sw.) DC at Middle Awash Rift Valley Area,Northeastern Ethiopia. Journal of Arid Environments 58/135.154.

[9]   Taherzadeh MJ, Eklund R, Gustafsson L, Niklasson C, Lide´n G. 1997. Characterization and fermentation of dilute-acid hydrolyzates from wood. Ind Eng Chem Res 36:4659–4665.

[10]   Taye, T., Ameha, T., Adefiris, W. and Getu, E. (2004). Biological Impact Asssessment on selected IAS Plants on Native Species Biodiversity, Report submitted to EARO, Ethiopia, tree species in semi-arid Senegal. Forest Ecology and Management 176: 253-264.

[11]   US Congress, 1984, Commercial Biotechnology: An International Analysis, report OTA-BA-218, US Congress, Office of Technology Assessment, Washington DC USA.

[12]   USDOE, 2003, Advanced bioethanol technology - website: www.ott.doe.gov/biofuels/, US Department of Energy, Office of Energy Efficiency and Renewable Energy, Office of Transportation Technologies, Washington DC USA.

[13]   Wyman, C.E., Dale, B.E., Elander, R.T., Holtzapple, M., Ladisch, M.R. and Lee, Y.Y., (2005) Coordinated development of leading biomass pretreatment technologies. *Bioresour. Technol.* 96 1959-1966).

[14]   Wheals AE, Basso LC, Alves DMG, Amorim HV. Fuel ethanol after 25 years. Trends Biotechnol 1999;17:482–7.

[15]   Wooley R, Ruth M, Sheehan J, Ibsen K, Majdeski H and Galvez A, 1999, Lignocellulosic biomass to ethanol - Process design and economics utilizing co-current dilute acid prehydrolysis and enzymatic hyrolysis - Current and futuristic scenarios, National Reneawable Energy Laboratory, Golden Colorade,USA.

[16]   Van Zyl, W., and Kargi, F., 2007, Consolidated Bioprocessing for bioethanol production using Saccharomyces cerevisiae Pages 205-235, Biofuels.

[17]   Vaccarino, C., Locurto, R., Tripodo, M. M., Patane, R., Lagana, G. and Ragno, A.(1989), "SCP from orange peel by fermentation with fungi–acid-treated peel," *Biol. Wastes* 30, 1-10.

[18]   Zhu JY, Pan XJ, Wang GS, Gleisner R (2009), "Sulfite pretreatment (SPORL) for Robust enzymatic saccharification of spruce and red pine". Bioresource Technology 100 (8): 2411–2418.doi:10.1016/j.biortech.2008.10.057.   PMID 19119005

7

Prediction of Smoke Propagation in a Big Multi-Story Building Using Fire Dynamics Simulator (*FDS*)

Ahmed Farouk Abdel Gawad, Hamza Ahmed Ghulman

Mechanical Engineering Department, College of Engineering and Islamic Architecture, Umm Al-Qura Univ., Makkah, Saudi Arabia

Email address:

afaroukg@yahoo.com (A. F. AbdelGawad)

Abstract: In the present work, the computational fluid dynamics (*CFD*) technique was used to predict the fire dynamics in a big three-story building. Important aspects of fire dynamics were investigated such as smoke propagation and temperature distribution. The study aims to decrease the fire hazards by computationally predicting the expected smoke movement in real-life conditions. Consequently, early evacuation plans can be established to save human lives by proper estimation of the smoke direction and density. Also, temperature rise has a potential effect on the safety of both humans and structures. Different factors were considered such as fire location, doors, and emergency openings. Important findings and notable conclusions are recorded.

Keywords: Fire Dynamics, Smoke Propagation, Computational Method, Unsteady Solution

1. Introduction

1.1. Importance

Due to the development of modern life, people may gather at the same time and place with intensive density. This situation may initiate series fires that lead to massive losses in human lives. Commonly, most of the deaths in fires are not due to direct fire, but because of suffocation with smoke, fumes and toxic gases. Usually, the lack of experience and awareness of individuals results in the increased risk and mortality rates.

Computational prediction of the most probable direction of smoke propagation assists to save human lives. Moreover, smoke-control schemes and evacuation plans can be established as part of the fire-safety strategy.

Examples of buildings where smoke prediction and control play a remarkable role include: holy and worship places, university campuses, shopping centers, big hotels, atrium buildings, large warehouse and industrial buildings, underground structures (car parks and tunnels), *etc.*

Usually, prediction and control of smoke flow within a building may cover one or more of the following objectives: (*i*) Assisting fire fighting, (*ii*) Guarantee safe flees for the occupants of the building, (*iii*) Protecting property.

1.2. Previous Investigations

The problem of fire dynamics simulation was investigated by many researchers. Men *et al.* [1] used large eddy simulations for studying fire-driven flows. Kashef *et al.* [2] carried out computational simulations of in-situ fire tests in road tunnels. Xin *et al.* [3] investigated computationally the turbulent buoyant flame using a mixture-fraction-based combustion model. Jahn *et al.* [4] concerned the effect of model parameters on the simulation of fire dynamics. Huo *et al.* [5] considered the locations of diffusers on air flow field in an office. Razdolsky [6] investigated mathematically the modeling of fire dynamics. Cheng and Hadjisophocleous [7] considered the dynamic modeling of fire spread in buildings. Ling and Kan [8] carried out numerical simulations on fire and analysis of the spread characteristics of smoke in a supermarket. Yang *et al.* [9] investigated both experimentally and numerically a storehouse fire accident. Zhang and Li [10] studied the thermal actions in localized fires in large enclosures. Sun *et al.* [11] investigated the progressive collapse analysis of steel structures under fire conditions. Wu and Chen [12] considered *3D* spatial information for fire-fighting search and rescue route analysis within buildings. Agarwal and Varma [13] studied the fire-induced progressive collapse of steel building structures.

Other investigators used fire dynamic simulator (*FDS*) code in their research work. He and Jiang [14] used *FDS* to assess effectiveness of air sampling-type detector for the protection of large open spaces. Webb [15] used *FDS* modeling for hot smoke testing in cinema and airport concourses. Smardz [16] validated *FDS* for forced and natural convection flows. Sun *et al.* [17] evaluated the fire-plume properties with *FDS* and the Clark coupled wildfire model. Coyle and Novozhilov [18] validated *FDS* using smoke management studies. Zhang *et al.* [19] assessed *FDS* predictions for heat flux and flame heights from fires in *SBI* tests.

Moreover, smoke propagation in buildings and structures was considered by many researchers. Wu *et al.* [20] proposed a distributed method for predicting building fires based on a two-layer zone model. Zhang and Wang [21] carried out a numerical simulation of smoke movement in vertical shafts during a high-rise building fire. Jiang *et al.* [22] modeled fire-induced radiative heat transfer in smoke-filled structural cavities. Yu *et al.* [23] studied the smoke control strategy due to fire in a high-rise building. Zhang *et al.* [24] extended the work of [21] using a modified network model. Bae *et al.* [25] developed a network-based program for unsteady smoke simulation in high-rise buildings.

Also, some investigators concerned the fire evacuation simulation. Tingyong *et al.* [26] studied the building fire evacuation based on continuous model of *FDS & EVAC* code. Tang and Ren [27] carried out *GIS*-based *3D* evacuation simulation for indoor fire. Zhang *et al.* [28] modeled and analyzed *3D* complex building interiors for effective evacuation simulations.

1.3. Present Investigation

The present study is based on the computational fluid dynamics (*CFD*) technique using Fire Dynamic Simulator (*FDS,v.5*). This code was developed and published by the National Institute of Standards and Technologies (*NIST*), U.S. Department of Commerce [29]. The study concerns the smoke propagation due to sample fires in a big three-story building. Different factors were considered such as fire location, doors, and emergency openings. Actually, this investigation is an extension of [30]. Smokeview [31-33] was used to represent the results of the present study as will be shown in the coming sections.

2. Governing Equations and *LES* Simulation

2.1. General Features of the Computational Modeling

FDS code [29] is a computational tool for the prediction of fire scenarios and smoke spread that are expected in almost all types of buildings. The code prediction depends on the architectural plans of the building in addition to the burning materials. The code is based on the solution of the governing equations of flow and combustion due to fire. The core algorithm of *FDS* is an explicit predictor-corrector scheme, second-order accurate in space and time. Turbulence is

treated by means of the Smagorinsky form of Large Eddy Simulation (*LES*). More details of the *FDS* code can be found in [34,35]. There is a big number of attempts for the validation of *FDS*. Some of them are illustrated in Sec. 1.2 and many others in [36]. The validation process may be carried out using the results of other *CFD* programs, codes and standards [29, 37-43], and/or experiments as shown in Fig. 1 that illustrates a typical hot-smoke test layout using smoke canister [15]. Generally, based on these validation investigations, the results of *FDS* can be trusted for almost all fire cases; providing a fine mesh is used to model the problem under-investigation.

2.2. Flow Governing Equations

FDS solves numerically a form of the Navier-Stokes equations appropriate for low-speed; thermally-driven flow with an emphasis on smoke and heat transport from fires [16] as follows:

Conservation of mass:

$$\frac{\partial \rho}{\partial t} + \nabla.\rho\, u_i = 0 \qquad (1)$$

Conservation of momentum:

$$\frac{\partial}{\partial t}(\rho\, u_i) + \nabla.\rho\, u_i\, u_j + \nabla p = \rho\, f + \nabla.\tau_{ij} \qquad (2)$$

Conservation of energy:

$$\frac{\partial}{\partial t}(\rho\, h) + \nabla.\rho\, h\, u_i = \frac{Dp}{Dt} + \dot{q}''' - \nabla.q + \Phi \qquad (3)$$

Equation of state for a perfect gas:

$$p = \rho\, R\, T \qquad (4)$$

In terms of the mass fractions of the individual gaseous species, the mass conservation equation can be written as:

$$\frac{\partial}{\partial t}(\rho Y_i) + \nabla.\rho\, Y_i\, u = \nabla.\rho\, D_i\, \nabla Y_i + \dot{m}_i''' \qquad (5)$$

Where, u_i is velocity in *i*-direction, i =1, 2, 3, ρ is fluid density, f is summation of external forces, τ_{ij} is shear stresses, p is pressure, h is enthalpy, $\dot{q}'''$ is heat release rate per unit volume (*HRRPUV*), q is the heat transfer, Φ is any heat source, and T is the temperature. Y_i is the mass fraction.

2.3. Large Eddy Simulations (LES) and Sub-Grid Scale Models

Large eddy simulation resolves large scales of the flow field solution allowing better fidelity than alternative approaches such as Reynolds-averaged Navier-Stokes (*RANS*) methods. It also models the smallest scales of the solution, rather than resolving them as direct numerical simulation (*DNS*) does.

For incompressible flow, the continuity equation and

Navier-Stokes equations are filtered, yielding the filtered incompressible continuity equation,

$$\frac{\partial \overline{u}_i}{\partial x_i} = 0 \qquad (6)$$

and the filtered Navier-Stokes equations,

$$\frac{\partial \overline{u}_i}{\partial t} + \frac{\partial}{\partial x_j}(\overline{u}_i\,\overline{u}_j) = -\frac{1}{\rho}\frac{\partial \overline{p}}{\partial x_i} + v\frac{\partial^2 \partial \overline{u}_i}{\partial x_j \partial x_j} - \frac{\partial}{\partial x_j}\tau_{ij} \qquad (7)$$

Where, $\overline{p}$ is the filtered pressure field and $\tau_{ij} = \overline{u_i u_j} - \overline{u}_i\,\overline{u}_j$ is the subgrid-scale stress tensor. τ_{ij} is found by an eddy viscosity representation for small scales as [44]:

$$\tau_{ij} = \frac{1}{3}\tau_{kk}\,\delta_{ij} = -2\,\upsilon_T\,\overline{S}_{ij} \qquad (8)$$

Where, δ_{ij} is the Kronecker's delta. To find τ_{ij}, the Smagorinsky-Lilly sub-grid scale SGS model, which was developed by Smagorinsky [45] and used in the first LES simulation by Deardorff [46], is used.

The eddy viscosity is modeled as:

$$v_T = (C_s\,\varDelta_g)^2\sqrt{2\,\overline{S}_{ij}\,\overline{S}_{ij}} = (C_s\,\varDelta_g)^2\,|S| \qquad (9)$$

Where, $\varDelta_g$ is the filter width that is calculated as:

$$\varDelta_g = (\varDelta_x \varDelta_y \varDelta_z)^{1/3} \qquad (10)$$

$\varDelta_x$, $\varDelta_y$ and $\varDelta_z$ are the grid sizes in the three Cartesian coordinates x, y and z, respectively. C_s is a modeling constant that is problem-dependent. The magnitude of the large-scale strain rate tensor is defined as:

$$\overline{S}_{ij} = \frac{1}{2}(\frac{\partial \overline{u}_i}{\partial x_j} + \frac{\partial \overline{u}_j}{\partial x_i}) \qquad (11)$$

2.4. Combustion Model and Radiation Transport

FDS uses the mixture fraction model as the default combustion model [34]. The mixture fraction is a conserved scalar quantity. It is defined as the fraction of gas at a given point in the flow field that originated as fuel, as follows:

$$Z = \frac{sY_F - (Y_{O2} - Y_{O_2}^\infty)}{sY_F^I + Y_{O_2}^\infty} \;;\; s = \frac{v_{O_2}\,W_{O_2}}{v_F\,W_F} \;;\; v_F = 1 \qquad (12)$$

Where, Y is the mass fraction. Subscripts F and O_2 refer to fuel and oxygen, respectively. Y_F^I is the fuel mass fraction in fuel stream. Superscript ∞ refers to "far away from the fire". v is the stoichiometric coefficient. W is the molecular weight of gas. By design, mixture fraction varies from $Z=1$ in a region containing only fuel to $Z=0$ in regions (typically far away from the fire) where only ambient air with un-depleted oxygen is present.

Radiative heat transfer is included in the model via the solution of the radiation transport equation for a non-scattering grey gas, and in some limited cases using a wide-band model. The equation is solved using a technique similar to finite-volume methods for convective transport, thus the name given to it is the Finite-Volume Method (FVM) [16].

3. Building Description and Computational Aspects

3.1. Building Description

The present model is a three-story building, Fig. 2, with overall plan dimensions of about 41×17 m^2. The overall height is 10 m. The offices and facility are concentrated in an area of 18×13 m^2. A central-rectangular hollow-section extends from the first floor to the roof of the third floor with a cross-section of 4.4×3.2 m^2. The main stairs are at the right of the building. There is a stair door at each floor, Figs. 2a,f,g, with dimensions of 2.3×0.9 m^2. The main door (entrance) is located at the rear of the first floor, Fig. 2h, with dimensions of 2.7×2.4 m^2. Some furniture samples appear in the third floor, Figs. 2a,f.

The source of fire is a wooden disk that was altered vertically between the three floors according to the fire case, Fig. 2a.

3.2. Computational Mesh and Domain

The governing equations were approximated on a rectilinear mesh (grid). A computational mesh of $202\times85\times50$ cells was used. Thus, the cells were almost cubic with dimensions of $0.2\times0.2\times0.2$ m^3. Figure 3 shows some horizontal and vertical sections that illustrate the cells of the computational mesh. As can be seen, the mesh was very fine. Thus, the mesh was capable of capturing the features of both the flow and thermal fields.

As can be seen in Fig. 2, the computational domain was extended above the roof of the third floor and behind the building rear wall by about 0.5 m. These two extensions were intended to facilitate smoke exit from the upper vent (emergency opening) and the main door, respectively.

3.3. Boundary Conditions

Concerning the flow field, no-penetration and no-slip conditions are applied on the solid surfaces. Flow speed is determined at the openings/vents. All solid surfaces are assigned thermal boundary conditions, plus information about the burning behavior of the material. Heat and mass transfers to and from solid surfaces are handled with empirical correlations. Also, material properties of solids may be prescribed as a function of temperature [16]. For all the present building cases, the fire power was set suitable to such applications [35,47]. The normal temperature (without fire) in the building was taken as 20^oC.

3.4. Investigated Cases

To investigate the effect of different possible real-life situations, fifty-seven cases were considered, table 1. These cases cover the location of fire source, the opening/closing of the stair doors and main door, and the operation of the emergency opening (vent).

The actual situation of the building has no ceiling opening (vent). The authors of the present work propose an idea to reduce fire/smoke hazards by considering an active outlet vent in the ceiling of the third floor. This emergency vent operates automatically as the fire emerges depending on the signal of heat detectors. The vent is located in the geometric center of the central-rectangular hollow-section, Fig. 4a, with dimensions of $1.0 \times 1.0 \ m^2$. The vent opens (activates) when the temperature reaches 40^oC. For simplicity, a heat detector was placed just above the fire source, Fig. 4b. The heat detector is moved from the ceiling of one floor to another following the fire source. The vent is equipped by a fan that may operate at three different modes, namely: (i) no operation (zero velocity), (ii) outlet velocity of 1 m/s, (iii) outlet velocity of 5 m/s.

The cases of table 1 cover the fire location at the three floors. Symbols "F", "S", and "T" refer to the location of the fire source in the first, second, and third floors, respectively. Symbol "v" refers to vent operation. In the coming sections, the stair doors will be referred as "door-1", "door-2", and "door-3" for the first, second, and third floors, respectively.

4. Results and Discussions

The presentation of the results considers three main times after the fire ignition, namely: 60s (1 minute), 300s (5 minutes), and final period. Actually, 60s was chosen as a suitable time for preliminary quick evacuation of the building after fire ignition with proper alarming. Moreover, 300s was considered as a suitable time for complete evacuation of the building. Final period is the time at which the smoke pattern reaches its steady (constant) shape within the building without further change with time.

4.1. Fire Source at the First Floor

Figure 5 shows the results of the smoke propagation for different cases. As it can be seen in Fig. 5a, after 60s of fire ignition, the smoke patterns are much similar to each others with small differences. Very small amount of smoke enters the stair area due to the open stair doors, Fig. 5a(ii). The opening of the emergency vent draws the smoke from the back walking corridor near the stair area, Figs. 5a(iii),5a(iv). Some smoke gathers in the back corridor near the closed door-3 in case F6v, Fig. 5a(iv).

Figure 5b shows the smoke patterns after 300s. The smoke propagates in the three floors of the building especially the third floor for all cases without the emergency vent, Figs. 5b(i-vi). Smoke fills the stair area when door-1 is open, Fig. 5b(i). Of course, no smoke enters the stair area when all stair doors are closed, Fig. 5b(ii). The upper part of the stair area

is filled with smoke when door-3 is open, Fig. 5b(iii) while the rest of the stair area has no smoke. Small amount of smoke gathers at the upper part of the stair area when door-2 is open, Fig. 5b(iv). The stair area is partially filled with smoke when two of the stair doors are open, Figs. 5b(v,vi).

When the emergency vent is open, smoke is concentrated in the third floor, with low density in the other two floors, Figs. 5b(vii-ix). There is no smoke in the stair area when the three doors are closed, Figs 5b(viii,ix). When the fan of the emergency vent works with full capacity (5 m/s), the smoke density reduces in the third floor, Fig. 5b(ix).

Figure 5c shows the smoke propagation at the final period. Table 2 shows the time in seconds of the final period for each case of Fig. 5c. The maximum period of 1800s occurs when door-1 is closed (cases F8 and F12). It is clear that, by the final period, the smoke completely fills the three floors, Figs. 5c(i,ii). The upper part of the stair area is filled with smoke when door-3 is open, Figs. 5c(iii,iv). The stair area is free of smoke when the three stair doors are closed, Figs. 5c(ii,v,vi). Thus, some of the occupants can survive in the stair area until the fire fighters arrive to rescue them providing that the stair doors are well-protected against smoke leakage.

When the emergency vent is open, the smoke is drawn to the third floor, which has less smoke density comparing to other cases without the emergency vent, Figs. 5c(iv-vi). Even in the first floor, which has the fire source, the smoke is concentrated near the ceiling and thus leaving space for occupants to move with lowering their heads. When the fan of the emergency vent works with full capacity (5 m/s), the smoke density reduces in all floors, Fig. 5c(vi), especially the second floor which becomes almost empty of smoke. Hence, the occupants can survive in the second floor until being rescued.

As seen in Table 2, for all cases, the temperature distribution is almost the same with maximum temperature of 170^oC, which is located in the fire area. Generally, temperature increase in the second and third floors is small. There is no temperature increase in the stair area. It is clear that the main door, whether open or closed, has a negligible effect on the smoke propagation.

4.2. Fire Source at the Second Floor

Figure 6 shows the results of the smoke propagation for different cases. Generally, as it can be seen in Fig. 6a, after 60s of fire ignition, similar behavior to that of Fig.5a is noticed. There are small differences between the smoke patterns. Very small amount of smoke enters the stair area due to the open stair doors, Fig. 6a(ii). The opening of the emergency vent draws the smoke from the back walking corridor near the stair area, Figs. 6a(iii,iv). Some smoke gathers in the back corridor near the closed door-3 in case S6v, Fig. 6a(iv).

Figure 6b shows the smoke patterns after 300s. Mainly, the smoke propagates in the two upper floors of the building especially the third floor for all cases without the emergency vent, Figs. 6b(i-iv). Smoke fills the stair area when two or more stair doors are open, Figs. 6b(i,iv). Of course, no smoke enters the stair area when all stair doors are closed, Fig. 6b(*ii*). The upper part of the stair area is filled with smoke when *door-3* is open, Fig. 6b(*iii*) while the rest of the stair area has no smoke.

When the emergency vent is open, smoke is concentrated in the third floor, with no smoke at all in the first floor, Figs. 6b(*v-vii*). Smoke gathers in the upper part of the second floor below the ceiling. Thus, occupants can move in the second floor and leave to the first floor by lowering their heads. There is no smoke in the stair area when the three doors are closed, Figs. 6b(*vi,vii*). When the fan of the emergency vent works with full capacity (5 *m/s*) and all stair doors are closed, the smoke density reduces in the third floor, Fig. 6b(*vii*).

Figure 6c shows the smoke propagation at the final period. Table 3 shows the time in seconds of the final period for each case of Fig. 6c. The maximum period of 1800*s* occurs for almost all cases, without the emergency vent working, except case *S7* (1300*s*) when the two doors; *door-1* and *door-3*, are open. It is clear that, by the final period, the smoke completely fills the three floors, Figs. 6c(*i-v*). The only exception is case *S8*, when *door-1* is closed, the smoke density in the first floor is low in the back corridor, Fig. 6c(*vi*). Thus, the first floor is a good resort for late evacuation when securing *door-1*. The stair area is free of smoke when the three stair doors are closed, Fig. 6c(*ii*).

When the emergency vent is open, the smoke is drawn to the third floor, which has less smoke density comparing to other cases without the emergency vent, Figs. 6c(*vii-ix*). Even in the second floor, which has the fire source, the smoke is concentrated near the ceiling and thus leaving space for occupants to move with lowering their heads. The worst case (*S6v*) happens when the three stair doors are closed, Fig. 6c(*ix*). Generally, it seems that the operating speed of the fan of the emergency vent has very low effect on the smoke density.

As seen in Table 3, for all cases, the temperature distribution is almost the same with maximum temperature of 170°C, which is located in the fire area. Generally, temperature increases considerably in the second floor with partial increase in the third floor. There is small temperature increase in the stair area. There is no temperature increase in the first floor. It is clear that the main door, whether open or closed, has a negligible effect on the smoke propagation.

4.3. Fire Source at the Third Floor

The results of this section reveal that temperature rises to very high values (around 1000°C), which leads to sudden flashover through the building. Moreover, the building structure starts to burn gradually, which may lead to a building collapse eventually. The operation of the emergency vent prevents completely this flashover. Thus, the cases are divided into two sections; one for flashover and the other for operation of the emergency vent.

4.3.1. Cases of Flashover

Whether or not "flashover" occurs during the course of a fire is one of the most important outcomes of a fire

calculation. Flashover is characterized by the rapid transition to fire behavior from localized burning source to the involvement of all combustibles in the enclosure. High radiation heat transfer levels from the original burning item, the flame and plume directly above it, and the hot smoke layer spreading across the ceiling are all considered to be responsible for the heating of the other items, leading to their ignition. Factors affecting flashover include enclosure size, ceiling and wall conductivity and flammability, and heat- and smoke-producing quality of enclosure contents [48].

In the present study, warning signs of flashover were noticed just before the actual occurrence of flashover. These signs include heat build-up and "rollover". Rollover means small, sporadic flashes of flame that appear near ceiling level or at the top open doorways of smoke-filled enclosures [48].

This section covers all the investigated cases (*T1-T13*) without the operation of the emergency vent. Figure 7 shows the smoke propagation due to the fire source in the third floor. After 60*s*, the smoke pattern is exactly the same for all cases, Fig. 7a. Smoke is concentrated in the upper portion of the third floor, while the other two floors are completely free of smoke. Just before flashover, Fig. 7b, the smoke completely fills the third floor. When the three stair doors are open or *door-1* and *door-3* are open, smoke fills completely the stair area and sneaks partially to the first and second floors, Fig. 7b(*i*). When the three stair doors are closed or only one door is open, smoke fills completely the third floor, and partially the second floor, whereas the first floor is approximately free of smoke, Fig. 7b(*ii*). When *door-3* is open, smoke partially fills the upper half of the stair area and the second floor, Fig. 7b(*iii*). It is noticed that, in all cases, the first floor is completely or partially free of smoke, which represents good resort for occupants to get out of the building through the main door.

Figure 7c shows the smoke propagation at time of flashover for all cases (*T1-T13*). Both smoke and fire propagate in the whole building. Figure 7d shows the smoke propagation after time of flashover without structure burning for many cases. It is clear thus fire is decaying. Figure 7e illustrates the smoke propagation after time of structure burning for some cases (*T4, T6, T7, T10, T13*). Some parts of the structure are vanished due to flashover and high temperature rise. As can be seen in table 4, when the structure starts to burn, the maximum temperature is constant at 1000°C. It is clear from table 4 that the time of flashover is case-dependent and the maximum temperature is around 1000°C.

Figure 8 demonstrates the temperature distribution in cases of flashover. It is clear that sudden temperature rise occurs at the time of flashover in the whole building, Fig. 8b. However, in cases of structure burning, the temperature lowers after its sudden rise, Fig. 8c.

Figure 9 shows the development of the structure burning with time for case *T6* as an example.

4.3.2. Cases of Operation of Emergency Vent

For all cases of operation of emergency vent (*T1v-T6v*),

there is no flashover and structure burning. Figure 10 illustrates the patterns of smoke propagation with operation of the emergency vents (without flashover). After 60s, the smoke pattern is the same for all cases, Fig. 10a. Smoke is concentrated in the upper portion of the central area of the third floor. Smoke is sucked out the building through the emergency vent.

After 300s, the same pattern is kept as that after of 60s when the fan of the emergency vent operates at full capacity (5 m/s) with the three stair doors are open, Fig. 10b(ii). In other cases, Figs. 10(i,iii,iv), the smoke propagates in other parts of the third floor but restricted to the upper portion. The two other floors are completely free of smoke.

After 1800s (the final period), the minimum density of the smoke is noticed when the fan of the emergency vent operates at full capacity (5 m/s), Figs. 10c(ii,v). Maximum smoke density is seen in the case ($T2v$) of the closed stair doors and the emergency vent is open with no fan working (0 m/s), Fig. 10c(iii). In all cases, Fig. 10c(i-v), the first and second floors as well as the stair area are completely free of smoke.

Fig. 11 demonstrates the temperature distribution in all the six cases. It is obvious that the operation of the emergency vent reduces the maximum temperature to 170^oC which is much less and safer in comparison to the cases without emergency vent when maximum temperatures becomes about 1000^oC, table 4.

4.4. Smoke Exit of the Main Door

The smoke exit of the main door is an important factor in the evacuation plans. Table 5 illustrates the time at which the smoke starts to exit from the main door for all cases. As can be seen in table 5, the time of smoke exit, from the main door, increases considerably when the fire source moves from the first floor to the third floor. Considering the fire source in the first floor, the operation of the emergency vent approximately doubles the time required for the smoke to start exiting from the main door. However, for the fire source in the second or third floors, the operation of the emergency vent prevents completely the exit of smoke from the main door. Thus, the operation of the emergency vent helps greatly in the evacuation of the occupants from the main door of the building.

Figure 12 shows the progress of smoke exit of the main door with time of case $F1$ as an example. The smoke exiting from the main door is restricted to the upper portion of the main door till 180s (3 $minutes$). Thus, occupants can easily and safely leave the main door.

Fig. 1. *Typical hot-smoke test layout using smoke canister [15].*

(a) Front view.

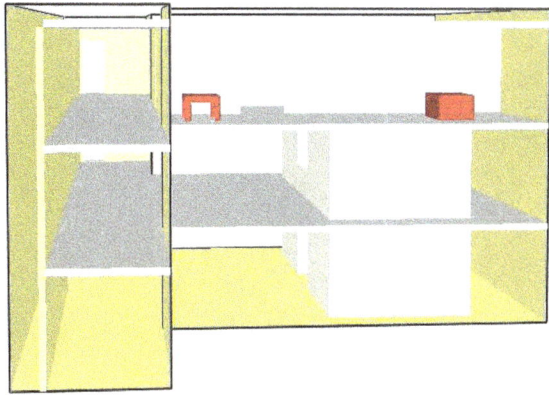

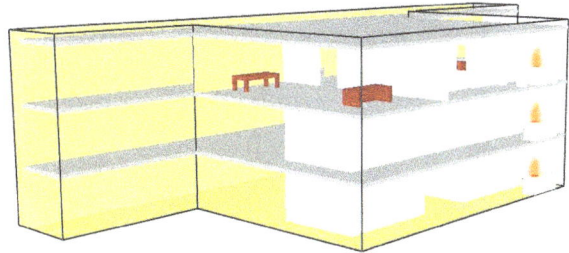

(b) Side view.

(c) Inclined view.

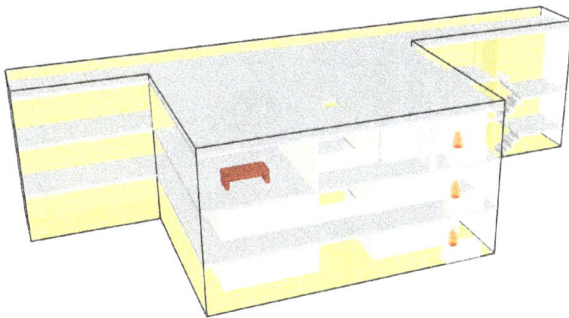

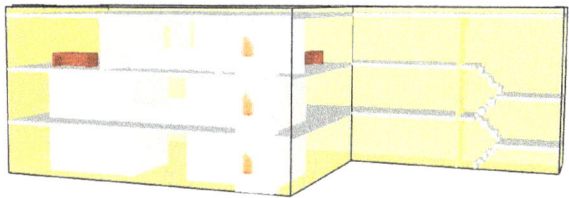

(d) Inclined view.

(e) Inclined view.

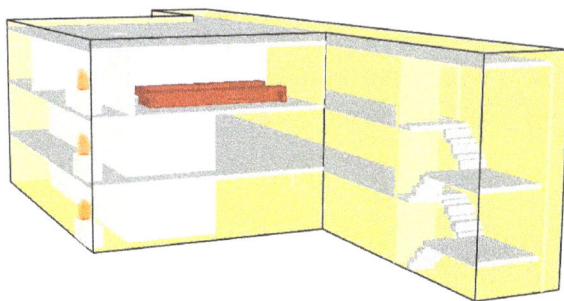

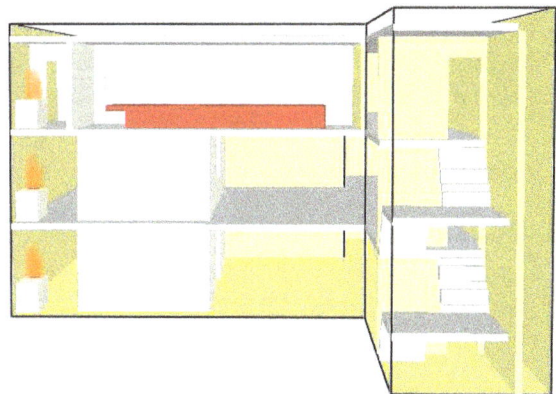

(f) Inclined view.

(g) Side view.

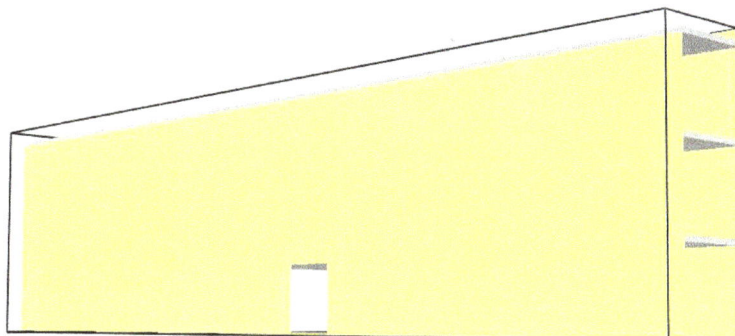

(h) Rear view.

Fig. 2. *Views showing the details of the three floors of the building.*

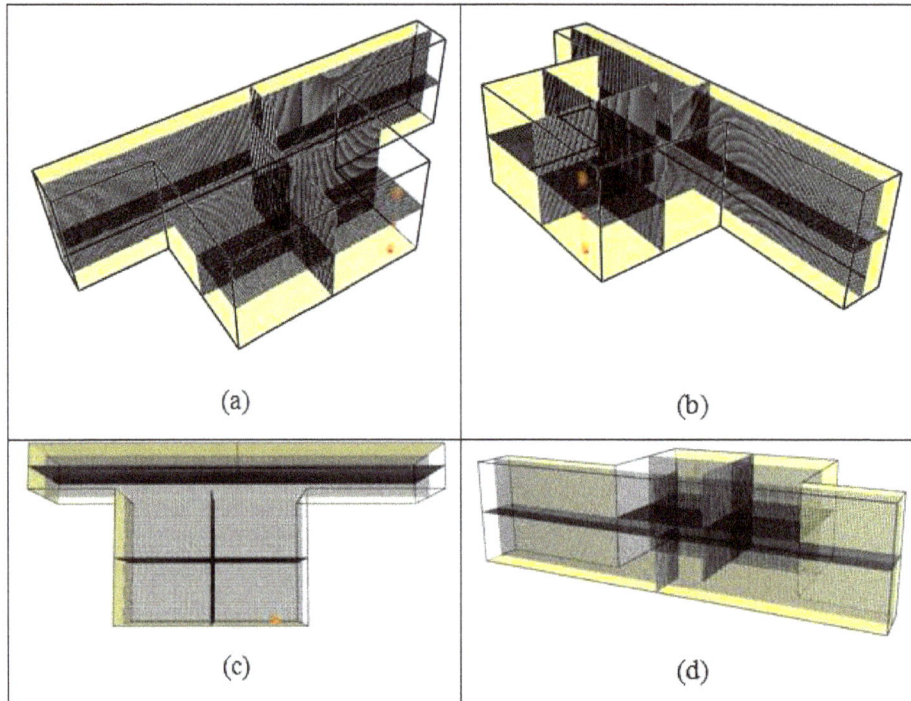

Fig. 3. *Views of the elements (cells) of the computational mesh.*

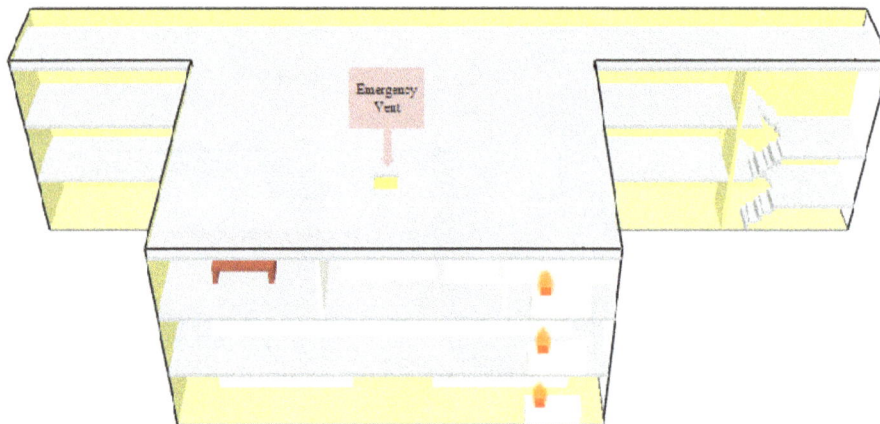

(a) Top view showing the location of the proposed opening (emergency vent).

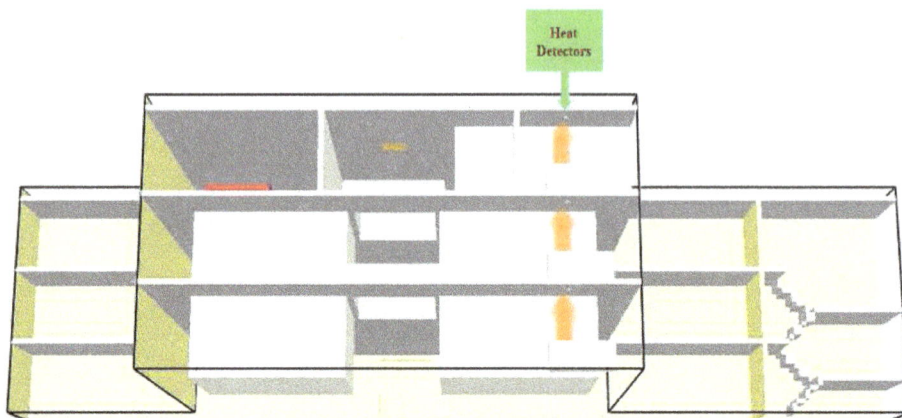

(b) Locations of the three heat detectors.

Fig. 4. *Views of the smoke detectors and emergency vent.*

Table 1. Cases of the present study.

No.	First Floor Cases	Second Floor Cases	Third Floor Cases	Main door	First floor stair door	Second floor stair door	Third floor stair door	Emergency vent
1	F1	S1	T1	Open	Open	Open	Open	Closed
2	F2	S2	T2	Open	Closed	Closed	Closed	Closed
3	F3	S3	T3	Open	Closed	Closed	Open	Closed
4	F4	S4	T4	Open	Closed	Open	Closed	Closed
5	F5	S5	T5	Open	Open	Closed	Closed	Closed
6	F6	S6	T6	Open	Open	Open	Closed	Closed
7	F7	S7	T7	Open	Open	Closed	Open	Closed
8	F8	S8	T8	Open	Closed	Open	Open	Closed
9	F9	S9	T9	Closed	Open	Open	Open	Closed
10	F10	S10	T10	Closed	Closed	Closed	Closed	Closed
11	F11	S11	T11	Closed	Closed	Closed	Open	Closed
12	F12	S12	T12	Closed	Closed	Open	Closed	Closed
13	F13	S13	T13	Closed	Open	Closed	Closed	Closed
14	F1v	S1v	T1v	Open	Open	Open	Open	Open (0 m/s)
15	F2v	S2v	T2v	Open	Closed	Closed	Closed	Open (0 m/s)
16	F3v	S3v	T3v	Open	Open	Open	Open	Open (1 m/s)
17	F4v	S4v	T4v	Open	Closed	Closed	Closed	Open (1 m/s)
18	F5v	S5v	T5v	Open	Open	Open	Open	Open (5 m/s)
19	F6v	S6v	T6v	Open	Closed	Closed	Closed	Open (5 m/s)

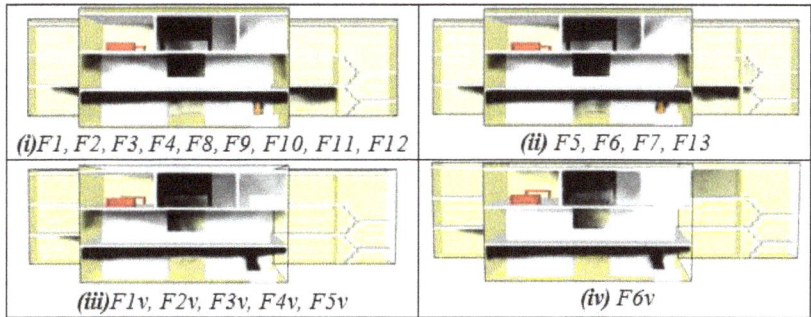

(i)F1, F2, F3, F4, F8, F9, F10, F11, F12　　(ii) F5, F6, F7, F13

(iii)F1v, F2v, F3v, F4v, F5v　　(iv) F6v

a. Smoke propagation after 60s.

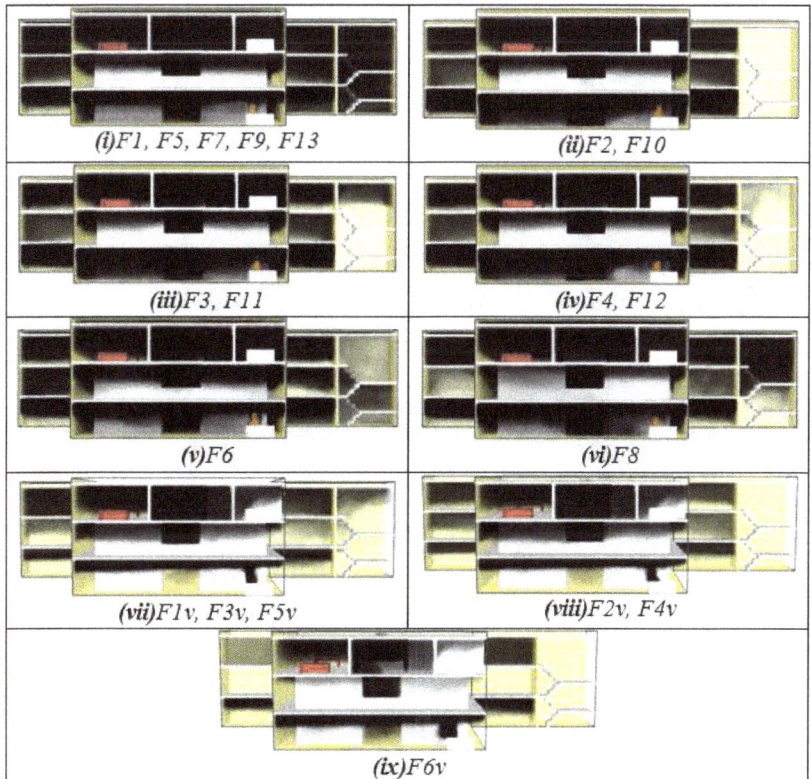

(i)F1, F5, F7, F9, F13　　(ii)F2, F10

(iii)F3, F11　　(iv)F4, F12

(v)F6　　(vi)F8

(vii)F1v, F3v, F5v　　(viii)F2v, F4v

(ix)F6v

b. Smoke propagation after 300s.

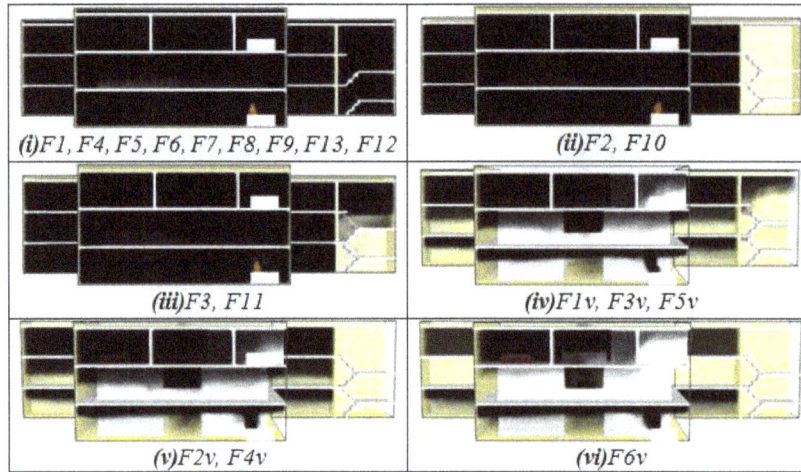

c. Smoke propagation at final period.

Fig. 5. *Views of the smoke propagation due to fire source in the first floor.*

Table 2. *Final period and maximum temperature due to fire source in the first floor for different cases.*

Case	Final time	Maximum Temperature = 170°C
F7, F9, F13	900s	
F10, F11	1000s	
F1, F3, F5, F6, F2, F4, F1v, F3v, F5v, F2v, F4v, F6v	1200s	
F8, F12	1800s	

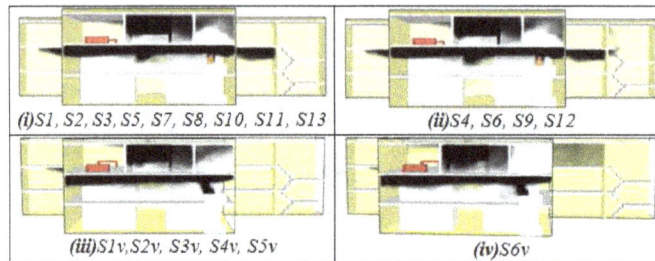

a. Smoke propagation after 60s.

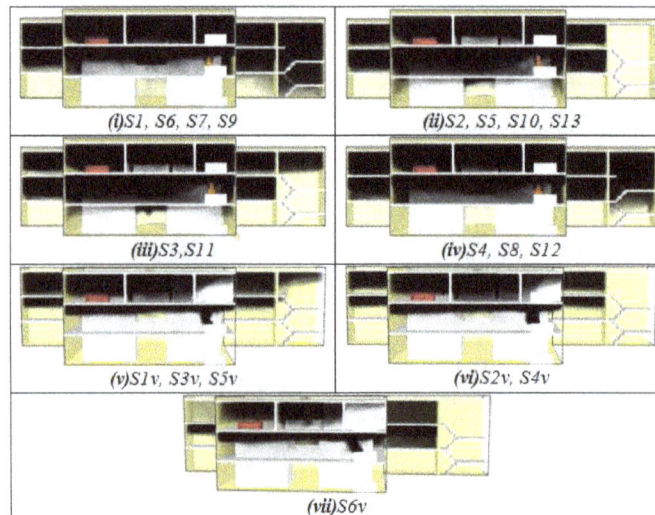

b. Smoke propagation after 300s.

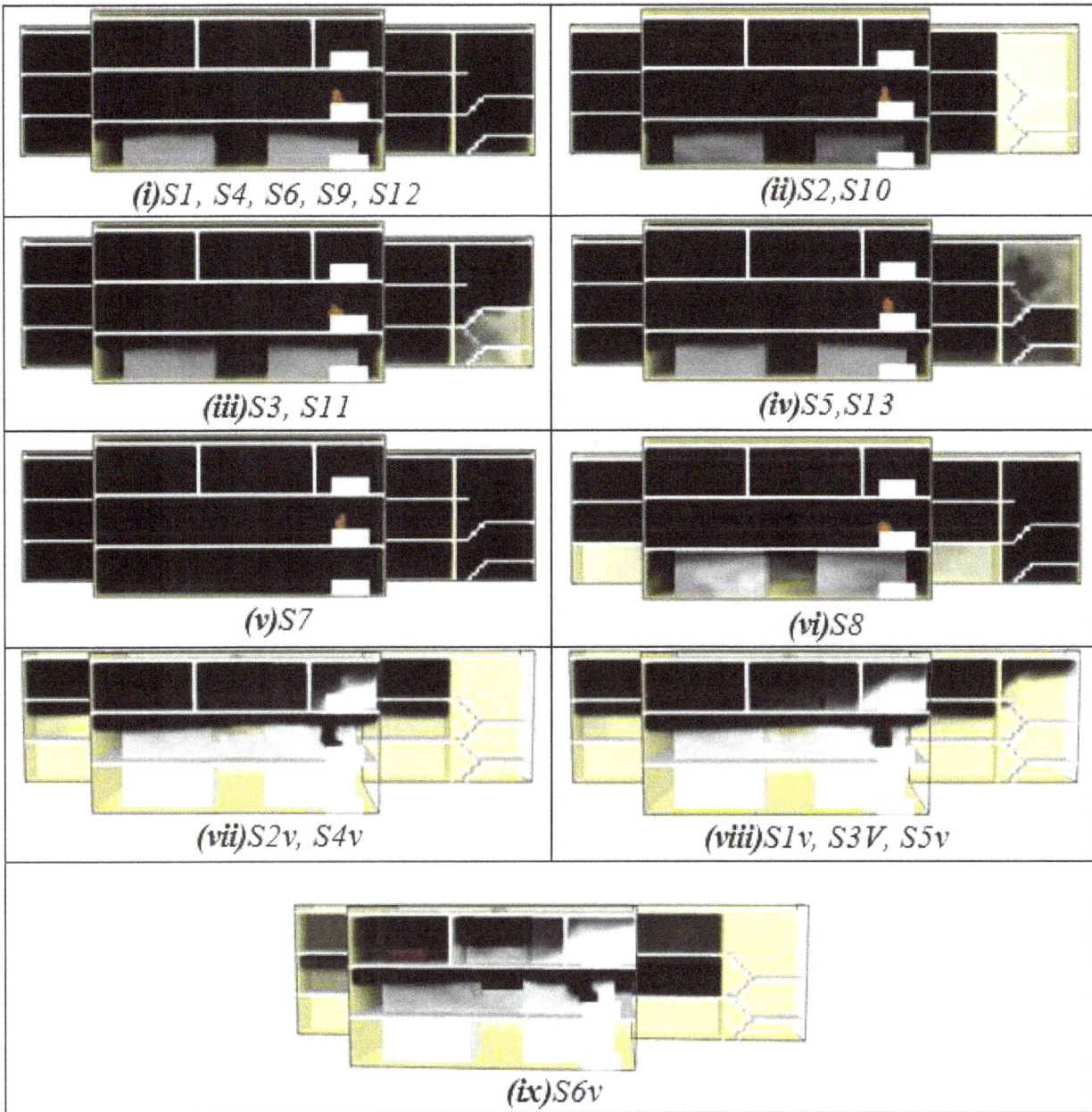

c. Smoke propagation at final period.

Fig. 6. *Views of the smoke propagation due to fire source in the second floor.*

Table 3. *Final period and maximum temperature due to fire source in the second floor for different cases.*

Case	Final time	Maximum Temperature = 170°C
S1v, S3V, S5v, S2v, S4v, S6v	1200s	
S7	1300s	
S1, S4, S6, S9, S12, S2,S10, S3, S11, S5,S13, S8	1800s	

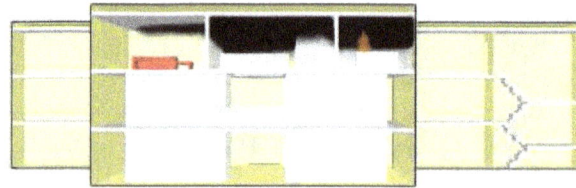

T1,T2, T3, T4, T5, T6, T7, T8, T9, T10, T11, T12, T13
a. Smoke propagation after 60s.

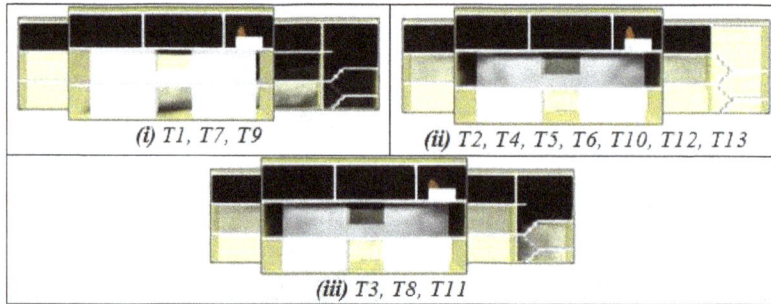

(i) T1, T7, T9   *(ii) T2, T4, T5, T6, T10, T12, T13*

(iii) T3, T8, T11

b. Smoke propagation before flashover.

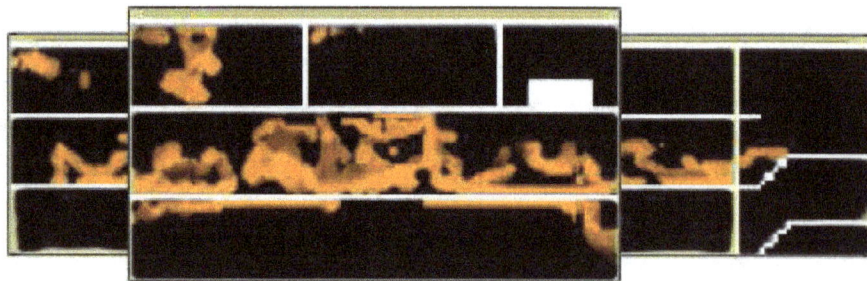

T1, T2, T3, T4, T5, T6, T7, T8, T9, T10, T11, T12, T13
c. Smoke propagation at time of flashover.

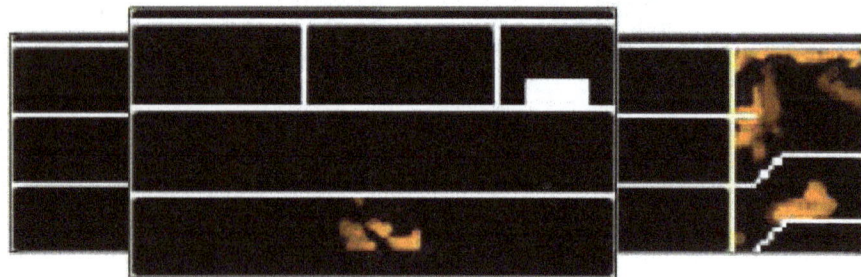

T1, T2, T3, T5, T8, T9, T11, T12
d. Smoke propagation after time of flashover without structure burning.

T4, T6, T7, T10, T13
e. Smoke propagation after time of structure burning.

Fig. 7. *Views of the smoke propagation due to fire source in the third floor.*

Table 4. *Time of flashover and maximum temperature.*

Case	Time of flashover (s)	Max. Temperature (°C)
T1	1365	1020
T2	1255	970
T3	1348	1020
T4	1280	1000
T5	1274	1020
T6	1253	1000
T7	1370	1000
T8	1384	970
T9	1387	1020
T10	1267	1000
T11	1357	970
T12	1264	1020
T13	1255	1000

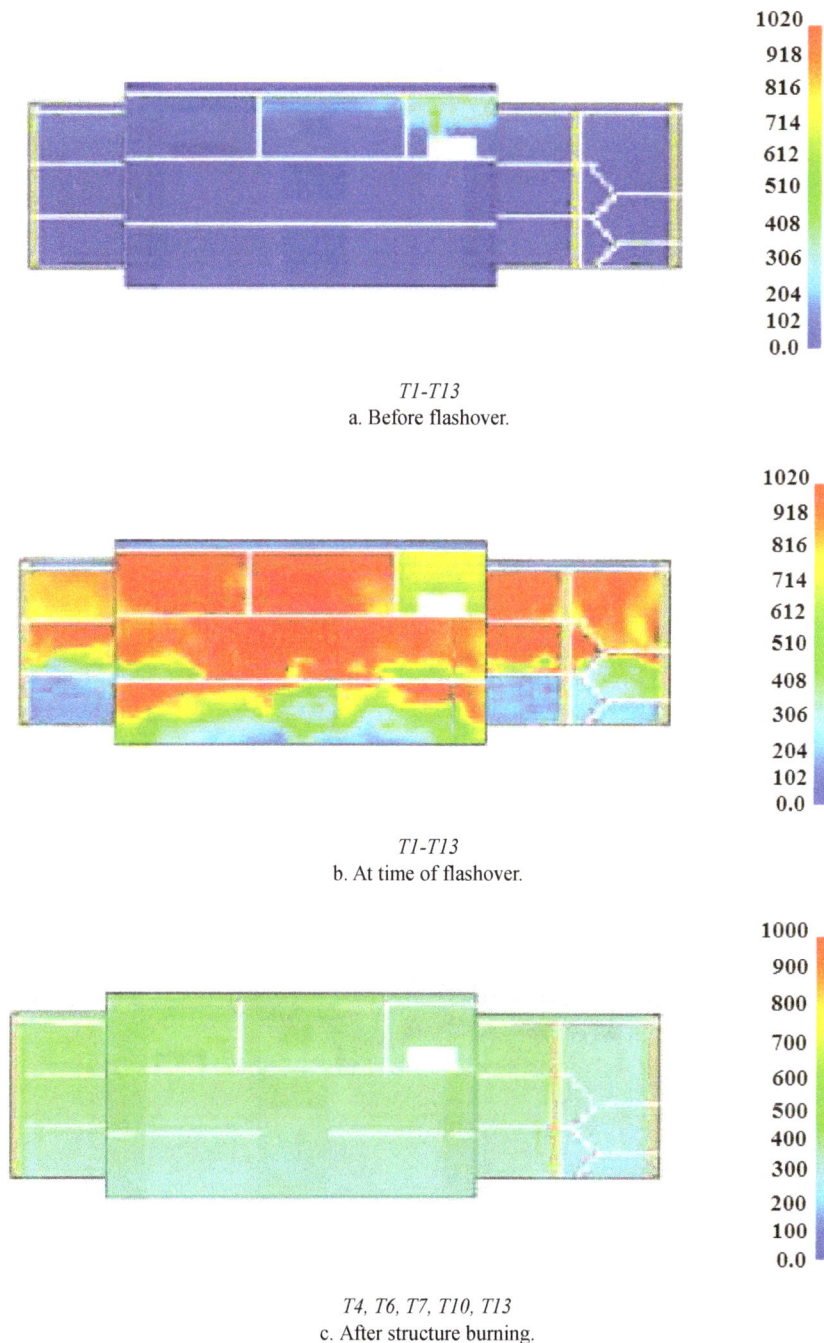

T1-T13
a. Before flashover.

T1-T13
b. At time of flashover.

T4, T6, T7, T10, T13
c. After structure burning.

Fig. 8. *Temperature distribution due to fire source in the third floor in cases of flashover.*

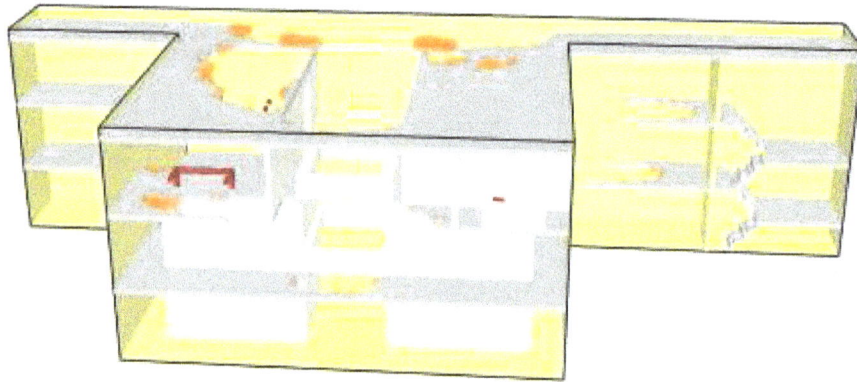

(a) Time = 1332*s*

(b) Time = 1400*s*

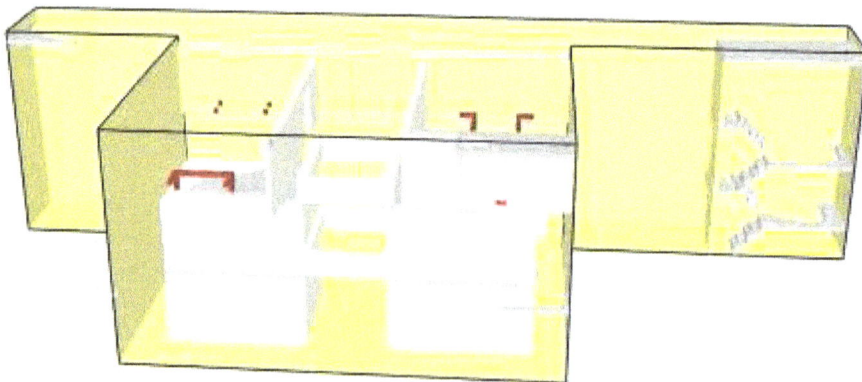

(c) Time = 1800*s*

Fig. 9. *Views of the structure burning due to fire source in the third floor at different time periods (Case T6 as an example).*

T1v, T2v, T3v, T4v, T5v, T6v
a. Smoke propagation after 60*s*.

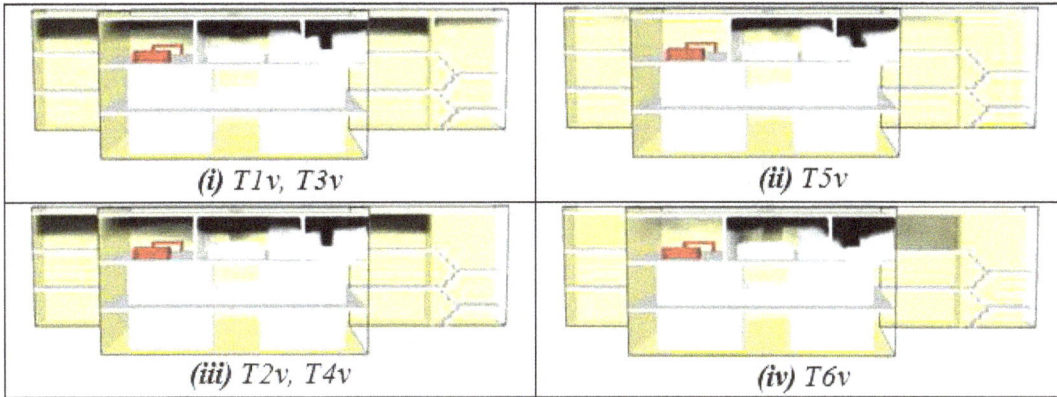

b. Smoke propagation after 300*s*.

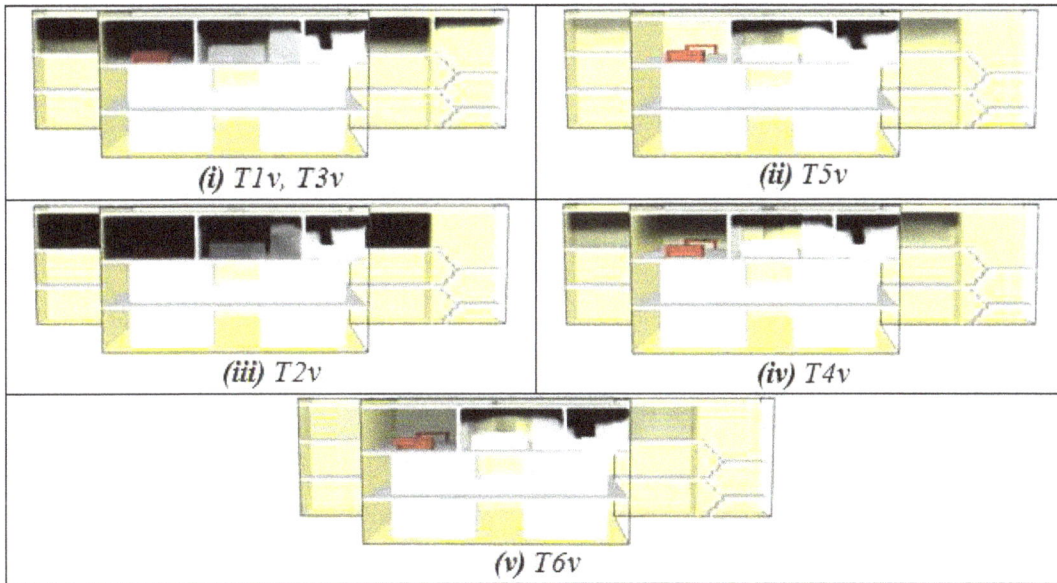

c. Smoke propagation at final period (1800*s*).

Fig. 10. *Views of the smoke propagation due to fire source in the third floor without flashover (operation of emergency vent).*

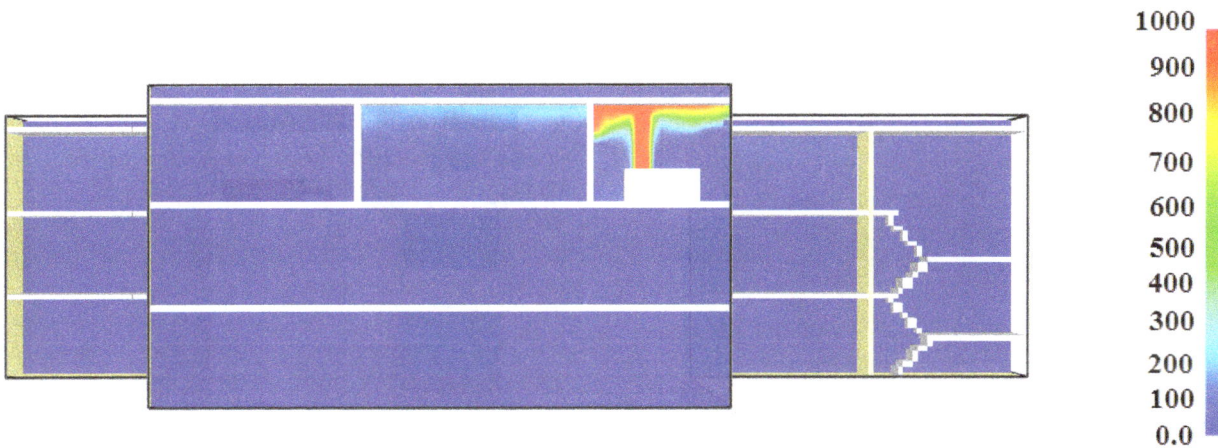

T1v, T2v, T3v, T4v, T5v, T6v (Max Temperature = 170°C)

Fig. 11. *Temperature distribution due to fire source in the third floor without flashover.*

Table 5. *Time of smoke exit out of the main door.*

First floor cases	Time (s)	Second floor cases	Time (s)	Third floor cases	Time (s)
F1	30	S1	240	T1	1125
F2	30	S2	270	T2	1250
F3	30	S3	220	T3	1345
F4	30	S4	220	T4	1272
F5	30	S5	215	T5	1268
F6	30	S6	485	T6	1247
F7	30	S7	230	T7	1357
F8	30	S8	675	T8	1382
F9	Closed	S9	Closed	T9	Closed
F10	Closed	S10	Closed	T10	Closed
F11	Closed	S11	Closed	T11	Closed
F12	Closed	S12	Closed	T12	Closed
F13	Closed	S13	Closed	T13	Closed
F1v	60	S1v	No exit	T1v	No exit
F2v	60	S2v	No exit	T2v	No exit
F3v	57	S3v	No exit	T3v	No exit
F4v	57	S4v	No exit	T4v	No exit
F5v	57	S5v	No exit	T5v	No exit
F6v	57	S6v	No exit	T6v	No exit

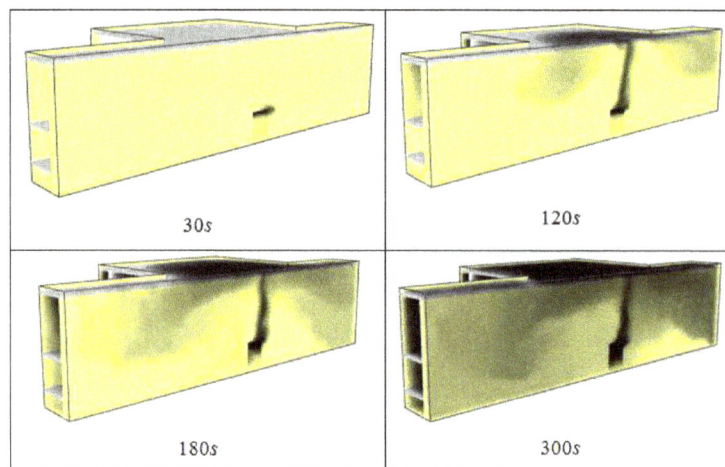

Fig. 12. *Progress of smoke exit of the main door with time (Case F1 as an example).*

5. Conclusions

Based on the above results and discussions, the following concluding points can be stated:

(i) Temperature may rise to unexpected very high levels and flashover occurs when the smoke does not find its way out of the building. This is typically happened in the present study for the cases of fire source in the third floor. Depending on the material of the structure, this may lead to structure burning and/or building collapse.

(ii) Distribution of temperature and its maximum value are dependent on the fire location and the outlet openings of the smoke.

(iii) Emergency vents in the roof of the top floor, which operate in time of fire based on the signal of heat/smoke detectors, play an outstanding job in sucking the smoke outside the building. Thus, flashover is prevented and evacuation of the occupants becomes much easier and safer.

(iv) In some of the investigated cases, the emergency vents may cause the first and/or the second floors to be fully or partially free of smoke for a relatively large period of time (20 or 30 *minutes*).

(v) Naturally, the increase of the speed of the fan of the emergency vent draws more smoke out of the building. However, in certain cases, just the vent opening without fan operation helps greatly in sucking the smoke out of the building.

(vi) The location of the fire source, relative to the building floors, affects significantly the smoke propagation and density.

(vii) The inner openings between different floors facilitate the smoke propagation from one floor to another. This situation is very tricky to the occupants and may cause serious injuries.

(viii) The condition of the stair doors; open or closed, has an important effect on the smoke propagation. Smoke may move from the upper floor to a lower floor through the open doors of the stair area.

(ix) In certain cases, the stair area may be fully or partially free of smoke for a considerably long period of time. Thus, the stair area becomes a good resort for occupants until the arrival of fire fighters to rescue them providing that the stair doors are well-insulated against smoke leakage.

(x) Generally, the condition of the main door; open or closed, has a little effect on the smoke propagation. However, it is essential to study the smoke movement out of it for proper evacuation plans.

(xi) Occupants should be trained to obey the emergency and evacuation plans especially lowering their heads below the smoke layer to survive. In many of the investigated cases, the smoke is constrained to the upper portion of the floor just below the ceiling.

Acknowledgement

The authors would like to acknowledge Engs. B. S. Qattan, O. S. Yamani, H. H. Bahha, S. A. Gubali, B. A. Al-Sobhi, and N. M. Al-Harbi for their efforts to complete the present study.

Nomenclature

C_s	= modeling constant.
F	= summation of external forces.
H	= enthalpy.
Q	= heat transfer.
$\dot{q}'''$	= heat release rate per unit volume ($HRRPUV$).
P	= pressure.
$\bar{p}$	= filtered pressure field.
$\overline{S_{ij}}$	= magnitude of the large-scale strain rate tensor.
T	= temperature.
u_i	= velocity in i-direction, i =1, 2, 3.
$\overline{u_i u_j}$	= nonlinear filtered advection term.
W	= molecular weight of gas.
Y	= mass fraction.
Y_F^I	= fuel mass fraction in fuel stream.
Z	= mixture fraction.

Greek

δ_{ij}	= Kronecker's delta.
Δ_g	= filter width.
Δ_x, Δ_y and Δ_z	= grid sizes in the Cartesian coordinates x, y and z, respectively.
Φ	= any heat source.
v	= stoichiometric coefficient.
v_T	= Turbulent eddy viscosity.
ρ	= fluid density.
τ_{ij}	= subgrid-scale stress tensor.

Superscripts and Subscripts

∞	= refers to "far away from the fire".
F	= refers to "fuel".
O_2	= refers to "oxygen".

Abbreviations

ASTM	= American Society for Testing and Materials.
BFST	= Bureau of Fire Standards and Training.
CFD	= Computational Fluid Dynamics.
DNS	= Direct Numerical Simulation.
FDS	= Fire Dynamic Simulator.
FPA	= Fire Protection Association Australia.
FVM	= Finite-Volume Method.
HRRPUV	= Heat Release Rate per Unit Volume.
LES	= Large Eddy Simulation.
NFPA	= National Fire Protection Association.
NIST	= National Institute of Standards and Technologies.
RANS	= Reynolds-Averaged Navier-Stokes.
SBI	= Single Burning Item.
SGS	= Sub-Grid Scale.

References

[1] W. Men, K. B. Mcgrattan, and H. R. Baum, "Large Eddy Simulations of Fire-Driven Flows", *ASME* National Heat Transfer Conference, Vol. 2, 1995.

[2] A. Kashef, N. Bénichou, G. Lougheed, and A. Debs, "Computational Fluid Dynamics Simulations of in-Situ Fire Tests in Road Tunnels", *5th* International Conference-Tunnels Fires, London, *UK*, pp. 185-196, Oct. 25-27, 2004.

[3] Y. Xin, J. P. Gore, K. B. McGrattan, R. G. Rehm, and H. R. Baum, "Fire Dynamics Simulation of a Turbulent Buoyant Flame Using a Mixture-Fraction-Based Combustion Model", Combustion and Flame J., Vol. 141, pp. 329-335, 2005.

[4] W. Jahn, G. Rein, and J. L. Torero, "The Effect of Model Parameters on the Simulation of Fire Dynamics", Fire Safety Science, Vol. 9, pp. 1341-1352, 2008.

[5] Y. Huo, Y. Gao1, and W. Chow, "Locations of Diffusers on Air Flow Field in an Office," The Seventh Asia-Pacific Conference on Wind Engineering, Taipei, Taiwan, November 8-12, 2009.

[6] L. Razdolsky, "Mathematical Modeling of Fire Dynamics," Proceedings of the World Congress on Engineering 2009, London, *U.K.*, Vol. II, July 1 - 3, 2009.

[7] H. Cheng, and G. V. Hadjisophocleous, "Dynamic Modeling of Fire Spread in Building", Fire Safety Journal, Vol. 46, No. 4, pp. 211-224, 2011.

[8] D. Ling, and K. Kan, "Numerical Simulations on Fire and Analysis of the Spread Characteristics of Smoke in Supermarket", In Advanced Research on Computer Education, Simulation and Modeling, pp. 7-13, Springer Berlin Heidelberg, 2011.

[9] P. Yang, X. Tan, and W. Xin, "Experimental Study and Numerical Simulation for a Storehouse Fire Accident", Building and Environment, Vol. 46, No. 7, pp. 1445-1459, 2011.

[10] C. Zhang, and G. Q. Li, "Fire Dynamic Simulation on Thermal Actions in Localized Fires in Large Enclosure", Advanced Steel Construction, Vol. 8, pp. 124-136, 2012.

[11] R. Sun, Z. Huang, and I. W. Burgess, "Progressive Collapse Analysis of Steel Structures under Fire Conditions", Engineering Structures, Vol. 34, pp. 400-413, 2012.

[12] A. H. Wu, and L. C. Chen, "3D Spatial Information for Fire-fighting Search and Rescue Route Analysis within Buildings", Fire Safety Journal, Vol. 48, pp. 21-29, 2012.

[13] A. Agarwal, and A. H. Varma, "Fire Induced Progressive Collapse of Steel Building Structures: The Role of Interior Gravity Columns", Engineering Structures, Vol. 58, pp. 129-140, 2014.

[14] M. He, and Y. Jiang, "Use FDS to Assess Effectiveness of Air Sampling-Type Detector for Large Open Spaces Protection", Vision Fire & Security, 2005.

[15] A. Webb, "FDS Modelling of Hot Smoke Testing, Cinema and Airport Concourse", Thesis of Master of Science, The Faculty of the Worcester Polytechnic Institute, USA, 2006.

[16] P. Smardz, "Validation of Fire Dynamics Simulator (FDS) for Forced and Natural Convection Flows", Master of Science in Fire Safety Engineering, University of Ulster, 2006.

[17] R. Sun, M. A. Jenkins, S. K. Krueger, W. Mell, and J. J. Charney, "An Evaluation of Fire-Plume Properties Simulated with the Fire Dynamics Simulator (FDS) and the Clark Coupled Wildfire Model," Can. J. for Res., Vol. 36, pp. 2894-2908, 2006.

[18] P. Coyle, and V. Novozhilov, "Further Validation of Fire Dynamics Simulator Using Smoke Management Studies", International Journal on Engineering Performance-Based Fire Codes, Vol. 9, No. 1, pp.7-30, 2007.

[19] J. Zhang, M. Delichatsios, and M. Colobert, "Assessment of Fire Dynamics Simulator for Heat Flux and Flame Heights Predictions from Fires in SBI Tests", Fire Technology, Vol. 46, pp. 291–306, 2010.

[20] N. Wu, R. Yang, and H. Zhang, "A Distributed Method for Predicting Building Fires Based on a Two-Layer Zone Model", ASME 2013 International Mechanical Engineering Congress and Exposition, 2013.

[21] X. T. Zhang, and S. L. Wang, "Numerical Simulation of Smoke Movement in Vertical Shafts during a High-Rise Building Fire", Applied Mechanics and Materials, Vol. 438, pp. 1824-1829, 2013.

[22] Y. Jiang, G. Rein, S. Welch, and A. Usmani, "Modeling Fire-Induced Radiative Heat Transfer in Smoke-Filled Structural Cavities", International Journal of Thermal Sciences, Vol. 66, pp. 24-33, 2013.

[23] Y. Yu, Y. Y. Chu, and D. Liang, "Study on Smoke Control Strategy in a High-rise Building Fire", Procedia Engineering, Vol. 71, pp. 145-152, 2014.

[24] X. Zhang, S. Wang, and J. Wang, "Numerical Simulation of Smoke Movement in Vertical Shafts during High-Rise Fires Using a Modified Network Model", Journal of Chemical & Pharmaceutical Research, Vol. 6, No. 6, 2014.

[25] S. Bae, H. J. Shin, and H. S. Ryou, "Development of CAU_USCOP, A Network-Based Unsteady Smoke Simulation Program for High-Rise Buildings. Building Simulation, Vol. 7, No. 5, pp. 503-510, 2014.

[26] F. Tingyong, X. Jun, Y. Jufen, and W. Bangben, "Study of Building Fire Evacuation Based on Continuous Model of FDS & EVAC", In Computer Distributed Control and Intelligent Environmental Monitoring (CDCIEM), IEEE Conference, pp. 1331-1334, 2011.

[27] F. Tang, and A. Ren, "GIS-based 3D evacuation simulation for indoor fire", Building and Environment, Vol. 49, pp. 193-202, 2012.

[28] L. Zhang, Y. Wang, H. Shi, and L. Zhang, "Modeling and analyzing 3D complex building interiors for effective evacuation simulations", Fire Safety Journal, Vol. 53, pp. 1-12, 2012.

[29] http://www.fire.nist.gov/fds/, http://www.nist.gov/index.html

[30] A. F. Abdel-Gawad, and H. A. Ghulman, "Fire Dynamics Simulation of Large Multi-story Buildings, Case Study: Umm Al-Qura University Campus", International Conference on Energy and Environment 2013 (ICEE2013), Universiti Tenaga Nasional, Putrajaya Campus, Selangor, Malaysia, 5-6 March 2013. [Institute of Physics (IOP) Conference Series: Earth and Environmental Science, Vol. 16, issue 1, 2013, doi:10.1088/1755-1315/16/1/012040].

[31] G. P. Forney, Smokeview (Version 5)-A Tool for Visualizing Fire Dynamics Simulation Data - Volume I: User's Guide, NIST Special Publication 1017-1, 2010.

[32] G. P. Forney, Smokeview (Version 5)-A Tool for Visualizing Fire Dynamics Simulation Data - Volume II: Technical Reference Guide, NIST Special Publication 1017-2, 2010.

[33] G. P. Forney, Smokeview (Version 5)-A Tool for Visualizing Fire Dynamics Simulation Data - Volume III: Verification Guide, NIST Special Publication 1017-3, 2010.

[34] K. McGrattan, S. Hostikka, J. Floyd, H. Baum, R. Rehm, W. Mell, and R. McDermott, Fire Dynamics Simulator (Version 5)-Technical Reference Guide-Volume 1: Mathematical Model, NIST Special Publication 1018-5, 2010.

[35] K. McGrattan, R. McDermott, S. Hostikka, and J. Floyd, Fire Dynamics Simulator (Version 5)-User's Guide, NIST Special Publication 1019-5, 2010.

[36] K. McGrattan, S. Hostikka, J. Floyd, and B. Klein, Fire Dynamics Simulator (Version 5) Technical Reference Guide - Volume 3: Validation, NIST Special Publication 1018-5, 2010.

[37] National Fire Protection Association (NFPA): The American authority on fire, electrical, and building safety: http://www.nfpa.org

[38] British Standards: http://shop.bsigroup.com

[39] The Bureau of Fire Standards and Training (BFST), Division of State Fire Marshal, Florida, USA:http://www.myfloridacfo.com/sfm/bfst/bfst_index.htm

[40] Fire Protection Association Australia (FPA), Australia: http://www.fpaa.com.au

[41] ASTM International, formerly known as the American Society for Testing and Materials (ASTM), USA: http://www.astm.org/Standards/fire-and-flammability-standards.html

[42] Alaska Fire Standards Council, Alaska, USA http://dps.alaska.gov/AFSC/

[43] Fire Commissioner of Canada Standards: http://www.hrsdc.gc.ca/eng/labour/fire_protection/policies_sta ndards

[44] A. Leonard, "Energy Cascade in Large-Eddy Simulations of Turbulent Fluid Flows", Advances in Geophysics A, Vol. 18, pp. 237–248, 1974.

[45] J. Smagorinsky, "General Circulation Experiments with the Primitive Equations", Monthly Weather Review, Vol. 91 (3), pp. 99–164, 1963.

[46] J. Deardorff, "A Numerical Study of Three-Dimensional Turbulent Channel Flow at Large Reynolds Numbers", Journal of Fluid Mechanics, Vol. 41 (2), pp. 453–480, 1970.

[47] Z.-C. Grigoraş, and D. Diaconu-Şotropa, "Establishing the Design Fire Parameters for Buildings", Bul. Inst. Polit. Iaşi, t. LIX (LXIII), f. 5, pp. 133-141, 2013.

[48] H.-J. Kim, and D. G. Lilley, "Heat Release Rates of Burning Items in Fires", 38th Aerospace Sciences Meeting & Exhibit, Reno, Nevada, USA, 10-13 January 2000, AIAA 2000-0722.

8

Ethics, Communication, and Propaganda About Energetic and Environmental Topics

Francesca Marin[1], Alberto Mirandola[2]

[1]Department of Philosophy, Sociology, Education and Applied Psychology (FISPPA), University of Padua, Padua, Italy
[2]Department of Industrial Engineering, University of Padua, Padua, Italy

Email address:
francesca.marin@unipd.it (F. Marin), alberto.mirandola@unipd.it (A. Mirandola)

Abstract: This paper aims at disentangling and explaining the meaning of communication, which should not be confused with the mere transmission of information. The focus will be put on scientific and technological communication, mainly in the field of energetic and environmental topics. Communication is a profession practiced by speakers and writers working within the mass media (newspapers, television, internet, etc.), but other people are involved in communication issues, particularly when communication deals with scientific and technological topics. Scientists, technicians and professionals, and engineers in particular, have a great responsibility when participating in the spread of technical information, which should not be confused with propaganda, whose meaning is explained in the paper. The paper will be developed in two stages. Firstly, by offering a conceptual framework, it will be argued that communication, rightly understood, is a special kind of action that is characterized by an *ethical commitment*, which should permeate our daily life, in particular the professional experience. On the contrary, propaganda cannot constitute an authentic communicative context because it involves senders and receivers, and not interlocutors. Indeed, propaganda generally aims at influencing opinions, attitudes and actions of a specific target audience on the basis of senders' personal interest or ideological thinking. Secondly, practical examples will be provided in the scientific and technical fields, with particular attention given to energetic and environmental issues. In fact, this is a critical context, because people are generally not prepared to deeply understand this matter and can easily be manipulated. Some examples will show how a given reality can be partially presented or misrepresented when speaking about the concept of sustainability, the evaluation of health or safety risks, the assessment of the potentiality of renewable energy sources, the difference between energy sources and energy carriers, the interpretation of climate changes, or the ideological opposition to industrial initiatives.

Keywords: Communication, Information, Ethics, Propaganda, Technical Writing, Media, Public Perception, News

1. Introduction

"Environmentalism began with environmental communication": this sentence effectively explains that communication has a large influence on the public opinion and on one's relationships with others. Communication is also a profession, generally practiced by journalists, speakers of mass media, etc.

A professional speaker or writer should promote better relations with people (colleagues, customers, clients, citizens, society). He must know the audience he addresses and adapt his speaking or writing to the characteristics and the needs of this audience, using a suitable language. He should be conscious of his responsibility towards the others.

Professional writing or speaking must create relations with the audience, aiming to promote knowledge and achieve the common good. This attitude of professional writers or speakers is an example of "ethical communication". In the present paper, the conceptual setting of professional communication will be stated, emphasizing and examining in depth the meaning and the differences between ethical communication and propaganda.

Particular attention will be given to communication regarding scientific and technological topics, which are often dealt with not only by scientists and technicians, but also by politicians and journalists. It will be pointed out that these topics should be faced by people who have proper knowledge of them; their writing or speaking should be objective, oriented to the improvement of the society and/or the environment, and not influenced by ideology, personal profit or for the benefit of a given group of people (a political party, association, etc.).

2. Propaganda and Communication

Nowadays communication flows easily from a wide variety of sources, and information can be gathered by everyone, at every time and everywhere thanks to the availability of different and widespread communication devices and channels. Although there are positive effects of these communication processes, such as availability, accessibility and affordability of information, the risk of deception and intentional control of information is higher than it was in past years. For example, the use of strategies of persuasion, manipulation and propaganda is common, often hidden, and can be found in all aspects of daily life.

For the purpose of this paper, we will focus on communication processes that characterize professional experience, emphasizing and analyzing the meaning and the differences between ethical communication and propaganda.

Generally, propaganda is the spread of ideas and information aimed at influencing opinions, attitudes and actions of a specific target audience.[1] This deliberate attempt to shape perceptions and direct behaviors arises from different motivations, such as private economic interests, political demagogy, and superficiality, and it is usually realized by presenting facts selectively (for example lying by omission) or in a distorted way. In this respect, a variety of approaches and devices are employed to circumvent or suppress the audience's ability to adequately judge the information. For example, an essential tool of propaganda is persuasion because its techniques and strategies aim to convince an audience and produce conviction. From this general definition, propaganda involves a communicative process, yet, can propaganda be totally included in the communicative field? To specify, can we really say that a professional who is making propaganda is communicating in the true sense of the word?

At first blush, we could provide an affirmative answer to such questions because propaganda is a dissemination of information, not objective and well-documented of course, but selective and distorted. Nowadays, when we think about communication we still attempt to describe it by the so-called "transmission model" or "standard view of communication". This model describes communication as a means of sending and receiving information, that is, as the mere transmission of a message. To exemplify, by communicating, information is sent as a message from a sender to a receiver, who is the target of the message. As a consequence, according to this model, communication is a linear and transactional process;

it is always unilateral, unidirectional and it is good or successful when the transmission of a message occurs effectively, removing anything that may slow down this transmission. In this respect, communication is regulated by the principles of efficacy and efficiency: to communicate effectively, information should be sent in a short time and with a minimal waste of resources. From a moral point of view, this model reduces the value of a communicative process to the proper functioning of a system, which is supposed to be effective and efficient.

Nevertheless, can communication be described as the mere transmission of a message from a sender to a receiver? Actually, the verb "to inform", and not "to communicate", suggests the act of sending and receiving information. To specify, communication may include the transmission of information, but cannot be reduced to it and this is due to the particular human interaction that characterizes every communication process – indeed, all the subjects involved in this process (speakers) are considered, from the beginning, as *interlocutors*. As a consequence, even when communication includes the transmission of information, those who receive the message are not deemed as mere receivers, but as interlocutors, that is, as speakers who cooperate within the communicative context. Rightly understood, communication is the *creation of a shared space*; in other words, to communicate means to disclose a shared space among interlocutors.[2]

These considerations are confirmed by the etymology of the word "communication" – indeed, this term derives from the Latin noun *communicatio*, which refers to the Latin verb *communicare* meaning "to make common", "to give to someone a share of something".[3] The notion of sharing or imparting is then intrinsic to the term "communication" because to communicate means to create a shared space of joint participation. As a consequence, involving interlocutors (and not merely senders and receivers), a communicative process is bidirectional.

On the contrary, propaganda is always unilateral and unidirectional and cannot constitute an authentic communicative context. Indeed, propaganda is the mere transmission of a message which has been intentionally selected and distorted by senders in order to influence receivers' opinions, attitudes and actions. In this way the receiver of the message is never considered as an interlocutor.

3. Considering Communication in the Category of Action

The previous analysis of the meaning and differences between propaganda and communication addresses a further relevant aspect of the latter: by involving interlocutors, the communicative field is characterized by an *ethical commitment* because it is oriented toward *reaching understanding*. As suggested by the German philosopher

---

1 It is difficult to define propaganda because on the one hand it can and does find its place in many fields such as advertising, entertainment, politics, professional experience, and on the other hand it assumes different forms, such as agitative and integrative propaganda, and white, grey and black propaganda. For all these distinctions, see Jowett, O'Donnell (20156), pp. 1-33; Brunello (2014), pp. 171-175.

2 See Fabris (2006).

3 See Peters (1999), p. 7.

Jürgen Habermas within his communicative action theory, communication cannot be reduced to merely describing the world because it is a special kind of action[4] which is of intentional character and fulfills a certain function, which is to reach a shared understanding.[5] The possible achievement of mutual understanding is thus included within the communicative practice. In this way, the communicative field is characterized by a normative background, that is, by moral principles which are implicitly presupposed by all interlocutors. To specify, for Habermas the communication between a speaker and a listener inherently involves the following universal validity claims: comprehensibility, truth, sincerity (or truthfulness) and rightness. In fact, it is only on the basis of the reference to these validity claims that we might accept or contest (reject as invalid) someone else's statement, considering it comprehensible or incomprehensible, true or false, fair or unfair and having or lacking a truthful attitude by the speaker. Quoting Habermas, «The concept of communicative action presupposes language as the medium for a kind of reaching understanding, in the course of which the participants, through relating to a world, reciprocally raise validity claims that can be accepted or contested».[6] These claims thus have intersubjective validity, are often raised only implicitly by the speakers, and are open to both criticism and justification.

According to this framework, ethics is intrinsically involved in the communicative field. In particular, as communication is oriented toward reaching understanding, all interlocutors are tacitly involved in the practice of giving reasons: when required, speakers may be called to explain and rationally justify their speech act as true, correct and authentic. In other words, rightly understood, communication is a field of moral choices and decisions: communicative processes are characterized by an ethical commitment and all participants are responsible for the fulfillment of a shared understanding.

On the basis of the previous considerations we could say that, in order to develop a good communicative practice, criteria of objectivity, rightness, honesty, truth, and sincerity should firstly be fulfilled. Secondly, all subjects involved in this practice should be recognized as interlocutors and be treated in terms of equality and parity. Thirdly, the speaker should know the content of what he is communicating, which requires an authentic commitment of data checking and updating. Finally, faced with a propagandistic dissemination of ideas and information, a good interlocutor does not passively adopt its content, but rather reads up on and carefully examines this information. Otherwise he would be reducing himself to a receiver.

4. Professional Experience and Communication

From our point of view, a professional should be aware of all aspects mentioned above for the following reasons: a profession is a particular working activity socially recognized and carried out by those who have specific competences and knowledge acquired through lengthy academic and practical training. A profession is then characterized by a *public commitment*: when a professional is asked for a certain professional service, he does not relate to only one client, but to the entire community. This aspect is confirmed by the etymology of the word "profession": indeed, this term derives from the Latin verb *profiteri*, which means "to declare aloud or publicly". In other words, those who practice a profession are involved in a particular human relationship and are making a professional commitment towards others that do not possess their knowledge and skills. As a consequence, the professional-client relationship is always asymmetrical, that is, characterized by an inequality of expertise among the subjects involved, and it is, by its very nature, a fiduciary relationship. Indeed, the professional holds the balance of power and the client is therefore forced to trust him.

The aspects of public commitment and of the fiduciary character of every professional-client relationship should thus be taken into account within the communicative processes that characterize professional experience. When professionals communicate, they are addressing the entire society and should be aware that those who do not possess their knowledge and skills will be forced in some ways to trust them. Professionals thus ought to be honest and accurate in all communication and adapt their speaking or writing to the characteristics and needs of the audience.

All of these aspects are addressed by professional codes, which are a form of self-regulation aimed at dealing with the problem of asymmetry in the relationship between professional and client and at avoiding any possible instrumentalization of the latter. For example, the Code of Ethics of Engineers promulgated by the American Society of Mechanical Engineers (ASME) states that in the fulfillment of their professional duties *"Engineers shall issue public statements only in an objective and truthful manner"*.[7] As a consequence, by disseminating propaganda, a professional intentionally selects and distorts a message in order to influence clients' opinions, attitudes and actions. In this way, an instrumentalization of clients, but even of the entire society, is at stake.

5. Technical and Scientific Communication

Communication regarding scientific or technical subjects is tricky, because the people it addresses are generally not

---

4 As argued by philosopher J.L. Austin within his speech act theory, many utterances are performative because these statements express the action character, meaning they perform an act by the fact of their being uttered. An example of performative utterance is promising because by uttering "I promise" I am not merely saying something, I am performing an act of promise directed towards other people. See Austin (1962).
5 Habermas (1984). For an analysis of Habermas' discourse-based morality, see Donald Moon (1995) and Rehg (2011).
6 Habermas (1984), p. 99.

7 ASME (1998).

prepared to fully understand this matter and may be swayed by misleading ideas if the speaker has a personal interest in influencing them. The ethical codes of some technical associations are based on concepts that are considered milestones. For example, the ASME Code of Ethics previously mentioned states that *"Engineers shall hold paramount the safety, health and welfare of the public in the performance of their professional duties"*.[8] However the professional duty of engineers is not only to design machines, instruments or processes, but also to inform the people correctly about what is really dangerous for health or safety and how to assess the cost/benefit ratio of a given technology. This means that engineers are not only builders and designers, but also have social responsibilities in communicating with policy makers and people about scientific and technical subjects.

In the mass media, the industrial society and its products (machinery, devices, chemical processes, plants) are often associated with many kinds of dangers, a lack of safety, environmental pollution, etc. According to this feeling, public opinion, generally oriented by certain "opinion makers", thinks that these negative features often prevail over the positive effects and are harmful for people's health and safety. To a certain extent this is true; but is only a part of the truth. In fact, technology has brought about the progress of medicine, the creation of products and systems for personal hygiene, the development of safety standards and corresponding equipment to reduce the effects of dangerous events; and, above all, the availability of a great number of resources which have made it possible to sustain a growing population and increase man's life span in the developed countries. Correct information must consider all these features: the natural and human systems are very complex and their equilibria are difficult to analyze, understand and explain.

Science and technology run together through the years and influence each other, causing an increase of knowledge and awareness about the development and interaction of human actions and natural events. Scientists and technicians have inside themselves the "culture of doubt", which promotes knowledge through research of new ways, processes and products. This culture is in contrast with the "fideistic" and ideological attitude of certain groups of persons who tend to simplify knowledge, are inclined to be self-confident in their own truth and refuse to modify their ideas. The information spread by these people, or people inspired by personal or political interest, is not reliable and can be misleading and sometimes dangerous.

Nowadays a lot of news and information about technical and scientific topics can be obtained from the web. However, everyone can introduce information into the web without any filter or controls. Therefore, on the web we can find both information and misinformation, communication and propaganda, truth and falsities. Certainly, this source of information can be very useful and contribute to knowledge when used in the right way by wise and qualified people, but on the Internet we can find everything and its opposite, in which case knowledge is reduced to mere opinion.

The scientific environment has also its communication problems. Scientific information is commonly spread via papers written by scientists and researchers. This is a good practice; however, some researchers write a large number of papers in order to improve their academic carrier[9] and the haste and anxiety of publishing often does not result in good quality. Even though these papers are generally submitted for peer review, these reviews are not always reliable. In fact, each outstanding reviewer is requested to examine many papers and sometimes they do not have enough time to elaborate upon their content: their review may sometimes be superficial. This is why the Impact Factor, which is generally considered a good indicator, is not always trustworthy. In any case, peer review is overall a good assessment system.

A practice used by scientific journalists when dealing with controversial topics is "balance treatment": the journalist presents the opposite opinions of scientists or technicians belonging to different currents of thought. The idea is to compare and then assess their different opinions, but if these opinions are well expressed and documented by apparently reliable arguments, where does the truth lie? Balance treatment can create more confusion than knowledge.

In conclusion, sometimes it is very difficult to distinguish between truth and falsity. Professional writers and speakers have a great responsibility towards the society when presenting scientific and technical topics.

6. Communication About Energetic and Environmental Topics

According to the ASME code, *"Engineers should consider environmental impact and sustainable development in the performance of their professional duties"*.[10] They should not only act accordingly in developing their work, but also inform people about the interaction between energy, economy and the environment; the meaning of sustainable development, which is connected to science, technology and economy, should be correctly explained. *Sustainability* does not simply refer to the impact of certain technologies on the environment, but is a more complex concept, because it also deals with:

- the resources needed to sustain human population (more than 7 billion in 2015 and still increasing);
- the social and economic impact;
- the constraints of nature and technology;
- the responsibility towards future generations.

These features are particularly challenging: technical and scientific communication should inform the public without falling into demagogy. For example, when it comes to resources and social organization, some opinion leaders of the so-called "green" people think that generally it would be

---

8 Ibidem.

9 In this respect, think about the academic "publish or perish" mandate (i.e. produce published work or you won't get tenure) which is clearly pervasive and can contribute to lots of bad writing.

10 Ibidem.

desirable *to let nature simply run with its own rhythms.* But they forget that, since the beginning of history, human beings have always interacted with nature and modified it, by abating forests and replacing them with farming, diverting rivers, breeding and eating animals, etc. They do not consider that nowadays our planet is very crowded; sometimes it is necessary to use the possibilities offered by technology, which is a valuable means that can prevent natural disasters or other dangerous events from occurring. Of course, technology must be used with moderation and wisdom. Energy conservation is a goal in all fields of activity: seeking high efficiency and rational organization of energy systems is paramount. Nevertheless, it is important to distinguish utopian projects and their effects from realistic ones.

As for renewable energy sources, their *real* potential must be explained to people, to avoid unrealistic reliance upon them. These sources have a very important positive characteristic: they are renewable for indefinite time and have a low environmental impact during operation. However, they also have some negative features:

a) Very low power density: the collection of significant amounts of power requires large surfaces (i.e. a huge request of areas for power plants) and hence a high consumption of materials, which must be extracted from the earth and treated by industrial manufacturers to obtain the requested products. Therefore, while the environmental impact during operation is negligible, the impact for the construction of the plants, their maintenance and final decommissioning and disposal is higher.

b) Unpredictable and variable availability, not consistent with the needs of the users. These features require the use of integrative and/or storage systems, often based on the exploitation of non-renewable resources.

As a consequence, it is utopian to think that renewable sources will be able to cover a large percentage of energy needs within a few decades. Consider the fact that in 2015 fossil fuels cover about 85% of the total energy supply in the world. Will it be possible to transform the world economy so rapidly in such a way? The authors think that spreading these ideas is misleading; but very often these concepts are developed by newspapers and other communication systems (Internet, television, etc.).

These issues are very complex: to face them, it is important to be able to "reason by systems", which is typical of modern scientists. In any case, the so-called "3 E's" (Energy, Environment, Economy) must always be considered, because these three aspects are closely connected to each other. There is a conflict between the need for resources in our crowed world, the environmental impact and the economy: finding the most acceptable compromise between these features is a very challenging problem for politicians, scientists, economists and engineers. These concepts must be correctly explained to the people.

The products delivered to the environment through the combustion of fossil fuels (gases, ashes, etc.) are polluting, of course: all the methods and systems able to abate and limit this pollution are paramount. Much progress has been made in this field: the pollution of engines, boilers, furnaces, etc. has been reduced year by year. People must be informed about this issue. Some incorrect information can be found regarding the pollution of our cities. While it is true that the concentration of products in the atmosphere often exceeds the limits imposed by regulations in some areas, these limits, stated in recent years, are very low. The concentration of products in some crowded areas, even though it may often go above the limits, is much lower than in the past decades (when regulations did not exist), thanks to the progress of technology. Can we remember the environmental situation in London or in some cities of the Po Valley in the 1950's or 1960's? This situation is much better now. Nevertheless, newspapers and television almost every day inform us about the overrunning of limits and its terrible effects, without saying anything about the improvements mentioned above. Incorrect information is not only telling lies, but also hiding part of the truth.

A recurring piece of information in the mass media is the number of deaths caused by pollution in a certain area. Is it really possible to make this assessment so precisely? And if this information is true, how can we explain that the length of human life in the cities of the developed countries has been continuously increasing?

A different question is the situation in some developing areas, where rampant industrialization brings huge problems. These countries should learn some lessons from the history of the "old" industrialized economies; it does not make sense that the new industrial countries should try to follow the path of the old ones. The recent methods now offered via scientific and technological knowledge can help foster more sustainable development, without making the same mistakes of the past.

A topic which gives rise to continuous controversy is the disposal or treatment of rubbish. The amount of rubbish produced in developed countries is very high; too high certainly. An obvious way to face this problem is through a significant reduction of this production. But the effects of this policy can be reached in the long term through suitable organization. In the meantime, the rubbish must be treated in some way. The classical solution in the past was to dispose of it into dumping grounds. Of course, this is not a good solution. A rational policy is the combination of:

- waste separation;
- recycling of some of the rubbish;
- incineration of a suitable part of the rubbish with energy conversion;
- disposal of the remaining part and incineration residues into well-managed dumping grounds.

Generally, incineration provokes heated discussions and is obstructed in every way, claiming that this technology is highly polluting and is harmful for public health. First of all, it is obvious that the combustion of every substance causes the release of products into the environment. This is why many technological systems have been introduced to minimize these emissions and are applied in all the plants burning fuels, fossil or not. A typical statement opponents make is that incineration causes the emission of unacceptable amount of macro- and micro-pollutants, mainly dioxin. If we consider the modern incinerators, this is not true: the emission of dioxin is close to zero; this result is obtained by the use of abatement systems and a continuous control of the combustion temperature, which must be kept between 850° and 950°C. Combustion is rigorously controlled on-line not only by plant operators, but also by public assessment boards. On the contrary, relatively high amounts of dioxin and other harmful products are emitted, for example, during violent demonstrations, when rubbish skips and other devices are burnt at low temperature in the streets among the people. The mass media generally gives glaring information about the (previously mentioned) opposition to incineration, but does not emphasize the emissions generated during those demonstrations.

The opposition to the incinerators is only one example of what can be observed whenever a project for a new industrial plant is submitted to a local administration: the initial reaction of the people is to strongly oppose it, a typical NIMBY reaction. Generally this attitude of the people is not rationally motivated, but is due to the influence of the mass media and small groups of individuals who are "specialized in opposition": they spread incorrect information about the consequences of the plants' operation on public health or safety. And they find a lot of followers: in fact, people are more attracted by negative than positive news, and when a danger is supposed to occur, they are prone to believe it. This is why it is very important to spread reliable information; scientists and technicians' responsibility in this matter is crucial. Facing a problem like this, the correct thinking is: if the plant is not built, what would the alternative realistic solution be? And would it be better or worse? And what about the cost and the social consequences of each alternative?

As a matter of fact, every human action includes a certain degree of risk and it is important to give this information to the people. Without risk, nothing would be done, no contribution to knowledge would be offered and each action would be bureaucratic and devoid of intelligence. The problem is the acceptability of an action or event having a given kind of risk, which mainly depends on three factors:[11] the probability of the occurrence of a danger, its consequences if the danger itself occurs, and the benefits coming from that action. These items must be carefully assessed and compared. Of course, this assessment should be performed by experts, weighting all the aspects and considering the general good of the community and the environment. Hence, the experts have the additional responsibility of informing the authorities and the people directly, or through mass media.

Generally, the people and also many politicians do not know what the difference is between "energy sources" and "energy carriers". One of the authors remembers that a certain politician, during an interview, said that "hydrogen will be the energy source of the future". This statement is not correct and creates confusion and false expectations among the audience: hydrogen is not a source, but, like electricity, is a transformation product, i.e. a carrier.

A question that must be carefully presented and discussed is that of the public incentives granted to encourage the start-up of a new technology, for example an innovative system exploiting renewable energy sources. In the authors' opinion, the incentives, paid by the community, can be granted for a limited time and should not be too high: in the long term, every technology should be self-sustaining, otherwise it would charge the community unacceptably. Correct information has to be spread regarding this matter.

7. Climate Changes and Information

The climate on the earth is the result of delicate and complex equilibria, which have been ruled by natural phenomena throughout the centuries since the origin of our planet.[12] Many different parameters influence the climate, mainly depending on the solar system: they determine the global and local temperatures, the composition of the atmosphere, the greenhouse effect, etc.

In the history of the world these parameters have seen remarkable variations; it may be said that the climate is continuously varying. Generally these variations have been very slow, but sometimes accelerations have been caused by specific causes. Human beings' curiosity about climate has pushed them to observe and study it since the prehistoric age, when the changes of the weather were mainly ascribed to the whims of gods.

In modern society, science and technology interact to give rational explanations to these phenomena. Many efforts are being made to assess the influence of the various parameters that are likely to determine the climate and the local weather; the possibility of forecasting short-term weather and the long-term trends of the climate is one of the main targets of these studies.

Understanding the mechanisms which regulate the climate is very challenging, because the interaction of multiple parameters can hardly be expressed by suitable systems of equations. Moreover, the values and the trends of these parameters from the past are not well known, because only in recent years have the measurement techniques, combined with proxy data, given reliable results.

This difficulty should encourage caution: scientists, aware of these difficulties, are continuously trying to improve their knowledge in this field and approach acceptable results step by step, i.e. slowly. However, a problem arises when politicians intervene and interfere with them. Generally politicians are not guided by the wish of obtaining knowledge, but by political interest and/or ideology. Therefore, the results of research are bent to a typical target of politicians: politics needs short-term solutions to get consensus, even though these solutions are not

11 European Commission (2000).

12 See Behringer (2007).

easy to reach. This attitude leads them to misrepresent the truth or hide part of it.

Returning to the question of climate, at present the prevailing current of thought is that nowadays the earth's climate is affected almost completely and exclusively by human activities, and particularly by the amount of greenhouse gases (mainly CO_2) emitted into the atmosphere.[13] The supporters of this statement are generally self-confident and self-referential, even if some other scientists[14] think that the influence of nature, which was the motor for the climate's changes throughout the whole history of our planet, cannot have become irrelevant in only a few decades. Atmospheric pollution must not be confused with climate; everyone accepts that preserving the environment and conserving natural resources are necessary, but this does not mean that political actions will be able to influence the climate significantly.

Once again, science should be cautious when dealing with such difficult problems, and communication should take this into account. On the contrary, the scientists who are not in line with the majority are often derided and deprived of research funds by "politically correct" lobbies or associations. At the current status of research, knowledge of past climate trends and causes is still uncertain because of the lack of reliable measurements, proxy data and testimonies. This uncertainty does not help us to completely understand the influence of countless parameters and events. In conclusion, these difficulties should result in more humility and caution in discussions of these issues, leaving room for doubt, which is the correct attitude held by the serious scientists and should also be that of the mass media.

8. Conclusion

Focusing on communication and propaganda about energetic and environmental topics, the aim of this paper was to address the ethical commitment that inherently characterizes communicative practice: involving interlocutors, and not merely senders and receivers, communication is the creation of a shared space and is oriented toward reaching understanding. As a consequence, communicative practice cannot be reduced to the transmission of information (although it might include this aspect) and propaganda is not an authentic communicative context because it involves senders and receivers (as noted, the former intentionally selects and distorts a message in order to influence the latter's opinions, attitudes and actions).

Communication is thus a field of moral choices and decisions, and all participants are responsible for the fulfillment of a shared understanding. Applied to scientific and technological communication, this means that scientists, technicians and professionals, engineers in particular, bear a great responsibility with respect to the spreading of technical information. Indeed, their communication should fulfill the criteria of objectivity, rightness, honesty, truth and sincerity in order to conveniently direct both citizens' actions and politicians' decisions. Furthermore people this communication addresses cannot passively accept its content: to be interlocutors, and not mere receivers, they should read up on and carefully examine this information.

In conclusion, the development of a good communicative practice requires that all participants acknowledge the ethical commitment intrinsic to communication. Nevertheless, this acknowledgement is a necessary, but not sufficient, condition: for example, someone might be aware of the ethical dimensions of communication yet decide to spread propaganda. Indeed, the promotion of a good communicative practice should move from the so-called "question of meaning", that is, *why* we should encourage good communication. In other words, why should we be interlocutors, and not mere senders and receivers? There are basically two closely related reasons. Firstly, a good communication safeguards the self and the other because it avoids the achievement of the former to the detriment of the latter and vice versa. Secondly, promoting good communication means setting up the conditions for the communication to continue itself: indeed, in the absence of interlocutors, other speech forms such as propaganda can always be more vociferously encouraged up to a point where the space for interlocution is completely closed.

References

[1]   American Society of Mechanical Engineers (ASME), *Code of Ethics of Engineers*, June 10, 1998 (http://web.mit.edu/2.009/www/resources/mediaAndArticles/ASME_ethics.pdf).

[2]   Austin, J.L. (1962), *How to Do Things with Words*, Oxford University Press, Oxford.

[3]   Behringer, W. (2007), *Kulturgeschichte des Klimas. Von der Eiszeit zur globalen Erwärmung* 5, aktualisierte Auflage, C.H. Beck, München (English Translation: *A Cultural History of Climate*, Polity Press, London 2009; Italian translation: *Storia culturale del clima*, Bollati-Boringhieri, Torino 2013).

[4]   Brunello, A.R. (2014), *A Moral Compass and Modern Propaganda? Charting Ethical and Political Discourse.* Review of History and Political Science, Vol. 2, Issue 2, pp. 169-197.

[5]   Donald Moon, J. (1995), *Practical Discourse and Communicative Ethics*, in S.K. White (ed. by), *The Cambridge Companion to Habermas*, Cambridge University Press, Cambridge, pp. 143-164.

[6]   European Commission, *First Report on the Harmonisation of Risk Assessment Procedures*, 26-27 October 2000 (http://ec.europa.eu).

[7]   Fabris, A. (2006), *Etica della comunicazione*, Carocci, Roma.

[8]   Fred Singer, S. (2008), *Nature, Not Human Activity, Rules the Climate*; Report of the NIPCC, The Heartland Institute, Chicago. (https://www.heartland.org/sites/all/modules/custom/heartland_migration/files/pdfs/22835.pdf).

---

13 See Intergovernmental Panel on Climate Change (IPCC).
14 In this respect, see Scafetta (2000) and Fred Singer (2008).

[9] Habermas, J. (1984), *The Theory of Communicative Action 1. Reason and the Rationalization of Society*, Polity Press, Cambridge.

[10] Intergovernmental Panel on Climate Change (IPCC): Assessment Reports AR1 (1990), AR2 (1995), AR3 (2001), AR4 (2007), AR5 (2013).

[11] Jowett, G.S., and O'Donnell, V. (2015^6), *Propaganda & Persuasion*, SAGE Publications, Singapore.

[12] Peters, J.D. (1999), *Speaking into the Air: A History of the Idea of Communication*, The University of Chicago Press, Chicago.

[13] Rehg, W. (2011), *Discourse Ethics*, in B. Fultner (ed. by), *Jürgen Habermas: Key Concepts*, Acument, Durham, pp. 115-139.

[14] Scafetta N., *Climate Change and Its Causes. A Discussion about some Key Issues*, Science & Public Policy Institute, March 18 2000. http://scienceandpublicpolicy.org/images/stories/papers/originals/climate_change_cause.pdf.

9

Investigation of Unsteady Mixed Convection Flow near the Stagnation Point of a Heated Vertical Plate embedded in a Nanofluid-Saturated Porous Medium by Self-Similar Technique

A. A. Abdullah[1], F. S. Ibrahim[2], A. F. Abdel Gawad[3], A. Batyyb[3]

[1]Department of Mathematical Sciences, Umm Al-Qura University, Makkah, Saudi Arabia
[2]Department of Mathematics, University College, Umm Al-Qura University, Makkah, Saudi Arabia
[3]Mech. Eng. Dept., College of Engineering and Islamic Architecture, Umm Al-Qura University, Makkah, Saudi Arabia

Email address:
batyyb@hotmail.com (A. Batyyb)

Abstract: This paper aims to study the problem of unsteady mixed convection in a stagnation flow on a heated vertical surface embedded in a nanofluid-saturated porous medium. The employed mathematical model for the nanofluid takes into account the effects of Brownian motion and thermophoresis. The presence of a solid matrix, which exerts first and second resistance parameters, is considered in this study. The self-similar solutions for the system of equations governing the problem are obtained. The resulting system of ordinary differential equations that govern the flow is solved numerically using fourth-fifth order Runge-Kutta with shooting method. Numerical results for the dimensionless velocity, temperature and nanoparticle volume fraction as well as skin friction, Nusselt number and Sherwood number are produced for different values of the influence parameters.

Keywords: Unsteady Mixed Convection, Self-Similar Solution, Nanofluids, Stagnation, Porous Media

1. Introduction

The study of mixed convection flow has applications in several industrial and technical processes such as nuclear reactors cooled during emergency shutdown, solar central-receivers exposed to winds, electronic devices cooling by fans, heat exchanges placed in a low-velocity environment, etc. The mixed convection flows become important when the buoyancy forces, due to the temperature difference between the wall and the free stream, become large. The mixed convection flow in the stagnation flow region of a vertical plate has been investigated by Ramachandra et al. [1].

When there is an impulsive change in the velocity field, the inviscid flow develops instantaneously, but the flow in the viscous layer near the wall develops slowly which becomes fully-developed steady flow after sometime. For a small period of time, the flow is dominated by the viscous forces and the unsteady acceleration, but for a large period of time, it is dominated by the viscous forces, the pressure gradient and the convective acceleration. Scshadri et al. [2] studied the unsteady mixed convection flow in the stagnation region of a heated vertical plate due to impulsive motion. The boundary-layer flow development of a viscous fluid on a semi-infinite flat plate due to impulsive motion of the free stream was investigated by Hall [3], Dennis [4] and Watkins [5]. The corresponding problem over a wedge was studied by Smith [6], Nanbu [7], and Williams and Rhyne [8].

Kumari [9] examined the temporal development of momentum and thermal boundary layers on an impulsively-started wedge with a magnetic field and has obtained the solution numerically starting from the initial steady state to the final steady state. The flow development of the laminar boundary on an impulsively-started translating and spinning rotational symmetric body was considered by Ece [10]. Qzturk and Ece [11] studied the unsteady forced convection heat transfer from a translating and spinning body. Brown and Riley [12] presented an analysis that covered three distinct phases in the temporal development of the free convection flow past a suddenly heated semi-infinite vertical plate. The unsteadiness in the flow field arises due to the step change in wall temperature. Ingham [13] has considered essentially the same problem as that of Brown and Riley [12], but instead of taking the step change in wall temperature, the wall temperature T_∞ is suddenly raised to $T_w = T_\infty + Ax^m$, where A is a positive constant, m is a constant and x is the distance measured from the leading edge of the plate.

The mixed convection flow at a two-dimensional stagnation point was investigated by Amin and Riley [14]. The forced flow is a stagnation point flow and the free convection part is due to a pressure gradient that is induced by temperature variations along the boundary. Hassanien, et al. [15] analyzed the problem of unsteady free convection flow in the stagnation-point region of a rotating sphere embedded in a porous medium. The unsteady flow and heat transfer of a viscous fluid in the stagnation region of a three-dimensional body embedded in a porous medium was investigated by Hassanien, et al. [16]. Hassanien and Al-Arab [17] studied the problem of thermal radiation and variable viscosity effects on unsteady mixed convection flow in the stagnation region on a vertical surface embedded in a porous medium with surface heat flux

The research topic of nanofluids has received considerable interest worldwide. Inherently low thermal conductivity is a primary limitation in developing energy-efficient heat transfer fluids that are required for ultrahigh-performance cooling. A very small amount of guest nanoparticles, when dispersed uniformly and suspended stably in host fluids, can provide dramatic improvements in the thermal properties of the host fluids. According to Yacob et al. [18], nanofluids are produced by dispersing the nanometer-scale solid particles into base liquids with low thermal conductivity such as water and ethylene glycol. Nanoparticles are usually made of metal, metal oxide, carbide, nitride and even immiscible nano-scale liquid droplets. Congedo et al. [19] compared different models of nanofluid (regarded as a single phase) to investigate the density, specific heat, viscosity and thermal conductivity and discussed the water–Al2O3 nanofluid in details by using CFD. Hamad et al. [20] introduced a one-parameter group to represent similarity reductions for the problem of magnetic field effects on free-convective nanofluid flow past a semi-infinite vertical flat plate following a nanofluid model proposed by Buongiorno [21]. Hamad [22] obtained the analytical solutions for convective flow and heat transfer of a viscous incompressible nanofluid past a semi-infinite vertical stretching sheet in the presence of magnetic field. Khan and Pop [23] obtained similarity solutions depending on Prandtl, Lewis, Brownian motion and thermophoresis numbers on the steady boundary-layer flow, heat and mass over a stretching surface in its plane. Further, Abu-Nada and Chamkha [24] presented the natural convection heat transfer characteristics in a differentially-heated enclosure filled with a CuO–EG–water nanofluid for different variable thermal conductivity and variable viscosity models. For more information, see also Das et al. [25], and

Kakaç and Pramuanjaroenkij [26]. Muthtamilselvan et al. [27] claimed that it is difficult to have a precise idea on how nanoparticles enhance the heat transfer characteristics of nanofluids.

In this study, our main objective is to analyze mixed convection in stagnation flow on a heated vertical surface embedded in a nanofluid-saturated porous medium. The effect of the presence of an isotropic solid matrix due to impulsive motion is considered. Moreover, we examine the combined effect of Brownian motion, thermophoresis parameters and nanoparticle fraction on boundary-layer flow and heat transfer and due to nanofluid. The governing boundary layer equations are transformed to a two-point boundary-value problem using similarity variables. These are numerically solved using fourth-fifth order Runge–Kutta method with shooting technique. The effects of governing parameters on fluid velocity, temperature and particle concentration are discussed and shown graphically and in tables as well.

2. Mathematical Analysis

Let us consider a semi-infinite vertical plate embedded in a saturated porous medium with temperature T_w and concentration ϕ_w. The ambient temperature and concentration, respectively are T_∞ and ϕ_∞. At $t = 0.0$ the ambient fluid is impulsively moved with a velocity U_e and at the same time the surface temperature is suddenly raised. The flow field is over a heated vertical surface where the upper half of the field is assisted by the buoyancy force. However, the buoyancy force opposes the lower part. The surface of the plate is assumed to have arbitrary temperature and concentration. The physical flow model and coordinate system is shown in Fig. 1.

Under above assumptions along with Boussinesq and boundary layer approximations, the governing equations of the conservation of mass, momentum, energy and nanoparticles volume fraction can be expressed as:

$$\frac{\partial u}{\partial x} + \frac{\partial v}{\partial y} = 0 ,\qquad (1)$$

$$\frac{\partial u}{\partial t} + u\frac{\partial u}{\partial x} + v\frac{\partial u}{\partial y} = \frac{\partial U_e}{\partial t} + U_e\frac{\partial U_e}{\partial x} + v\frac{\partial^2 u}{\partial y^2} + \frac{v}{K}(U_e - u)$$
$$+ \frac{\Gamma}{K^{1/2}}(U_e^2 - u^2) + [(1-\phi_\infty)\beta(T - T_\infty) \qquad (2)$$
$$- (\rho_p - \rho_f)(\phi - \phi_\infty)/\rho_f]g$$

$$\frac{\partial T}{\partial t} + u\frac{\partial T}{\partial x} + v\frac{\partial T}{\partial y} = \alpha\frac{\partial^2 T}{\partial y^2} + \tau\left[D_B\frac{\partial T}{\partial y}\frac{\partial \phi}{\partial y} + \frac{D_T}{T_\infty}\left(\frac{\partial T}{\partial y}\right)^2\right], \qquad (3)$$

$$\frac{\partial \phi}{\partial t} + \left(u\frac{\partial \varphi}{\partial x} + v\frac{\partial \phi}{\partial y}\right) = D_B\frac{\partial^2 \phi}{\partial y^2} + \frac{D_T}{T_\infty}\left(\frac{\partial^2 T}{\partial y^2}\right), \qquad (4)$$

Where, x and y are the coordinates along and normal to the surface, respectively. v is the kinematic viscosity, time is denoted by t, u and v are the velocity components along the x and y directions, respectively. K is the permeability of the porous medium and Γ is the empirical constant in the second-order resistance. T is the temperature, ϕ is the solid volume fraction, $\alpha = k / (\rho C)_f$ is the thermal diffusivity, D_B

is the Brownian diffusion coefficient, D_T is the thermophoretic diffusion coefficient, $\tau = (\rho C)_p / (\rho C)_f$ is the ratio of the effective heat capacity of the nanoparticle material to the heat capacity of the fluid, g is the acceleration due to gravity.

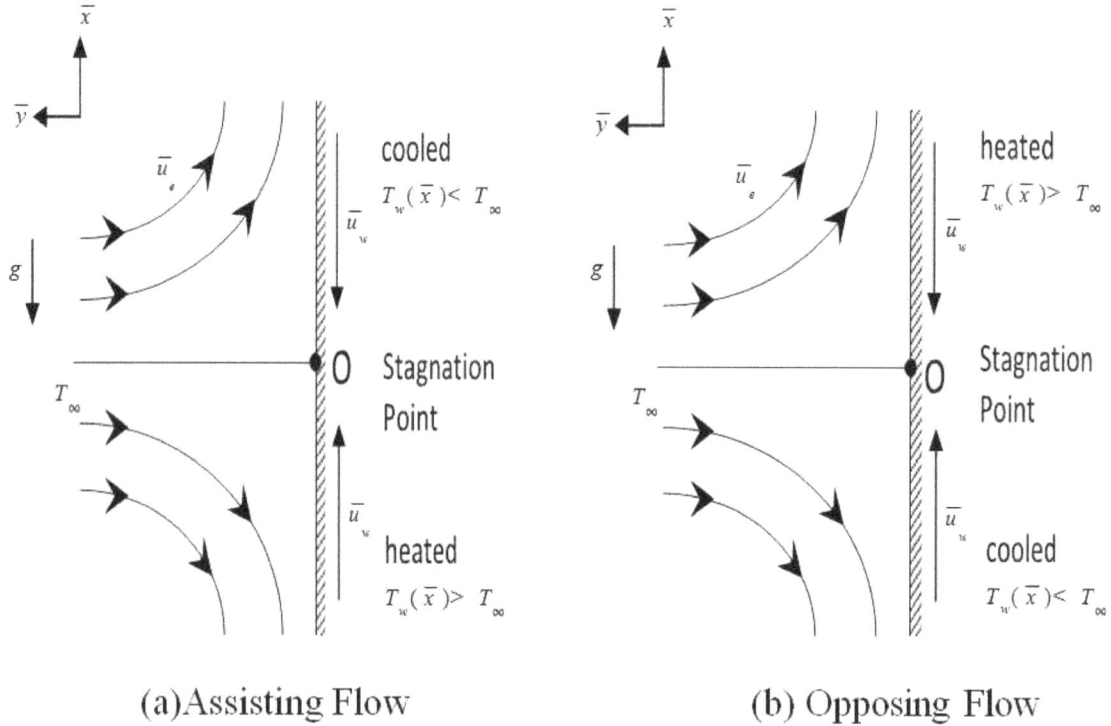

(a) Assisting Flow   (b) Opposing Flow

Fig. 1. *Physical model and coordinate system.*

The proposed initial conditions are

$$u(x,y,t) = v(x,y,t) = 0, \quad T(x,y,t) = T_\infty, \quad \phi(x,y,t) = \phi_\infty \text{ for}$$
$$t < 0. \tag{5}$$

The proposed boundary conditions for $t \geq 0$ are

$$u(t,x,0) = v(t,x,0) = 0, \quad u(t,x,\infty) = U_e = ax/(1-ct), \quad a > 0,$$
$$T(t,x,0) = T_w(t,x) = T_\infty + bx/(1-ct)^2, \quad T(t,x,\infty) = T_\infty, \tag{6}$$
$$\phi(t,x,0) = \phi_w(t,x) = \phi_\infty + bx/(1-ct)^2, \quad \phi(t,x,\infty) = \phi_\infty, \quad b,c > 0,$$

where a and c are constants (with a > 0 and c ≥ 0, where ct < 1), and both have dimension time^{-1}, while b is a constant and has dimension temperature/length, with b > 0 and b < 0 corresponding to the assisting and opposing flows, respectively, and b = 0 is for forced convection limit (absence of buoyancy force).

To transform equations (1-4) into a set of ordinary differential equations, we now introduce the following dimensionless quantities; the mathematical analysis of the problem is simplified by introducing the following similarity transforms:

$$\eta = \sqrt{\frac{a}{\upsilon(1-ct)}} y, \quad \psi = \sqrt{\frac{a\upsilon}{1-ct}} x f(\eta),$$
$$\theta(\eta) = \frac{T-T_\infty}{T_w - T_\infty}, \quad \varphi(\eta) = \frac{\phi - \phi_\infty}{\phi_w - \phi_\infty} \tag{7}$$

The equation of continuity is satisfied if we choose a stream function $\psi(x,y)$ such that $u = \dfrac{\partial \psi}{\partial y}$, $v = -\dfrac{\partial \psi}{\partial x}$.

Using the similarity transformation quantities (7), the governing Eqs. (2–6) are transformed to the ordinary differential equation as follows:

$$f''' + 1 + A(1 - f' + \frac{1}{2}\eta f'') - f'^2 + f f'' + \lambda(\theta - N_r \varphi)$$
$$+ \gamma(1 - f') + \Delta(1 - f'^2) = 0 \tag{8}$$

$$\frac{1}{Pr}\theta'' - A(2\theta + \frac{1}{2}\eta\theta') - f'\theta + f\theta' + N_b\theta'\varphi' + N_t\theta'^2 = 0 \tag{9}$$

$$\frac{1}{Le}\varphi'' - A(2\varphi + \frac{1}{2}\eta\varphi') - f'\varphi + f\varphi' + \frac{1}{Le}\frac{N_t}{N_b}\theta'' = 0 \tag{10}$$

With the boundary conditions:

$$f'(\eta) = 0, \quad f(\eta) = 0, \quad \theta(\eta) = 0, \quad \varphi(\eta) = 0,$$
$$for \ t < 0,$$
(11)

and proposed boundary conditions for $t \geq 0$ are

$$f(0) = 0, \quad f'(0) = 0, \quad \theta(0) = 1, \quad \varphi(0) = 1,$$
$$f'(\infty) = 1, \quad \theta(\infty) = 0, \quad \varphi(\infty) = 0,$$
(12)

Where the governing parameters are defined as:

$$A = \frac{c}{a}, \quad Pr = \frac{v}{\alpha}, \quad Le = \frac{v}{D_B}, \quad \lambda = \frac{Gr_x}{Re_x^2},$$

$$\gamma = \frac{1}{Da_x Re_x}, \quad \Delta = \frac{\Gamma}{Da_x^{1/2}}, \quad N_b = \frac{\tau}{v} D_B (\phi_w - \phi_\infty),$$
(13)

$$N_t = \frac{\tau}{v}\left(\frac{D_T}{T_\infty}\right)(T_w - T_\infty), \quad N_r = \frac{(\rho_p - \rho_f)(\phi_w - \phi_\infty)}{\rho_f \beta_T (1 - \phi_\infty)(T_w - T_\infty)}$$

Where f', θ and φ are the dimensionless velocity, temperature and particle concentration, respectively. The prime denotes differentiation with respect to the similarity variable η. $Gr_x = \frac{g\beta(T_w - T_\infty)x^3}{v^2}$, $Re_x = \frac{U_e x}{v}$ and $Da_x = \frac{K}{x^2}$ are Grashof, Reynolds and Darcy numbers, respectively. A, λ, Pr, Le, N_T, N_B, N_r, γ and Δ denotes the unsteadiness parameter, mixed convection parameter, Prandtl number, Lewis number, thermophoresis parameter, Brownian motion parameter, nanofluid buoyancy ratio, first resistant parameter and second resistant parameter, respectively.

The important physical quantities of interest in this problem are the skin friction coefficient (wall shear stress) C_f, local Nusselt number Nu_x and the local Sherwood number Sh_x that are defined as:

$$C_f = \frac{\mu}{\rho U_\infty^2}\left(\frac{\partial u}{\partial y}\right)_{y=0}, \quad Nu_x = \frac{x}{(T_w - T_\infty)}\left(\frac{\partial \theta}{\partial y}\right)_{y=0},$$
$$Sh_x = \frac{x}{(\phi_w - \phi_\infty)}\left(\frac{\partial \phi}{\partial y}\right)_{y=0}$$
(14)

Substituting Eq. (7) into Eq. (18), we get

$$C_f = Re_x^{-1/2} f''(0), \quad Nu_x = -Re_x^{-1/2} \theta'(0),$$
$$Sh_x = -Re_x^{-1/2} \varphi'(0)$$
(15)

3. Results and Discussion

The resulting differential systems (8)–(10) subjected to the boundary conditions (12) are solved numerically through fourth-fifth order Runge–Kutta method (RK45) using a shooting technique. The values of the governing parameters are chosen arbitrary. However, the numerical results are presented for some representative values of these governing parameters. In order to see the physical insight, the numerical values of velocity $f'(\eta)$, temperature $\theta(\eta)$, and nanoparticle volume fraction $\varphi(\eta)$ with the boundary layer have been computed for different parameters as unsteadiness parameter A, mixed convection parameter λ, nanofluid buoyancy ratio parameter N_r, thermophoresis parameter N_t, Brownian motion parameter N_b, first resistant parameter γ, second resistant parameter Δ. Prandtl number Pr and Lewis number Le.

Table 1 indicates results for wall values for the gradients of velocity, temperature and volume fraction functions which are proportional to the friction factor, Nusselt number and Sherwood number, respectively. From this table, we notice that as N_r increases, the friction factor increases, the heat transfer rate (Nusselt number) and mass transfer rate (Sherwood number) decrease. As N_t and N_b increase, the friction factor and surface mass transfer rates increase whereas the surface heat transfer rate decreases

Table (1). Effect of Nt, Nb and Nr on f`(0), θ`(0) and φ`(0) with A=0.5, λ=1, γ=0.5, Δ =0.5, Pr=10 and Le=10

N_b	N_t	$N_r=0.5$			$N_r=1$			$N_r=3$		
		f``(0)	-θ`(0)	-φ`(0)	f``(0)	-θ`(0)	-φ`(0)	f``(0)	-θ`(0)	-φ`(0)
	0.1	1.86358	1.98266	2.99959	1.75281	1.97640	2.97942	1.29405	1.94971	2.89209
	0.3	1.87689	1.46341	3.00885	1.74724	1.46030	2.96609	1.20280	1.44720	2.77490
0.1	0.5	1.89459	1.18916	3.30288	1.75777	1.18808	3.23693	1.17626	1.18396	2.93645
	0.7	1.91206	1.02290	3.60614	1.77309	1.02289	3.52129	1.17597	1.02359	3.12943
	0.9	1.92805	0.91043	3.88186	1.78917	0.91098	3.78222	1.18647	0.91421	3.31690
	0.1	1.91563	1.10471	3.50357	1.82008	1.10377	3.48831	1.42727	1.09997	3.42351
	0.3	1.93714	0.89987	3.66024	1.83864	0.89972	3.64082	1.43308	0.89932	3.55799
0.3	0.5	1.95564	0.78110	3.80811	1.85644	0.78133	3.78544	1.44766	0.78250	3.68860
	0.7	1.97137	0.70268	3.93471	1.87243	0.70308	3.90958	1.46449	0.70499	3.80212
	0.9	1.98482	0.64593	4.04428	1.88657	0.64642	4.01722	1.48123	0.64867	3.90151
	0.1	1.94644	0.69021	3.53118	1.85358	0.69053	3.51722	1.47246	0.69207	3.45816
	0.3	1.96535	0.60599	3.66488	1.87203	0.60649	3.64911	1.48897	0.60871	3.58235
0.5	0.5	1.98126	0.55157	3.77327	1.88824	0.55213	3.75618	1.50642	0.55460	3.68382
	0.7	1.99473	0.51253	3.86302	1.90230	0.51310	3.84492	1.52299	0.51563	3.76836
	0.9	2.00627	0.48247	3.93986	1.91454	0.48304	3.92098	1.53822	0.48556	3.84111

	0.1	1.96866	0.48162	3.52373	1.87681	0.48224	3.51037	1.50008	0.48497	3.45396
	0.3	1.98501	0.44299	3.63003	1.89347	0.44363	3.61569	1.51811	0.44640	3.55514
0.7	0.5	1.99883	0.41555	3.71507	1.90787	0.41617	3.70001	1.53507	0.41889	3.63650
	0.7	2.01059	0.39448	3.78574	1.92032	0.39508	3.77015	1.55044	0.39771	3.70439
	0.9	2.02073	0.37744	3.84655	1.93115	0.37802	3.83051	1.56425	0.38055	3.76297

The effects of the first and second resistances of the porous medium on the gradients of velocity, temperature and volume fraction functions are been illustrated in Table 2. From this table we conclude that both the first and second resistances enhance the wall shear stress and the mass transfer rate and reduce the heat transfer rate.

Table (2). *Effects of γ and Δ on $f\,'(0)$, $-\theta`(0)$ and $-\phi`(0)$ with A=0.5, λ=1, N_T=0.5, N_B=0.5, N_R=1, Pr=10 and Le=10*

γ	Δ	$f\,'(0)$	$-\theta(0)$	$-\phi(0)$
	0.0	1.74114	0.55465	3.71225
	2.0	2.42814	0.54777	3.86815
0.0	4.0	2.94365	0.54481	3.96362
	6.0	3.37615	0.54328	4.03332
	0.0	2.27894	0.54976	3.82839
	2.0	2.82588	0.54591	3.93648
2.0	4.0	3.27555	0.54397	4.01254
	6.0	3.66738	0.54292	4.07146
	0.0	2.69411	0.54729	3.90496
	2.0	3.16479	0.54480	3.98909
4.0	4.0	3.56984	0.54347	4.05272
	6.0	3.93132	0.54275	4.10394
	0.0	3.05265	0.54579	3.96396
	2.0	3.47213	0.54411	4.03302
6.0	4.0	3.84350	0.54318	4.08770
	6.0	4.18059	0.54269	4.13297

We computed the solutions for the dimensionless velocity, temperature and nanoparticle volume fraction in Figs. 2–10. The effects of all the parameters governing the problem are discussed.

The variation of the non-dimensional velocity, temperature and nanoparticle concentration for $N_t = 0.5, N_b = 0.5, N_r = 1.0, \lambda = 1.0, \text{Pr} = 10, Le = 10, \gamma = 0.5, \Delta = 0.5$ with unsteadiness parameter A is illustrated in Fig. 2. It can be observed from Fig. 2(a) that first for velocity distribution, there is a special point ($\eta \approx 1.4$) called 'crossing over point' and the velocity profiles have completely conflicting behavior before and after that point. The value of the velocity profile for fixed η increases before that point and slightly decreases after that. Thus, due to the increase of unsteadiness parameter, A, the velocity initially enhances but ultimately it increases the thickness of momentum boundary layer. On the other hand the temperature and nanoparticle volume fraction profiles of Figs. 2 (b, c), show that the unsteadiness controls the heat and mass transfer. The control of heat and mass transfer is of great practical significance. For the increase of A, the temperature and nanoparticle volume fraction at a point decreases because the thermal boundary-layer thickness rapidly decreases due to increase of unsteadiness.

The effect of mixed convection parameter λ on the non-dimensional velocity, temperature and nanoparticle volume fraction is illustrated in Fig. 3. From these figures, it is

observed that the velocity is increased as λ increase, but both the temperature and nanoparticle fraction decrease with increasing λ.

The velocity profile decreases and both the temperature and nanoparticle volume fraction increases as nanofluid buoyancy ratio parameter N_r increases as showed in in Fig. 4.

Figure5 presents the effect of thermophoresis N_t on the velocity, temperature and volume fraction distributions. It is observed that the momentum boundary-layer thickness increases with an increase of N_t. As the parameter N_t increases, the thermal and nanoparticle volume fraction boundary-layer thickness increase for the specified conditions.

Furthermore, increasing the value of the Brownian motion N_b causes a thickening of momentum and thermal boundary layers, whereas thinning of the nanoparticle volume fraction boundary layer as shown in Fig.6. Physically, it is true due to the fact that the large values of the Brownian motion parameter impacts a large extent of the fluid. It results in the thickening of the momentum and thermal boundary layers. Hence, the present analysis shows that the flow field is appreciably influenced by the Brownian motion N_b.

The effect of both first resistant parameter γ and second resistant parameter Δ is showed in Fig. 7 and Fig. 8, respectively. From these figures, it is observed that both γ and Δ have the same behavior. The velocity profile increases but the temperature profile and volume fraction profile decreases as both γ and Δ increases.

The effect of Prandtl number Pr is illustrated in Fig. 9. From these figures, it is observed that as Pr increases the velocity and temperature profiles decrease. On the other hand, for the volume fraction profiles there is a crossing over point at ($\eta \approx 0.5$) where the volume fraction profile decreases before that point and slightly increases after that.

It is noticed, from Fig. 10 that an increase in the Lewis number Le results in an increase in the velocity, but in a decrease of the volume fraction within the boundary layer. In the temperature profile there is a crossing over point at ($\eta \approx 0.8$) where the temperature profiles have completely conflicting behavior before and after that point. The value of the temperature profile for fixed η decreases before that point and slightly increases after it. The present analysis shows that the flow field is appreciably influenced by the Lewis number Le.

(a)

$N_t=0.5, N_b=0.5, N_r=1.0, \lambda=1.0,$
$Pr=10. Le=10, \gamma=0.5, \Delta=0.5$

(b)

$N_t=0.5, N_b=0.5, N_r=1.0, \lambda=1.0,$
$Pr=10. Le=10, \gamma=0.5, \Delta=0.5$

A=0.5, 1, 2, 3

(c)

$N_t=0.5, N_b=0.5, H_r=1.0, \lambda=1.0,$
$Pr=10. Le=10, \gamma=0.5, \Delta=0.5$

A=0.5, 1, 2, 3

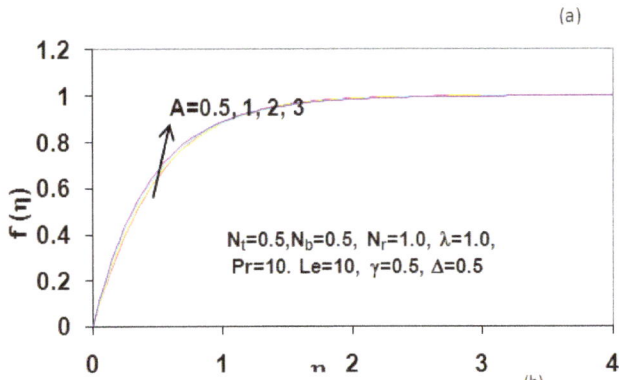

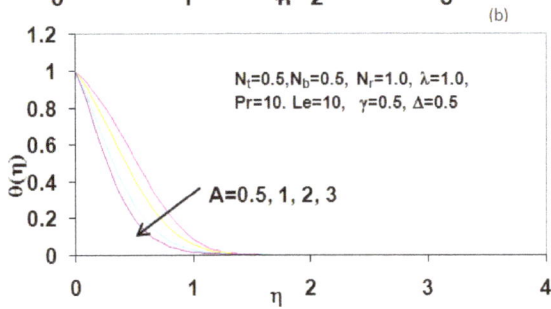

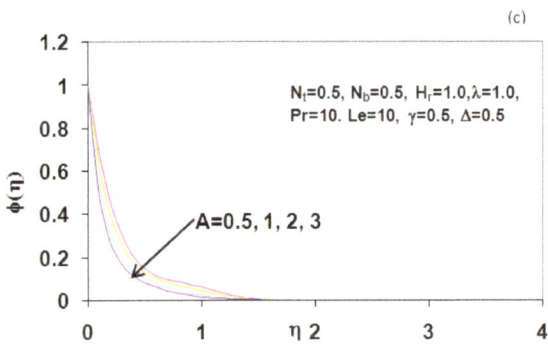

Fig. 2. *Effects of unsteadiness parameter* A *on (a) velocity, (b) Temperature and (c) Nanoparticle volume fraction profiles.*

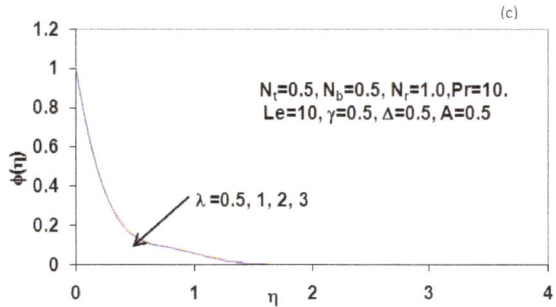

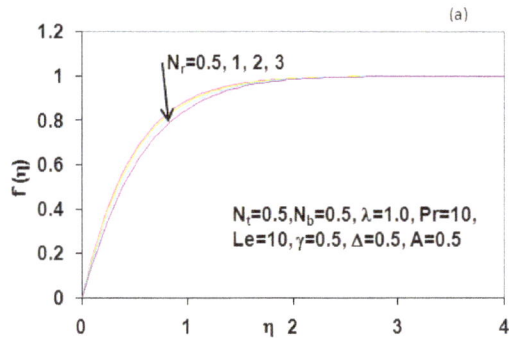

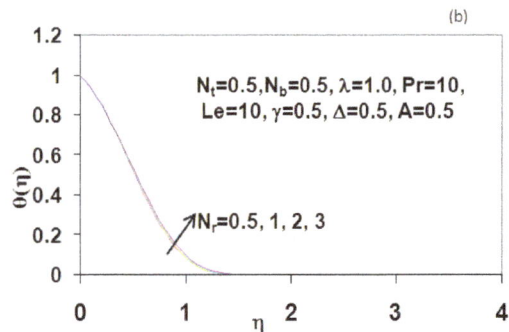

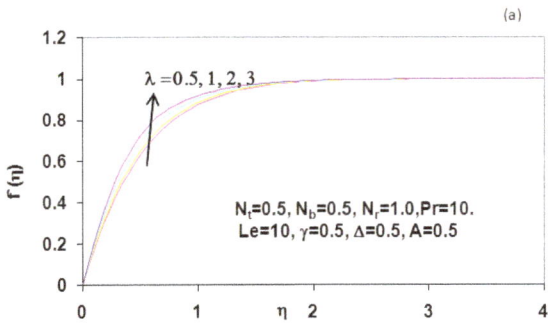

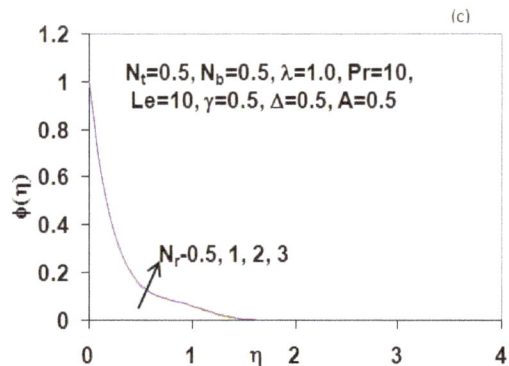

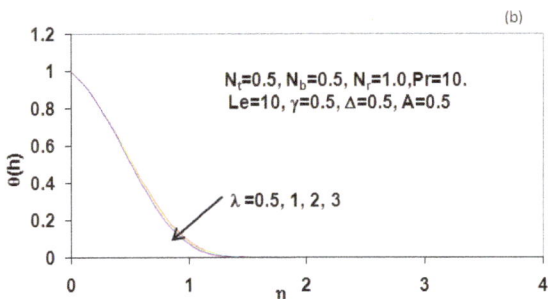

(c)

$N_t=0.5, N_b=0.5, N_r=1.0, Pr=10.$
$Le=10, \gamma=0.5, \Delta=0.5, A=0.5$

$\lambda=0.5, 1, 2, 3$

Fig. 3. *Effects of mixed convection parameter* λ *on (a) velocity, (b) Temperature and (c) Nanoparticle volume fraction profiles.*

(a)

$N_r=0.5, 1, 2, 3$

$N_t=0.5, N_b=0.5, \lambda=1.0, Pr=10,$
$Le=10, \gamma=0.5, \Delta=0.5, A=0.5$

(b)

$N_t=0.5, N_b=0.5, \lambda=1.0, Pr=10,$
$Le=10, \gamma=0.5, \Delta=0.5, A=0.5$

$N_r=0.5, 1, 2, 3$

(c)

$N_t=0.5, N_b=0.5, \lambda=1.0, Pr=10,$
$Le=10, \gamma=0.5, \Delta=0.5, A=0.5$

$N_r-0.5, 1, 2, 3$

Fig. 4 *Effects of nanofluid buoyancy ratio parameter* N_r *on (a) velocity, (b) Temperature and (c) Nanoparticle volume fraction profiles.*

(a)

$\lambda=0.5, 1, 2, 3$

$N_t=0.5, N_b=0.5, N_r=1.0, Pr=10.$
$Le=10, \gamma=0.5, \Delta=0.5, A=0.5$

(b)

$N_t=0.5, N_b=0.5, N_r=1.0, Pr=10.$
$Le=10, \gamma=0.5, \Delta=0.5, A=0.5$

$\lambda=0.5, 1, 2, 3$

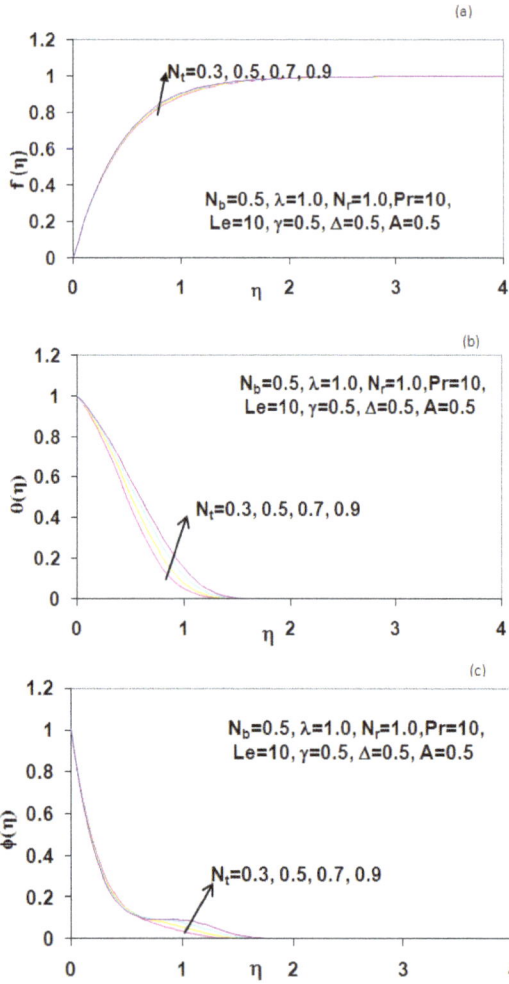

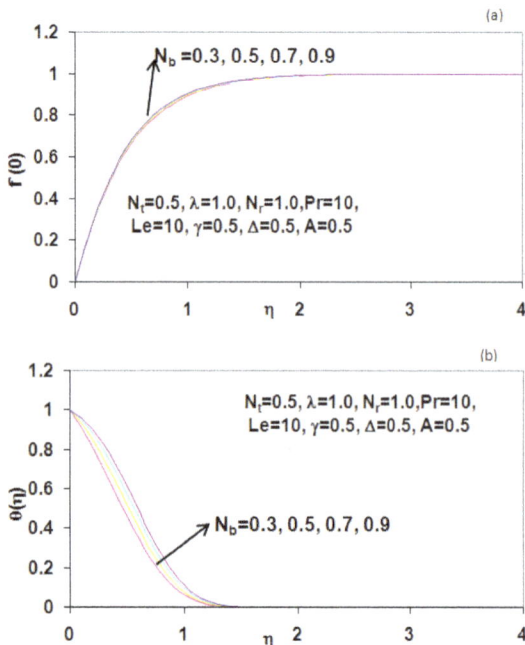

Fig. 5. *Effects of thermophoresis parameter N_t on (a) velocity, (b) Temperature and (c) Nanoparticle volume fraction profiles.*

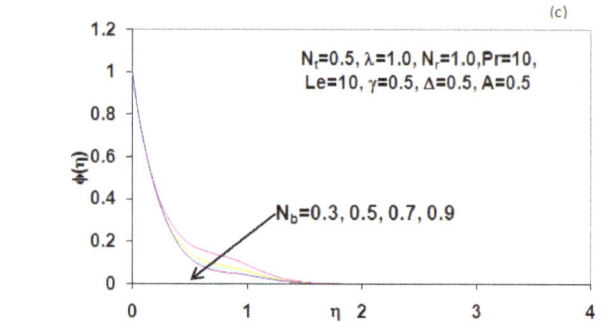

Fig. 6. *Effects of Brownian motion parameter N_b on (a) velocity, (b) Temperature and (c) Nanoparticle volume fraction profiles.*

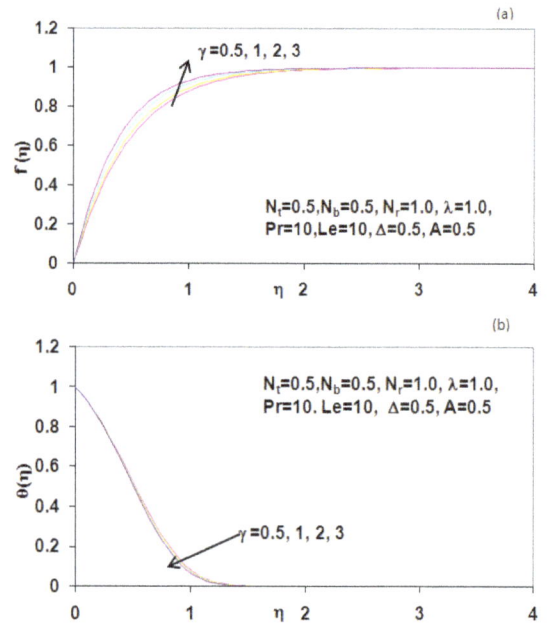

Fig. 7. *Effects of first resistant parameter γ on (a) velocity, (b) Temperature and (c) Nanoparticle volume fraction profiles.*

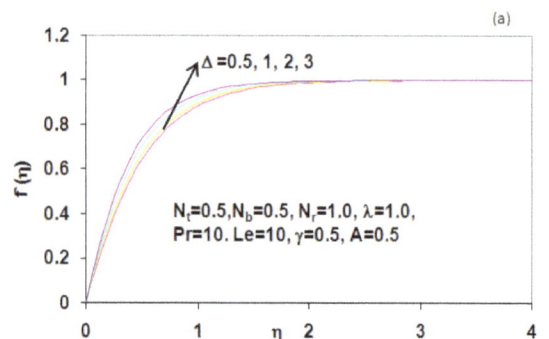

(b)

$N_t=0.5, N_b=0.5, N_r=1.0, \lambda=1.0,$
$Pr=10. Le=10, \gamma=0.5, A=0.5$

$\Delta=0.5, 1, 2, 3$

(c)

$N_t=0.5, N_b=0.5, H_r=1.0, \lambda=1.0,$
$Pr=10. Le=10, \gamma=0.5, A=0.5$

$\Delta=0.5, 1, 2, 3$

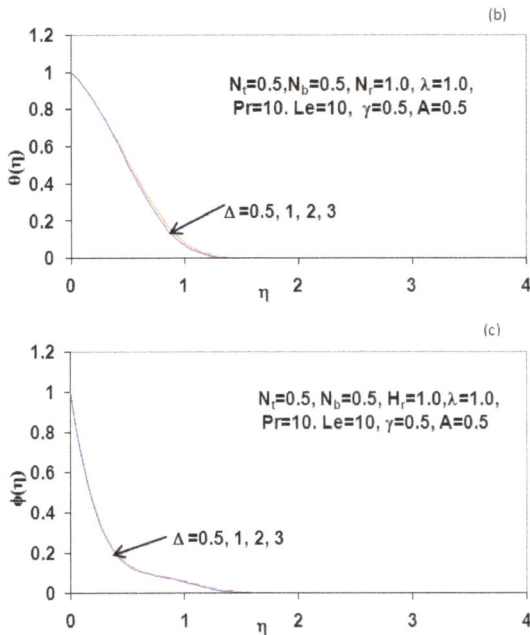

Fig. 8. *Effects of second resistant parameter* Δ *on (a) velocity, (b) Temperature and (c) Nanoparticle volume fraction profiles.*

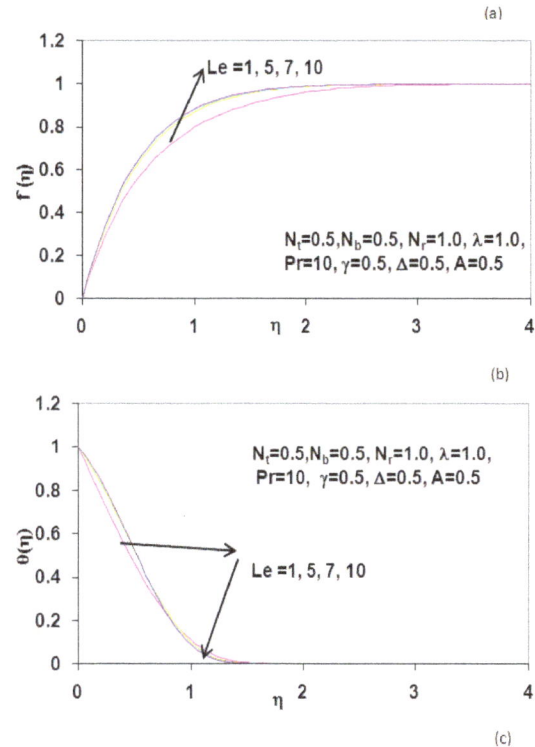

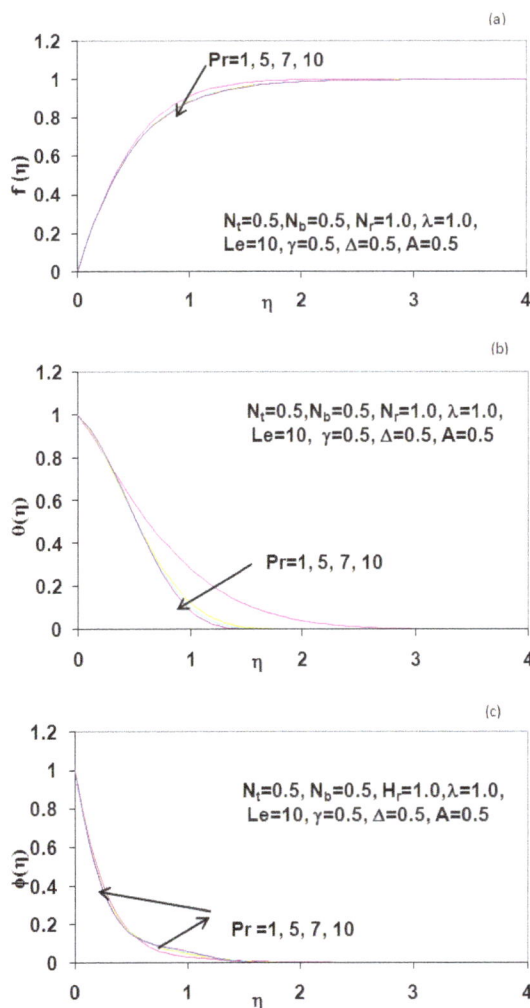

(a)

$Pr=1, 5, 7, 10$

$N_t=0.5, N_b=0.5, N_r=1.0, \lambda=1.0,$
$Le=10, \gamma=0.5, \Delta=0.5, A=0.5$

(b)

$N_t=0.5, N_b=0.5, N_r=1.0, \lambda=1.0,$
$Le=10, \gamma=0.5, \Delta=0.5, A=0.5$

$Pr=1, 5, 7, 10$

(c)

$N_t=0.5, N_b=0.5, H_r=1.0, \lambda=1.0,$
$Le=10, \gamma=0.5, \Delta=0.5, A=0.5$

$Pr=1, 5, 7, 10$

Fig. 9. *Effects of Prandtl number* Pr *on (a) velocity, (b) Temperature and (c) Nanoparticle volume fraction profiles.*

(a)

$Le=1, 5, 7, 10$

$N_t=0.5, N_b=0.5, N_r=1.0, \lambda=1.0,$
$Pr=10, \gamma=0.5, \Delta=0.5, A=0.5$

(b)

$N_t=0.5, N_b=0.5, N_r=1.0, \lambda=1.0,$
$Pr=10, \gamma=0.5, \Delta=0.5, A=0.5$

$Le=1, 5, 7, 10$

(c)

$N_t=0.5, N_b=0.5, H_r=1.0, \lambda=1.0,$

$Le=1, 5, 7, 10$

Fig. 10. *Effects of Lewis number* Le *on (a) velocity, (b) Temperature and (c) Nanoparticle volume fraction profiles.*

4. Conclusions

In the present work, we studied theoretically the problem of unsteady mixed convection boundary-layer flow near a stagnation point of a heated vertical surface embedded in a nanofluid-saturated porous medium. The resulting system of nonlinear partial differential equations is transformed to a system of ordinary differential equations by the means of self-similar solution. The obtained system is solved numerically using an efficient numerical shooting technique with a fourth-fifth order Runge–Kutta method scheme (MATLAB package). The solutions for the flow, heat and mass transfer characteristics are evaluated numerically for various values of the governing parameters, namely unsteadiness parameter A, mixed convection parameter λ, nanofluid buoyancy ratio parameter N_r, thermophoresis parameter N_t, Brownian motion parameter N_b, first resistant parameter γ, second resistant parameter Δ. Prandtl number Pr and Lewis number Le.

The following are brief summary conclusions drawn from

the analysis:

1. The magnitude of the skin friction coefficient $f''(0)$ increases, while the local Nusselt number decreases with increasing the nanofluid parameters N_t, N_b and N_r.

2. The local Sherwood number increases as both N_t and N_b parameter increases. However, it decreases as N_r increases.

3. The thickness of momentum boundary layer decreases with an increase in nanofluid buoyancy ratio parameter N_r and Prandtl number Pr. However, it increases with increasing all other parameters.

4. The thickness of thermal boundary layer increases with an increase in the nanofluid parameters N_t, N_b and N_r.

5. The nanoparticle volume fraction boundary-layer thickness increases with an increase in both N_t and N_r parameters and decreases with an increase in N_b parameter.

6. Both the first and second resistance parameters γ and Δ enhanced the momentum boundary-layer thickness and reduced both the thermal and nanoparticle volume fraction boundary-layer thickness

Acknowledgement

The authors would like to thank National Science, Technology and Innovation Plan (NSTIP) at Kingdom of Saudi Arabia (Project ID 12-MAT2296-10) for the financial support.

Nomenclature

A	Unsteadiness parameter
a, b, c	Constants in Eq. (6)
C_f	Skin friction coefficient
D_a	Darcy number
D_B	Brownian diffusion coefficient
D_T	Thermophoresis diffusion coefficient
f	dimensionless stream function
G	Acceleration due to gravity
Gr	Grashof number
Le	Lewis number
k	Thermal conductivity
K	Permeability
N_b	Brownian motion parameter
N_r	Nanofluid buoyancy ratio parameter
N_t	Thermophoresis parameter
Pr	Prandtl number
Re	Local Reynolds number
Sh	Sherwood number
t	Time
T	Fluid temperature
T_w	Temperature at the surface
T_∞	Ambient temperature as y tends to infinity

u, v	Velocity components along x and y directions, respectively
x, y	Distances along and normal to the surface

Greek symbols

α	Thermal diffusivity
β	Coefficient of volumetric thermal expansion
φ	Dimensionless nanoparticle volume fraction
Γ	Empirical constant
η	Pseudo similarity variable
θ	Dimensionless temperature
λ	Buoyancy parameter
μ	Coefficient of viscosity
γ	First order resistance
$\dfrac{\partial u}{\partial x} + \dfrac{\partial v}{\partial y} = 0$	Second order resistance
v	Kinematic viscosity

Subscripts

e, w, ∞	Conditions at the edge of the boundary layer, at the surface and in the free stream

References

[1] Ramachandra N., Chen T. and Armaly B, 1988. Mixed convection in the stagnation flows adjacent to vertical surface, J. Heat Transfer Vol. 110, pp. 173-177.

[2] Seshadri R., Sreeshylan N. and Nath G., 2002. Unsteady mixed convection flow in the stagnation region of a heated vertical plate due to impulsive motion, Int. J. Heat and Mass Transfer Vol. 45 pp. 1345-1352.

[3] Hall M., 1969. The boundary layer over an impulsively started flat plate, Proc. R. Soc. Vol. 310A, pp. 401-414.

[4] Dennis S., 1972. The motion of a viscous fluid past an impulsively started semi-infinite flat plate, J. Inst. Math. It's Appl. Vol. 10, pp. 105-117.

[5] Watkins C., 1975. Heat transfer in the boundary layer over an impulsively started flat plate, J. Heat Transfer Vol. 97, pp. 492-484.

[6] Smith S., 1967. The impulsive motion of a wedge in a viscous fluid, Z. Angew. Math. Phys. Vol. 18 , pp. 508-522.

[7] Nanbu K., 1971. Unsteady Falkner Skan flow, Z. Angew. Math. Phys. Vol. 22, pp. 1167-1172.

[8] Williams J. and Rhyne T.,1980. Boundary layer development on a wedge impulsively set into motion, SIAM J. Appl. Math. Vol. 38, pp. 215-224.

[9] Kumari M., 1997. Development of flow and heat transfer on a wedge with a magnetic field, Arch. Mech. Vol. 49, pp. 977-990.

[10] Ece M., 1992. An initial boundary layer flow past a translating and spinning rotational symmetric body, J. Eng. Math. Vol. 26, pp. 415-428.

[11] Ozturk A. and Ece M., 1995. Unsteady forced convection heat transfer from a translating and spinning body, J. Energy Resour. Technol. Vol. 117, pp. 318-323.

[12] Brown S. and Riley N., 1973. Flow past a suddenly heated vertical plate, J. Fluid Mech. Vol. 59, pp. 225-237.

[13] Ingham D., 1985. Flow past a suddenly heated vertical plate, Proc. R. soc. Vol. 402A, pp. 109-134.

[14] Amin N. and Riley N., 1995. Mixed convection at a stagnation point, Quart. J. Mech. Vol. 48, pp. 111-121.

[15] Hassanien I., Ibrahim, F. and Omer Gh.,2004. Unsteady free convection flow in the stagnation-point region of a rotating sphere embedded in a porous medium, Mech.Mech. Eng. Vol. 7, pp. 89-98.

[16] Hassanien, I., Ibrahim, F. and Omer Gh.,2006. Unsteady flow and heat transfer of a viscous fluid in the stagnation region of a three-dimensional body embedded in a porous medium, J.Porous Media, Vol. 9, pp. 357-372.

[17] Hassanien, I. and Al-Arabi, T.,2008. Thermal Radiation and variable viscosity effects on unsteady mixed convection flow in the stagnation region on a vertical surface embedded in a porous medium with surface heat flux Vol. 29, pp. 187 – 207.

[18] Yacob N., Ishak A. and Pop I., 2011. Falkner–Skan problem for a static or moving wedge in nanofluids. Int J Thermal Sci. Vol 50, pp. 133–139.

[19] Congedo P., Collura S. and Congedo P., 2009.Modeling and analysis of natural convection heat transfer in nanofluids. In: Proc ASME Summer Heat TransferConf. Vol. 3,pp. 569–579.

[20] Hamad M., Pop I. and Ismail A. 2011. Magnetic field effects on free convection flow of a nanofluid past a semi-infinite vertical flat plate. Nonlinear Analysis:Real World Appl. Vol. 12, pp. 1338–1346.

[21] Buongiorno J., 2006. Convective transport in nanofluids. ASME J Heat Transfer.Vol. 128, pp. 240–250.

[22] Hamad M., 2011. Analytical solution of natural convection flow of a nanofluid over a linearly stretching sheet in the presence of magnetic field. Int. Commun. Heat Mass Transfer. Vol. 38, pp. 487–492.

[23] Khan W and Pop I.,2010. Boundary-layer flow of a nanofluid past a stretching sheet. Int. J Heat Mass Transfer. Vol. 53, pp. 2477–2483.

[24] Abu-Nada E. and Chamkha A., 2010. Effect of nanofluid variable properties on natural convection in enclosures filled with a CuO–EG–water nanofluid. Int. J. Thermal Sci. Vol. 49, pp. 2339–2352.

[25] Das S., S Choi SU, Yu W. and Pradeep T., 2007. Nanofluids: Science and Technology. New Jersey: Wiley.

[26] Kakaç S. and Pramuanjaroenkij A., 2009. Review of convective heat transfer enhancement with nanofluids. Int. J Heat Mass Transfer. Vol. 52, pp. 3187–3196.

[27] Muthtamilselvan M., Kandaswamy P. and Lee J., 2010. Heat transfer enhancement of copper–water nanofluids in a lid-driven enclosure. Commun Nonlinear Sci. Numer. Simulat. Vol. 15, pp. 1501–1510.

[28] Pereyra V.,1978. PASVA3, an adaptive finite difference FORTRAN program for first order non-linear boundary value problems, in: Lecture Note in Computer Science, Vol. 76, Springer, Berlin.

[29] Seshadri R., Sreeshylan N. and Nath G., 2002. Unsteady mixed convection flow in the stagnation region of a heated vertical plate due to impulsive motion, Int. J. Heat and Mass Transfer Vol. 45 pp. 1345-1352.

10

Boiler Parametric Study of Thermal Power Plant to Opproach to Low Irreversibility

Sajjad Arefdehgani[1, *], Omid Karimi Sadaghiyani[2]

[1]Department of Mechanical engineering, Tabriz Branch, Islamic Azad University, Tabriz, Iran
[2]Department of Mechanical engineering, Urmia Branch, Urmia University, Urmia, Iran

Email address:

sajad.aref2008@gmail.com (S. Arefdehgani), st_o.sadaghiyani@urmia.ac.ir (O. K. Sadaghiyani)

Abstract: In this work, in order to reach to low irreversibilities, the energy and exergy have been analyzed in the boiler system of Tabriz power plant. First, it has been done to decrease the irreversibility of system. Second, flow and efficiencies of energy and exergy of the mentioned system have been studied. In the boiler, the energy efficiencies based on lower and higher heating value of fuel are 91.54% and 86.17% respectively. In other hand, the exergy efficiency is 43.98%. Comparing with Rosen and et al., it is demonstrated that, results have a logical agreement with experimental data. Accordingly, the EES used code, have been validated. Furthermore, the gas fired steam power plant efficiency has been increased by the use of irreversibilities reduction and diminution of excess combustion air and/ or the stack-gas temperature. Finally, these have been concluded, overall energy and exergy efficiencies of Tabriz power plant increase 0.497% and 0.46%, respectively when the fraction of excess combustion air decreases from 0.4 to 0.15.Also these efficiencies increase 2.196% nearly when, the stack-gas temperature decreases from 159 to 97 °C.

Keywords: Excess Air, Stack Gas, Exergy Efficiency, Energy Efficiency, Boiler, Tabriz Power Plant

1. Introduction

The leakage in energy supply and pollutant issue requires an improved using of energy sources. Thus, the anfractousity of power plant units has increased implicitly. Plant engineers are increasingly deigning an accurasly high performance. It will be done by thermodynamic calculations of high accuracy. Consequently, the time and cost of thermodynamic calculating during design and optimization have been rised significantly [2]. The most commonly-used analysis for measuring of an energy-conversion process efficiency is the first-law method. Also, by the second laws of thermodynamics, such as exergy (availability, available energy), entropy generation and irreversibility (exergy destruction) the efficiency has been evaluated. Exegetic analysis enriches thermodynamic evaluation of energy conservation, because it provides the tool for a clear difference between energy losses to the environment and internal irreversibilities in the process [3]. In the recent, exergy analysis has imprtant role to reach to the better understanding of the process, to measure sources of inefficiency, to realize quality of energy (or heat) used [2-5].

Exergy is consider as the maximum theoretical useful work (or maximum reversible work) gained by a system interacts with a steady state. Basicly Exergy is saved not conserved as energy but destructed in the system. Exergy destruction is as criteria for irreversibility. Therefore, an exergy analysis is applied to measure exergy destruction. Also it identifies the location, the magnitude and the source of thermodynamic inefficiencies in a thermal system [6].

Boiler efficiency has a significant effect great influence on heating- related energy savings. Thus, it is important to increase the heat transfer to the water and decrease the heat losses in the boiler. Heat can be lost from boilers by different way, including hot flue gas losses, radiation losses and, in the case of steam boilers, blow down losses [7]. For optimizing the operation of a boiler plant, it is necessary to identify where energy loss is likely to occur. A determined amount of energy is lost through flue gases as all the heat produced by the burning fuel cannot be transferred to water or steam in the boiler. Since most of the heat losses from the boiler exit with the flue gas, the recycling of this heat can result in substantial energy saving [8].

In a research, Exergetic analysis of pulp and paper production is presented. An exergy destruction as well as exergy efficiency relation is determined for each section of the system components and the whole system to indicate the largest exergy losses and possibilities of improvement. It is found that the largest exergy losses occurred in the steam plant and soda recovery and these sections are highly exergy inefficient. Further it is observed that with decrease of excess air and preheating of inlet air the exergy efficiency of boilers is increased [9].

An exergy destruction as well as exergy efficiency relation is determined for each section of pulp and paper production system components and the whole system to indicate the largest exergy losses and possibilities of improvement. It is found that the largest exergy losses occurred in the steam plant and soda recovery and these sections are highly exergy inefficient. Further it is observed that with decrease of excess air and preheating of inlet air the exergy efficiency of boilers is increased [10].

In the other research, the increase in boiler efficiency was obtained by utilizing existing combustion equipment, existing combustion controls system, and a new ZoloBOSS in-furnace laser based combustion measurement

system. This optimization improves the boiler efficiency by reducing the O2, improving and balancing the combustion, balancing the temperature and O2 distribution in the boiler and by continuous adaption to varying boiler conditions [11].

This demonstrates that there are huge savings potentials of a boiler energy savings by decreasing its losses. Recently, the technology involved in a boiler can be seen as having reached a plateau, with even marginal increase in efficiency ratherly hard to achieve [12].

Many investigator have helped to the fundamentals and performance of exergy analysis [13-19]. The history of exergy analysis was recently collected [20].

This work aims to identify and assess methods for increasing efficiencies of steam power plants, to provide options for improving their economic and environmental performance. In this study, several measures to improve efficiency, primarily based on exergy analysis, are considered. The modifications considered here, which increase efficiency by reducing the irreversibility rate in the steam generator, are decreasing the fraction of excess combustion air and/or decreasing the stack-gas temperature. The impact of implementing these measures on efficiencies and losses is investigated.

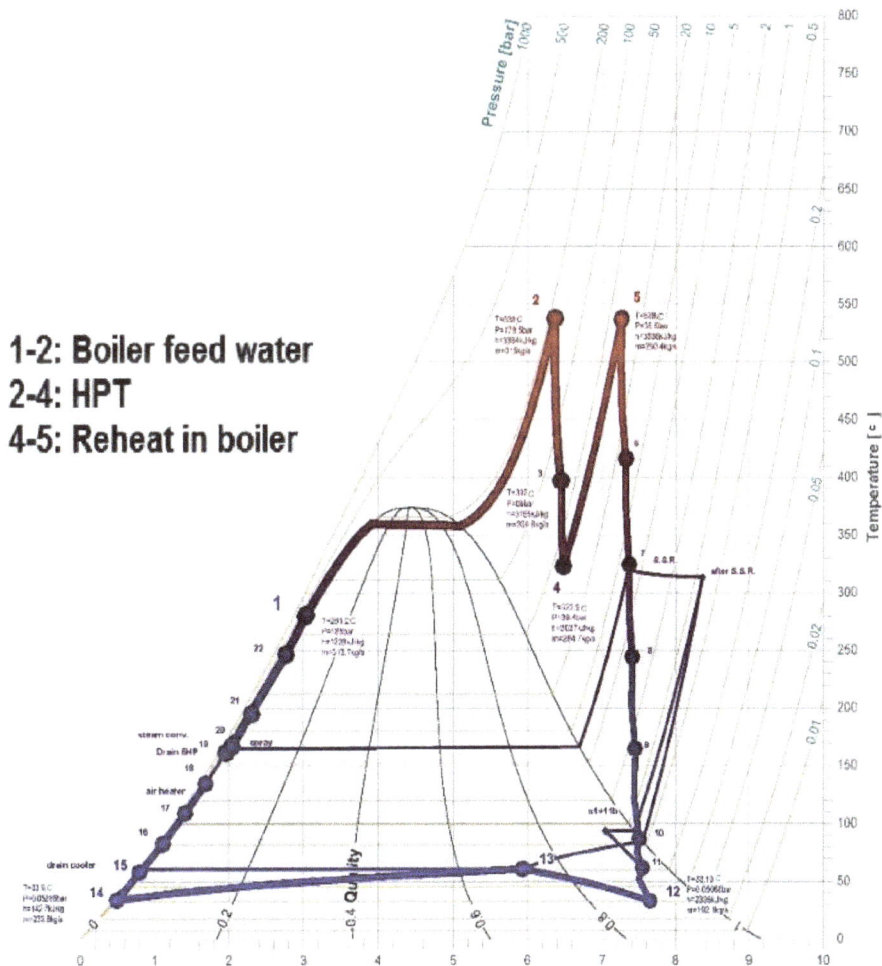

1-2: Boiler feed water
2-4: HPT
4-5: Reheat in boiler

(a)

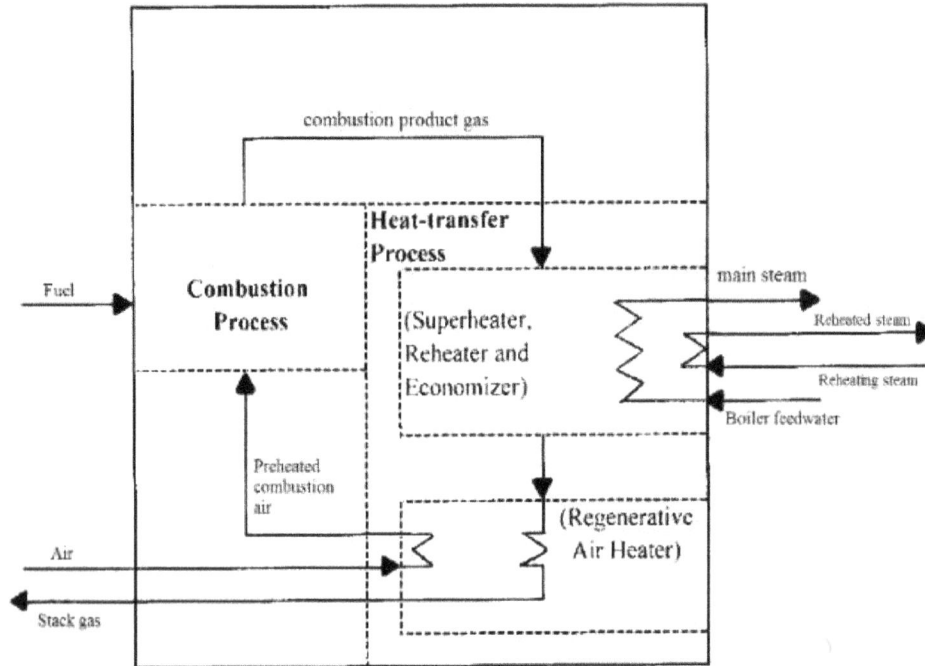

(b)

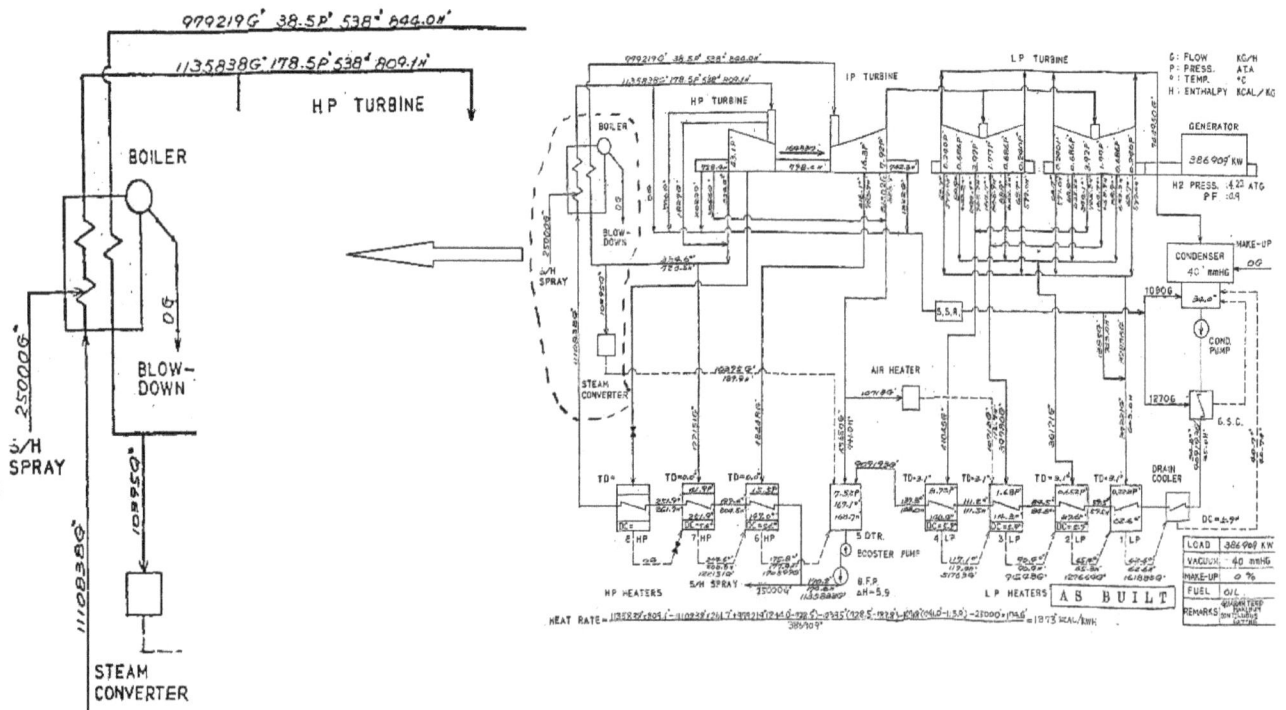

(c)

Figure 1. (a) State and process data on T-s chart of water, (b) Model of the Boiler and (c) Heat balance diagram for the Tabriz power plant.

2. Energy and Exergy Analysis

The state and process data on T-s chart of water and The Heat balance diagram for the boiler and power plant is shown in fig. 1 The following thermodynamic analysis of the power plant will consider the balances of mass, energy, entropy and exergy. The variation of kinetic and potential energies will be neglected and steady state flow will be considered. For a

steady state process, the mass balance for a control volume system in fig. 1 can be presented as

$$\sum_i \dot{m}_i = \sum_e \dot{m}_e \tag{1}$$

The energy balance for a control volume system is as:

$$\sum_i \dot{E}_i + \dot{Q} = \sum_e \dot{E}_e + \dot{W} \tag{2}$$

The entropy balance for a control volume system is as:

$$\sum_i \dot{S}_i + \sum_i \frac{\dot{Q}_i}{T} + \dot{S}_{gen} = \sum_e \dot{S}_e + \sum_e \frac{\dot{Q}_e}{T} \qquad (3)$$

Also, the exergy balance for a control volume system is written as

$$\sum_i Ex_i + \sum_i (1 - \frac{T}{T_k})\dot{Q}_k = \sum_e Ex_e + \dot{W} + \dot{I} \qquad (4)$$

Where the flue exergy rate is:

$$\dot{Ex} = \dot{m}(Ex) \qquad (5)$$

$$\dot{m}(Ex) = \dot{m}(Ex^{tm} + Ex^{ch}) \qquad (6)$$

The general form of exergy balance has been exhibited above. With calculating the chemical exergy of gas, the heat input will be included. The heat exergy term in Eq. (4) will be used to calculate the exergy loss associated with heat loss to the surroundings. The specific exergy is given by:

$$Ex^{tm} = (h - h_0) - T_0(s - s_0) \qquad (7)$$

3. Methods to Improve Plant Efficiency

In order to improve overall-plant efficiency, some methods

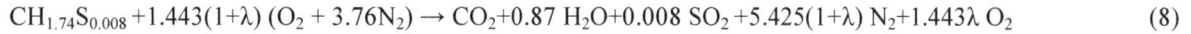

are considered here. These methods are as reducing excess combustion air and stack-gas temperature which are selected. It has been done to their potential benefits and application without excessive modifications to existing steam power plants.

4. Effect of Decreasing Excess Combustion Air

One method to reducing the exergy losses in the boiler is reducing of the air fraction in the combustion process. With decreasing of air fraction (while still remaining large enough to promote good fuel combustion), the temperature and the exergy of the combustion product gas increases. The heating of combustion air leads to increasing of product gas in the combustion process.

The plant optimization has been done in this analysis with decreasing the fraction of excess combustion air λ from 0.40 (the base-case value) to 0.15.

Modelling the oil used in the power plant as ($CH_{1.74}S_{0.008}$) and assuming complete combustion with excess air, the combustion reaction can be expressed as follows:

$$CH_{1.74}S_{0.008} + 1.443(1+\lambda)(O_2 + 3.76N_2) \rightarrow CO_2 + 0.87\,H_2O + 0.008\,SO_2 + 5.425(1+\lambda)\,N_2 + 1.443\lambda\,O_2 \qquad (8)$$

Where λ is the fraction of excess combustion air. Then, the air-fuel (AF) ratio can be written as

$$AF = \frac{\dot{m}_a}{\dot{m}_f} = \frac{\dot{n}_a M_a}{\dot{n}_f M_f} = \frac{1.443(1+\lambda)4.76M_a}{M_f} \qquad (9)$$

The chemical energy for the fuel can be written as

$$\text{Input Energy} = \dot{E}_f = \dot{n}_f \overline{HHV} \qquad (10)$$

Using eqs. (6) and (7) and noting that the thermomechanical exergy of fuel is zero at its assumed input conditions of $T_0 = 19°C$ and $P_0 = 1atm$, the fuel exergy can be written as

$$\dot{Ex}_f = \dot{n}_f \overline{Ex}^{ch} \qquad (11)$$

The energy flow rate of a gas flow can be written as the sum of the energy flow rates for its constituents:

$$\dot{E} = \sum_i \dot{n}_i [\bar{h} - \bar{h}_0]_i \qquad (12)$$

The exergy flow rate of a gas flow can be written with eqs. (6) and (7) as

$$\dot{Ex} = \sum_i \dot{n}_i [\bar{h} - \bar{h}_0 - T_0(\bar{s} - \bar{s}_0) + Ex^{ch}]_i \qquad (13)$$

The energy and exergy flow rate of combustion air can be written in terms of the mole flow rate of fuel n_f using eqs. (8), (12) and (13) and noting that the chemical exergy of air is zero, as

$$\dot{E}_a = \dot{n}_f 1.443(1+\lambda)[(\bar{h} - \bar{h}_0)_{O_2} + 3.76(\bar{h} - \bar{h}_0)_{N_2}] \qquad (14)$$

and

$$\dot{Ex}_a = \dot{n}_f 1.443(1+\lambda)\{[\bar{h} - \bar{h}_0 - T_0(\bar{s} - \bar{s}_0)]_{O_2} + 3.76(\bar{h} - \bar{h}_0 - T_0(\bar{s} - \bar{s}_0))_{N_2}\} \qquad (15)$$

It is useful to determine the hypothetical temperature of combustion gas in the boiler prior to any heat transfer (i.e., the adiabatic combustion temperature) to facilitate the evaluation of its energy and exergy and the breakdown of the

boiler irreversibility into portions related to combustion and heat transfer. The energy and exergy flow rates of the products of combustion can be written using eqs. (8), (12) and (13) as

$$\dot{E}_p = \dot{n}_f [(\bar{h} - \bar{h}_0)_{CO_2} + 0.87(\bar{h} - \bar{h}_0)_{H_2O} + 0.008(\bar{h} - \bar{h}_0)_{SO_2} + 5.425(1+\lambda)(\bar{h} - \bar{h}_0)_{N_2} + 1.443\lambda(\bar{h} - \bar{h}_0)_{H_2}] \qquad (16)$$

And

$$\dot{Ex}_p = \dot{n}_f \{ [\bar{h} - \bar{h}_0 - T_0(\bar{s} - \bar{s}_0) + Ex^{ch}]_{CO_2} + 0.87 [\bar{h} - \bar{h}_0 - T_0(\bar{s} - \bar{s}_0) + Ex^{ch}]_{H_2O}$$

$$+ 0.008 [\bar{h} - \bar{h}_0 - T_0(\bar{s} - \bar{s}_0) + Ex^{ch}]_{SO_2} + 5.425(1+\lambda) [\bar{h} - \bar{h}_0 - T_0(\bar{s} - \bar{s}_0) + Ex^{ch}]_{N_2} + 1.443\lambda [\bar{h} - \bar{h}_0 - T_0(\bar{s} - \bar{s}_0) + Ex^{ch}]_{O_2} \} \quad (17)$$

The adiabatic combustion temperature is determined using the energy balance in eq. (2) with $\dot{Q} = 0$ and $\dot{W} = 0$:

$$\dot{E}_f + \dot{E}_a = \dot{E}_p \quad (18)$$

$$\overline{HHV} + 1.443(1+\lambda) [(\bar{h} - \bar{h}_0)_{O_2} + 3.76(\bar{h} - \bar{h}_0)_{N_2}]_{T_a} = [(\bar{h} - \bar{h}_0)_{CO_2} + 0.87 (\bar{h} - \bar{h}_0)_{H_2O}$$

$$+ 0.008 (\bar{h} - \bar{h}_0)_{SO_2} + 5.425(1+\lambda) (\bar{h} - \bar{h}_0)_{N_2} + 1.443\lambda (\bar{h} - \bar{h}_0)_{O_2}]_{T_p} \quad (19)$$

Here T_p is evaluated using an iterative solution technique. Note that the flow rates of energy $\dot{E}_g$ and exergy $\dot{Ex}_g$ for the stack gas can then be evaluated using eqs. (16) and (17) because the composition of stack gas is same as that of the product gas.

An energy balance equation for the boiler can be written as

$$\dot{E}_f - \dot{E}_g = \dot{m}_{fw}(h_{fw,e} - h_{fw,i}) + \dot{m}_{re}(h_{re,e} - h_{re,i}) \quad (20)$$

Note that the exergy of combustion air at environment conditions is zero and, thus, not shown in eq. (7). The total irreversibility rate for the boiler $\dot{I}_{boiler}$ can then be expressed as

$$\dot{I}_{boiler} = \dot{I}_c + \dot{I}_{ht} \quad (23)$$

And the exergy efficiency Ψ_{boiler} as

$$\Psi_{boiler} = \frac{\dot{Ex}_{net,boiler}}{\dot{Ex}_f} \quad (24)$$

Where $\dot{Ex}_{net,boiler}$ represents the net exergy output rate for the H_2O that flows through the steam generator, i.e.

$$\dot{Ex}_{net,boiler} = (\dot{Ex}_{feed,e} - \dot{Ex}_{feed,i}) + (\dot{Ex}_{re,e} - \dot{Ex}_{re,i}) \quad (25)$$

The overall-plant thermal and exergy efficiencies are determined as follows:

$$\eta_{plant} = \frac{\dot{W}_{net}}{\dot{E}_f} \quad (26)$$

$$\Psi_{plant} = \frac{\dot{W}_{net}}{\dot{Ex}_f} \quad (27)$$

The variation with the fraction of excess combustion air λ of the product-gas temperature T_p and the irreversibility rates, are illustrated in Figs. 2 and 3. The variation with λ of the boiler exergy efficiency ψ_{boiler}, the overall-plant efficiencies based on exergy ψ_{plant} and energy η_{plant}, are illustrated in Fig. 5. Consider that, in Figs. 2, 3 and 7, the dotted line shows the estimated behavior of the mentioned parameters between $\lambda = 0$ (theoretical combustion air) and the point where calculations are made (i.e., $\lambda = 0.15$).

Substituting eqs. (10), (14) and (16) into eq. (18) and simplifying yields the following:

The irreversibility rates in the boiler associated with combustion $\dot{I}_c$ and heat transfer $\dot{I}_{ht}$ can be evaluated as follows:

$$\dot{I}_c = \dot{Ex}_f + \dot{Ex}_a - \dot{Ex}_p \quad (21)$$

$$\dot{I}_{ht} = \dot{Ex}_p + \dot{Ex}_{feed,i} + \dot{Ex}_{re,i} - \dot{Ex}_g - \dot{Ex}_{feed,e} - \dot{Ex}_{re,e} - \dot{Ex}_a \quad (22)$$

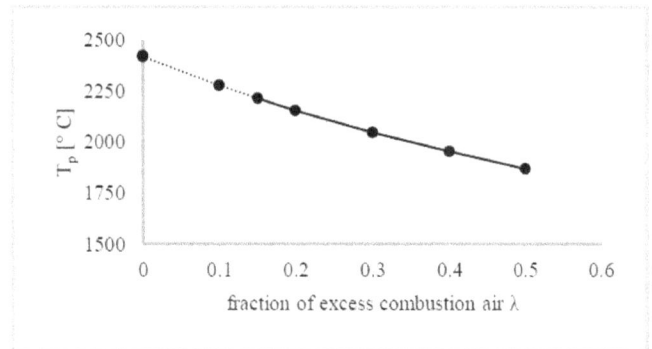

Figure 2. *The variation with fraction of excess combustion air λ of the combustion product-gas temperature T_p.*

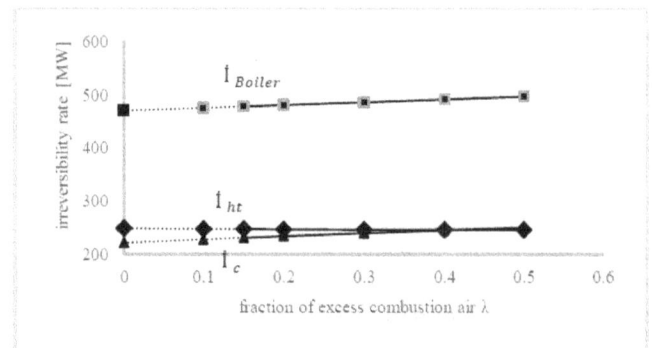

Figure 3. *The variation of irreversibility rates versus fraction of excess combustion air λ.*

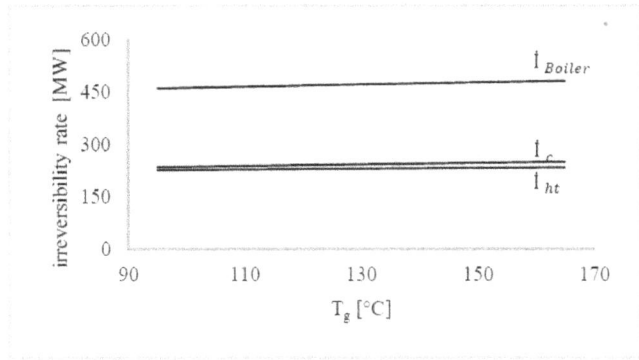

Figure 4. *The variation of irreversibility rates versus stack-gas temperature* T_g.

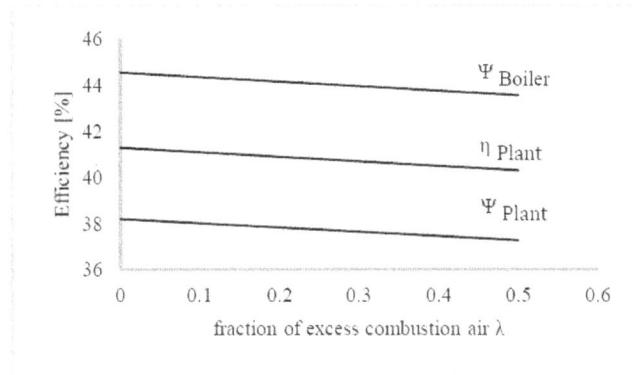

Figure 5. *The variation of efficiency versus fraction of excess combustion air* λ.

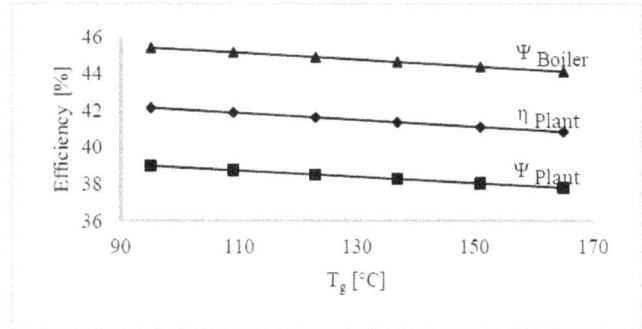

Figure 6. *The variation of efficiency versus stack-gas temperature* T_g.

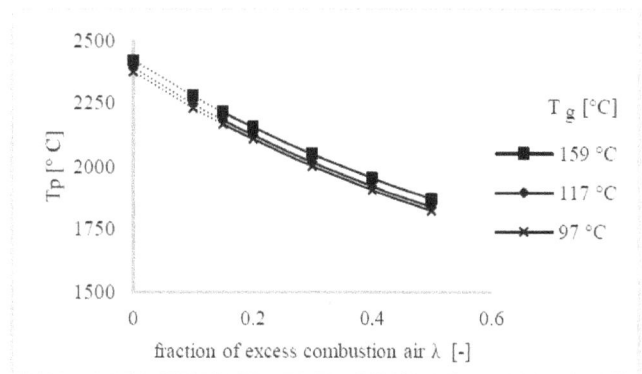

Figure 7. *The variation of the combustion product-gas temperature versus fraction of excess combustion air* λ, T_p *and stack-gas temperature* T_g.

Fig. 3 illustrated that with decreasing the fraction of excess combustion (λ), the combustion irreversibility decreases and, inrersely the heat transfer irreversibility increases consequently, the boiler irreversibility which is achieved by the summation of two above irreversibilities increases. It can concluded that with decreasing of λ the temperature of combustion products rises, and it leads to increasing of combustion products increasing. Also, the other conclusion is that, with decreasing of λ, the combustion products temperature increases, it leads to increasing in the difference of the fluid temerature with exhauoted gas totally, with increases of λ, $\dot{I}_{boiler}$ decreases and the consumption of fuel decreases consequently.

According to Fig. 5, with decreasing of λ, the boiler exergt efficiency (Ψ_{boiler}) and the overall plant efficiencies based on exergy Ψ_{plant} and energy ηplant increase. Because, with decreasing of λ, $\dot{I}_{boiler}$ decreases. And it leads to increasing of Ψ_{boiler}.

Also the decreasing of value of λ from 0.4 to 0.15, the energy and exergy efficiencies increase 0.497% and 0.46%, respectively. Forthermore if decreases form 0.4 to 0, two mentioned efficiencies increase 0.796% and 0.736%, respectively.

5. Effect of Decreasing Stack-Gas Temperature

Another method to diminish the irreversibility rate associated with combustion in the steam generator is reducing the stack-gas temperature T_g. When T_g decreases, more heat can be recovered in the regenerative air heater and can be used to rise the combustion-air temperature. Practical considerations provide lower limits for the temperature of the stack gas. By the use of a Teflon coating to safe materials from the corrosive acids (sulphuric and nitric) that condense out of stack gases at lower temperatures permits the stack-gas temperature Tg to be decreased by 56–83°C (100–150°F) below the nominal stack-gas temperature of 149°C (300°F) [21,22].

Also, two cases of reduced stack-gas temperature (relative to the base-case value of T_g = 159°C) are considered here: 117°C and 97°C. For each case, the energy recovery rate $\Delta\dot{E}$ from the regenerative air heater increases the energy (and temperature) of the combustion air. The energy flow rate of the preheated combustion air can be written as

$$\dot{E}_a = \Delta\dot{E} + \dot{n}_f 1.443(1 + \lambda)\left[(\bar{h} - \bar{h}_0)_{O_2} + 3.76(\bar{h} - \bar{h}_0)_{N_2}\right] \tag{28}$$

Where the term in square brackets is evaluated at the preheated combustion-air temperature for the base case.

In this work, the preheated combustion-air temperature T_a is evaluated frequently for different amounts of T_g, using eq.

(28). Then the combustion product-gas temperature T_p is obtained by eq. (19) and the fuel flow rate (n_f or m_f using eq. (20)). The exergy flow rates of the fuel, $\dot{Ex}_f$ the preheated combustion air, $\dot{Ex}_a$ the combustion product gas $\dot{Ex}_p$ and the stack gas $\dot{Ex}_g$ are determined as described in Section 4. The boiler irreversibilities ($\dot{I}_c$, $\dot{I}_{ht}$ and $\dot{I}_{boiler}$) and exergy efficiency Ψ_{boiler} and the overall-plant exergy efficiency Ψ_{plant} and thermal efficiency η_{plant}, are achived accordingly.

The variation with fraction of excess combustion air λ of stack-gas temperature T_g and several irreversibility rates are demonstrated in Fig. 8. The variation of the exergy efficiency of the boiler and the overall-plant exergy and thermal efficiencies, with stack-gas temperature T_g are demonstrated in Fig. 9.

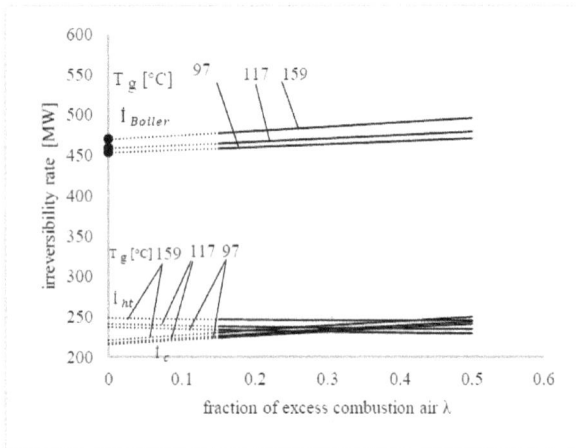

Figure 8. *The variation with fraction of excess combustion air λ of stack-gas temperature T_g and several irreversibility rates.*

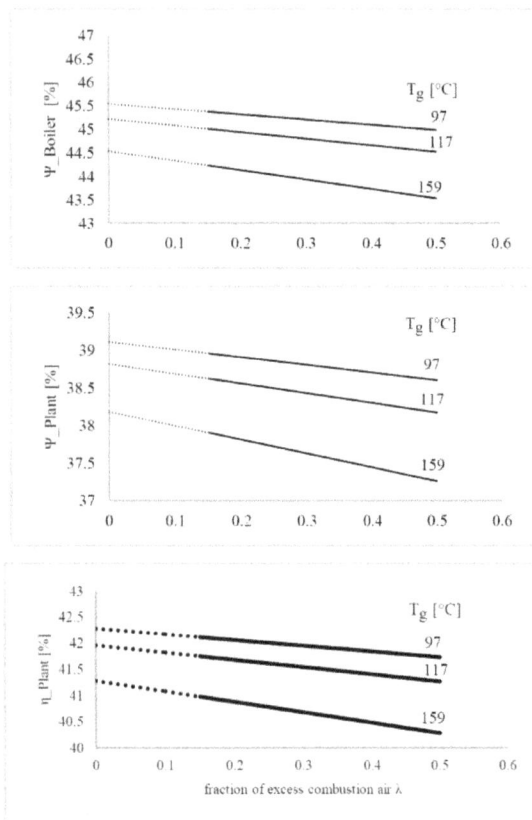

Figure 9. *The variation of the exergy efficiency of the boiler and the overall-plant exergy and thermal efficiencies, with stack-gas temperature T_g.*

According to fig. 8, $\dot{I}_{ht}$ and $\dot{I}_c$ decrease when, the stack-gas temperature (T_g), decreases. The decreasing of $\dot{I}_{ht}$ and $\dot{I}_c$ leads to decreasing of $\dot{I}_{boiler}$ two reasons, can be presented: first, with decrease of T_g, the temperature of pre-heated combustion air rises and the temperature of product gas will be increased consequently. The great value of product gas temperature makes the increasing in the difference between the temperatures of H_2O flow and exhausted gases.

Second, with decreasing in the value of T_g, two impacts can be shown as: significant decreasing in the $\dot{I}_{ht}$ and the slight decreasing in the $\dot{I}_c$. According the Fig. 9 it can be concluded that, with decreasing in the T_g, three parameters (Ψ_{boiler}, η_{plant} and Ψ_{plant}) will be rised.

The decreasing of Ψ_{boiler} leads to decreasing of $\dot{I}_{boiler}$. Also, with decreasing of T_g from 159°C to 97°C, Ψ_{plant} and η_{plant} increase 1.055% and 1.141% respectively.

6. Effect of Varying Excess Air and Stack-Gas Temperature Concurrently

The results of Sections 4 and 5 shows that the plant exergy efficiency can be rised by decreasing either the fraction of excess combustion air λ or the stack-gas temperature T_g. The other developments in plant exergy efficiency that may be achievable by changing λ and T_g concurrently are now analyzed. The parameter values considered are listed in Table 1.

Table 1. *Parameter values considered in investigating the effect of varying λ and T_g simultaneously.*

Parameter	Base-case value	Alternative values
Fraction of excess combustion air, λ	0.4	0.4, 0.3, 0.2, 0.15
Temperature of stack gas, T_g (°C)	159	117, 97

For the alternative values of λ and T_g, the mole flow rate of the fuel $\dot{n}_f$ is determined using eq. (20). The combustion-air temperature T_a, which changes only with T_g, is obtained by eq. (28). Then, the combustion product-gas temperature T_p is determined using eq. (19). Finally, exergy flow rates, such as $\dot{Ex}_f$ and, $\dot{Ex}_a$ are determined as described in Sections 4 and 5.

As λ and T_g are varied, the behaviours are illustrated for the combustion-gas temperature T_p (Fig. 7), several boiler irreversibility rates ($\dot{I}_c$, $\dot{I}_{ht}$ and $\dot{I}_{boiler}$) (fig. 4) and the boiler exergy efficiency Ψ_{boiler} well as the overall-plant exergy Ψ_{plant} and thermal η_{plant} efficiencies (figs. 5 and 6).

The concurrent affection of λ and T_g variation fig. 7. Exhibits the variation of T_p versus λ variation in the several amounts of T_g. fig. 4 shows the irreversiblities variations of boiler versus T_g variation.

According to fig. 8, in the all selected amounts of T_g decreasing of λ and $\dot{I}_{ht}$ increased and, $\dot{I}_c$ decrease inversly. Totally, $\dot{I}_{boiler}$ reduces. It is better to mention, the effect of $\dot{I}_{ht}$ is significantly more than $\dot{I}_c$ on the $\dot{I}_{boiler}$.

Accrding to fig. 9 three parameters of Ψ_{boiler}, η_{plant} and Ψ_{plant} are increased when, both λ and T_g are decreased. With decreasing of λ to 0.15 and T_g to 97°C, the energy and exergy efficiencies increases 1.36% and 1.268% respectively. With decreasing of λ to 0 and T_g to 97°C, the energy and exergy efficiencies increases 1.747% and 1.621% respectively.

7. Conclusions

With decreasing the fraction of excess combustion (λ), the combustion irreversibility decreases and, inrresely the heat transfer irreversibility increases consequently, the boiler irreversibility which is achieved by the summation of two above irreversibilities increases. It can concluded that with decreasing of λ the temperature of combustion products rises, and it leads to increasing of combustion products increasing. Also, the other conclusion is that, with decreasing of λ, the combustion products temperature increases, it leads to increasing in the difference of the fluid temerature with exhauoted gas totally, with increases of λ, $\dot{I}_{boiler}$ decreases and the consumption of fuel decreases consequently.

With decreasing of λ, the boiler exergt efficiency (Ψ_{boiler}) and the overall plant efficiencies based on exergy Ψ_{plant} and energy η_{plant} increase. Because, with decreasing of λ, $\dot{I}_{boiler}$ decreases. And it leads to increasing of Ψ_{boiler}.

Also the decreasing of value of λ from 0.4 to 0.15, the energy and exergy efficiencies increase 0.497% and 0.46%, respectively. Forthermore if decreases form 0.4 to 0, two mentioned efficiencies increase 0.796% and 0.736%, respectively.

$\dot{I}_{ht}$ and $\dot{I}_c$ decrease when, the stack-gas temperature (T_g), decreases. The decreasing of $\dot{I}_{ht}$ and $\dot{I}_c$ leads to decreasing of $\dot{I}_{boiler}$ two reasons, can be presented: first, with decrease of T_g, the temperature of pre-heated combustion air rises and the temperature of product gas will be increased consequently. The great value of product gas temperature makes the increasing in the difference between the temperatures of H_2O flow and exhausted gases.

Second, with decreasing in the value of T_g, two impacts can be shown as: significant decreasing in the $\dot{I}_{ht}$ and the slight decreasing in the $\dot{I}_c$. According the fig. 9 it can be concluded that, with decreasing in the T_g, three parameters (Ψ_{boiler}, η_{plant} and Ψ_{plant}) will be rised.

The decreasing of Ψ_{boiler} leads to decreasing of $\dot{I}_{boiler}$. Also, with decreasing of T_g from 159°C to 97°C, Ψ_{plant} and η_{plant} increase 1.055% and 1.141% respectively.

The concurrent affection of λ and T_g variation fig7. Exhibits the variation of T_p versus λ variation in the several amounts of T_g. fig. 4 shows the irreversiblities variations of boiler versus T_g variation.

In the all selected amounts of T_g decreasing of λ and $\dot{I}_{ht}$ increased and, $\dot{I}_c$ decrease inversly. Totally, $\dot{I}_{boiler}$ reduces. It is better to mention, the effect of $\dot{I}_{ht}$ is significantly more than $\dot{I}_c$ on the $\dot{I}_{boiler}$.

Three parameters of Ψ_{boiler}, η_{plant} and Ψ_{plant} are increased when, both λ and T_g are decreased. With decreasing of λ to 0.15 and T_g to 97°C, the energy and exergy efficiencies increases 1.36% and 1.268% respectively. With decreasing of λ to 0 and T_g to 97°C, the energy and exergy efficiencies increases 1.747% and 1.621% respectively.

Nomenclature

$\dot{Ex}$	Exergy rate [MW]
$\dot{ex}$	specific Exergy rate [MWkg^{-1}]
$\dot{E}$	Energy rate [MW]
h	specific enthalpy [Jkg^{-1}]
HHV	High Heat value[kjkg^{-1}]
$\dot{I}$	exergy destruction rate
LHV	low Heat value[kjkg^{-1}]
$\dot{m}$	mass flow rate [kgs^{-1}]
P	pressure [atm]
$\dot{Q}$	heat transfer rate [MW]
s	specific entropy [Jkg^{-1} K^{-1}] and [Jkmol^{-1} K^{-1}]
T	temperature [°C]
$\dot{W}$	work rate [MW]
$\dot{W}_s$	Isentropic work rate [MW]

Greek symbols

η	energy efficiency[%]
Ψ	exergy efficiency[%]

Subscripts

A	air
ch	chemical
c	Combustion
F	fuel
fw	Feed water
g	stack-gas
ht	Heat transfear
In	inlet
Out	outlet
p	Product gas Combustion
re	reheat
th	thermal
0	dead state conditions

References

[1]   Rosen, M.A., Tang, R., Improving steam power plant efficiency through exergy analysis: effects of altering excess combustion air and stack-gas temperature, Int. J. Exergy, 5 (2008), No. 1, pp. 31–51

[2]   Dincer, I., Al-Muslim, H., Thermodynamic analysis of reheat cycle steam power plants. Int.J. Energy Res, 25 (2001), pp. 727-739

[3]   Gallo, W.L.R., Milanez. L.F., Choice of a reference state for exergetic analysis, Energy, 15 (1990), pp. 113–121

[4]   Habib, M.A., Zubair, S.M., Second-law-based thermodynamic analysis of regenerative-reheat Rankine cycle power plants, Energy, 17 (1992), pp. 295–301

[5]   Cihan, A., et al., Energy-exergy analysis and modernization suggestions for a combined-cycle power plant. Int. J. Energy Res., 30 (2006), pp. 115–126

[6]   Song, T.W., et al., Exergybased performance analysis of the heavy-duty gas turbine in part-load operating conditions, Exergy, 2 (2002), 20, pp. 105–112

[7]   ERC, In: How to save energy and money in boilers and furnace systems. Energy Res. Centre (ERC) (2004), University of Cape Town, South Africa

[8]   Jayamaha, Lal, In: Energy Efficient Building Systems, Hardbook. Mcgraw Hill education, Europe, 2008.

[9]   Cownden, R., et al., Exergy analysis of a fuel cell power system for transportation applications, Exergy, An Int. J. 1 (2001), pp. 112–121

[10] Assari, M. R., et al., Exergy Modeling and Performance Evaluation of Pulp and Paper Production Process of Bagasse, a Case Study, Thermal Science, 18 (2014), 4, pp. 1399-1412

[11] Thavamani, S., et al., Increasing Boiler Efficiency through Intelligent Combustion Optimization, Proceedings, 20th Nuclear Engineering and the ASME 2012 Power Conf., California, USA, 2012, Vol. 4, pp. 753-761

[12] Sonia, Y., et al., a centurial history of technological change and learning curves for pulverized coal-fired utility boilers. Energy, 3 (2008), pp. 1996–2005

[13] Grimaldi, C.N., Bidini, G., Using exergy analysis on circulating fluidised bed boilers, International Journal of Energy, Environment, Economics, 2 (1992), No. 3, pp.205–213.

[14] Rosen, M.A., Dincer, I., Survey of thermodynamic methods to improve the efficiency of coal-fired electricity generation, Proc. Instn Mech. Engrs Part A: J. Power and Energy, 217 (2003a), pp.63–73.

[15] Sciubba, E., Su, T.M., Second law analysis of the steam turbine power cycle: a parametric study, Computer-Aided Engineering of Energy Systems, ASME, AES-Vol. 2–3 (1986), pp.151–165.

[16] [16] Rosen, M.A., Tang, R., Assessing and improving the efficiencies of a steam power plant using exergy analysis. Part 1: assessment, Int. J. Exergy, Vol. 3 (2006a), No. 4, pp.362–376.

[17] Rosen, M.A., Tang, R., Assessing and improving the efficiencies of a steam power plant using exergy analysis. Part 2: improvements from modifying reheat pressure, Int. J. Exergy, Vol. 3 (2006b), No. 4, pp. 377–390.

[18] Cengel, Y.A., Boles, M.A., Thermodynamics: An Engineering Approach, 5th ed., McGraw-Hill, Boston, 2006.

[19] Moran, M.J., Shapiro, H.N., Fundamentals of Engineering Thermodynamics, 5th ed., Wiley, New York, 2003.

[20] Rezac, P., Metghalchi, H., A brief note on the historical evolution and present state of exergy analysis, International Journal of Exergy, Vol. 1 (2004), pp. 426–437.

[21] Kamimura, F., Method and apparatus for treating flue gases from coal combustion using precoat agent with heat exchange, Fuel and Energy Abstracts 38 (1997), pp. 449.

[22] Kitto, J.B., Piepho, J.M., Making aging coal-fired boilers low-cost competitors, Power 139 (1995), No. 12, pp.21–26.

11

Performance investigation of multiple-tube ground heat exchangers for ground-source heat pump

Jalaluddin[1, *], Akio Miyara[2]

[1]Department of Mechanical Engineering, Hasanuddin University, Makassar, Indonesia
[2]Department of Mechanical Engineering, Saga University, Saga-shi, Japan

Email address:

jalaluddin_had@yahoo.com (Jalaluddin), miyara@me.saga-u.ac.jp (A. Miyara)

Abstract: The present study aims to investigate the performance of multiple-tube ground heat exchangers (GHEs). The multiple-tube GHEs with a number of pipes installed inside the borehole were simulated. Thermal interferences between the pipes and performance of multiple-tube GHEs are discussed. Increasing the number of inlet tube in the borehole increases the contact surface area and then leads to increase of heat exchange with the ground. However, ineffective of heat exchange in the outlet tube caused by thermal interferences from the inlet tube reduces the heat exchange rate for the GHEs. The GHE performances increase of 9.1 % for three-tube, of 13.6 % for four-tube, and of 20.1 % for multi-tube compared with that of the U-tube. The four-tube and multi-tube GHEs which consist of four pipes as heat exchange pipes where the multi-tube GHE provides better performance than that of the four-tube GHE. This fact indicates that thermal interferences between the pipes affect the performance. Thermal interferences between the pipes should be considered.

Keywords: Heat Exchange Rate, Thermal Interferences, Multiple-Tube GHEs

1. Introduction

The geothermal energy source is categorized based on ASHRAE [1] for using in high-temperature electric power production; > 150 °C, intermediate and low–temperature direct-use applications; < 150 °C, and Ground-source heat pump (GSHP) system applications; generally < 32 °C. The GSHP system has been widely used in engineering application for space heating and cooling. The GHEs used in the GSHP system are installed in either horizontal trenches or vertical boreholes. Short-term and long-term performances are important issues of the GSHP system. Both the short-term and long-term behavior of ground loop heat exchangers is critical to the design and energy analysis of ground-source heat pump systems [2]. Short-term analysis is required for detailed building energy analysis and the design of hybrid GSHP system [3, 4]. It helps to understand the effects of short duration peak loads on the ground response [5-9] and to establish the running control strategies for alternative operation modes in short time scales of operation such as for cooling, heating, and hot water heating according to different requirements [10]. The GHEs used in the GSHP system are installed in either horizontal trenches or vertical boreholes. Temperature distributions, energy and exergy performances

for two different horizontal GHEs [11, 12] and for three different vertical boreholes of 30, 60, and 90 m [13, 14] have been reported.

Several factors such as local conditions, ground heat exchanger (GHE) parameters, and operation conditions contribute significantly to the thermal performance of the GHE that used in the GSHP system to exchange heat with the ground. Analyzing the GHE performance in those conditions is needed to provide an accurate prediction of the performance in the GSHP system design. A number of studies have investigated the GHE performance in various backfilled materials, concrete pile foundations, and configuration shapes [15-18]. Heat exchange rate of the GHE was also evaluated by Jun et al. [19] with considering the effect of running time, shank spacing, depth of borehole, velocity in the pipe, thermal conductivity of grout, inlet temperature and soil type.

Experimental study of thermal performance of three types of GHEs including U-tube, double-tube, and multi-tube types installed in a steel pile foundation with 20 m of depth has been done [20]. The heat exchange rates of the GHEs in 24 hours of continuous operation with flow rates of 2, 4, and 8 l/min and the effect of increasing the flow rate have been discussed. The performance of the GHEs has been also investigated in different operation modes [21]. Operating the GHEs with different operation mode shows the different characteristic in their heat exchange rates.

This work investigates thermal interference and performance of multiple-tube ground heat exchangers (GHEs). The multiple-tube GHEs with a number of pipes installed inside the borehole were simulated in order to investigate the thermal interferences between the tubes and their performances.

2. Numerical Method

2.1. GHE Models

Three-dimensional unsteady-state models for multiple tubes of GHEs were built and simulated by using a commercial CFD code, FLUENT. Steel pipes, which are used as foundation pile for houses, were buried in the ground and used as boreholes for the GHEs. Multiple-tube GHEs including U-tube, multi-tube, three-tube, four-tube types where a number of pipes installed inside the boreholes at 20 m depth were simulated in order to investigate the thermal interferences between the tubes and their performances. The multiple tubes were inserted in the steel pile, and the gaps between the steel pile and tubes were grouted with silica-sand. In addition, the multi-tube consists of a central insulated-pipe as the outlet tube and four pipes as the inlet tubes placed around the central pipe. The three-tube and four-tube types were built with installing three and four tubes inside the boreholes which consists of two inlet and one outlet tubes for the three tube type and three inlet and one outlet tubes for the four-tube type, respectively. For all the GHEs, polyethylene is used as a tube material.

The ground around the GHEs is modeled of 5 m in radius. Fig. 1 shows the horizontal cross-sections of the multiple tube types of GHE models. Three-dimensional hybrid mesh generation was applied in the GHE models. All the related geometric parameters and material thermal properties for the GHEs are listed in Table 1.

Table 1. Related geometric parameters and material thermal properties of the GHEs.

Parameters	Value	Unit
U-tube, Tri-tube and Four-tube		
Inlet and outlet pipes of the multiple-tube GHEs (material: Polyethylene)		
Outer diameter, d_o	0.033	m
Inner diameter, d_i	0.026	m
Thermal conductivity, k_{PE}	0.35	W/(m K)
Multi-tube		
Inlet pipes of the multi-tube (material: Polyethylene)		
Outer diameter, d_o	0.025	m
Inner diameter, d_i	0.02	m
Thermal conductivity, k_{PVC}	0.35	W/(m K)
Outlet pipe of the multi-tube (material: Polyethylene)		
Outer diameter, d_o	0.02	m
Inner diameter, d_i	0.016	m
Thermal conductivity, k_{PVC}	0.35	W/(m K)
Pile foundation of the multiple-tube GHEs (material: Steel)		
Outer diameter, d_o	0.1398	m
Inner diameter, d_i	0.1298	m
Thermal conductivity, k_{Steel}	54	W/(m K)
Grout (material: Silica sand)		
Thermal conductivity, k_{grout}	1.4	W/(m K)

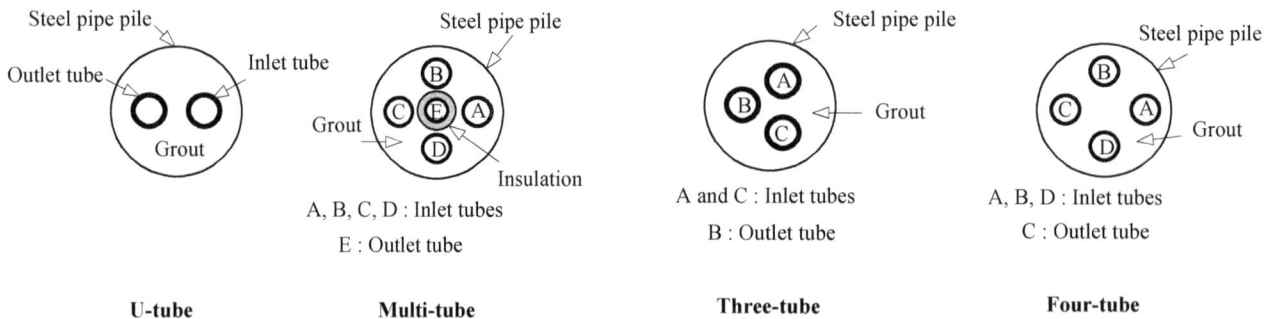

Figure 1. The horizontal cross-sections of the multiple-tube GHEs.

2.2. Boundary Conditions

The ground profiles around the borehole consist of Clay, sand, and Sandy-clay. This ground profiles are typical for Saga city, Japan [21]. The ground properties can be estimated using the values for similar ground profiles in this simulation.

The thermal characteristic parameters of the ground are:
- Clay (ρ = 1700 kg/m^3, k = 1.2 W/m K, c = 1800 J/kg K)
- Sand (ρ = 1510 kg/m^3, k = 1.1 W/m K, c = 1100 J/kg K)
- Sandy-clay (ρ = 1960 kg/m^3, k = 2.1 W/m K, c = 1200 J/kg K)

A constant and uniform temperature was applied to the top and bottom surfaces of the model. Variation of ground temperature near the surface due to ambient climate effect is negligible. Uniform initial ground temperature is assumed to be equal to the undisturbed ground temperature and constant of 17.7 °C. This value is based on recorded data of local ground temperature at Saga city, Japan [20]. Inlet temperature and flow rate of circulated water are specified as boundary conditions. The inlet water temperature was set to be constant of 27 °C. The flow rate of circulated water was set to 16 l/min. For the GHE models that used more than one inlet pipe, this flow rate is the total flow rate of the inlet pipes. k-epsilon two equation turbulence models were applied in the FLUENT simulation set-up. Scaled residuals for turbulence models were monitored.

3. Heat Transfer Model

Three-dimensional unsteady-state model used in simulation is:

$$k\left(\frac{\partial T^2}{\partial x^2} + \frac{\partial T^2}{\partial y^2} + \frac{\partial T^2}{\partial z^2}\right) = \rho c \frac{\partial T}{\partial t} \qquad (1)$$

where k is thermal conductivity (W/m K), T is temperature (K), ρ is density (kg/m³), c is specific heat (J/kg K), t is time and z is depth.

Temperature variation distribution of circulated water is simulated and the thermal performances of the GHEs were investigated by calculating their heat exchange rates through the water flow. The heat exchange rate is calculated by the following equation

$$Q = \dot{m} c_p \Delta T \qquad (2)$$

where $\dot{m}$ is flow rate, c_p is specific heat, and ΔT is the temperature difference between the inlet and outlet tubes of circulated water.

The heat exchange rate per unit borehole depth is defined as the following equation and it is used to express the performance of each GHEs.

$$\overline{Q} = Q / L \qquad (3)$$

where L is the depth of each GHE.

4. Results and Discussion

4.1. Thermal Interferences Effect

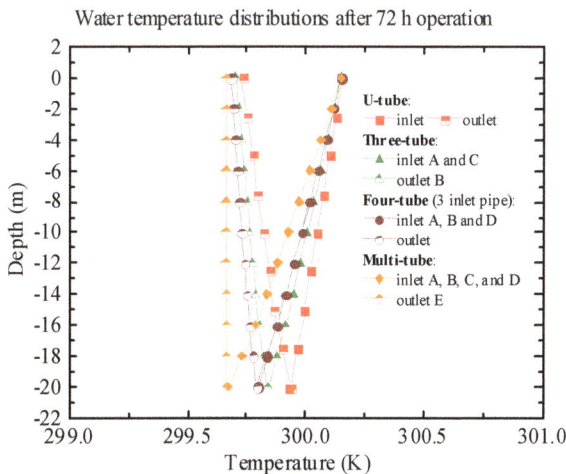

Figure 2. *Water temperature distributions of multiple- tube GHEs.*

Figure 2 shows the water temperature distribution of the multiple-tube GHEs after 72 h operation. The temperatures of water are measured at the central point of the tubes. Water temperature distributions decrease in the inlet and outlet tubes of the U-tube GHE. In the three-tube GHE, the water temperature distribution in the inlet tube decreases more than that of the U-tube but the water temperature changes in the

outlet tube is smaller than that of the U-tube. It indicates that effectiveness of the outlet tube decreases due to the thermal interferences from the two inlet tubes. The cross-sectional temperature distributions at 10 m depth after 72 h operation are shown in Fig. 3. The thermal interferences between the inlet and outlet tubes can be seen from these figures. In comparison with the U-tube GHE, the outlet tube of the three-tube GHE is affected by thermal interferences more significant than that of the U-tube GHE as shown in Fig. 3 (b). This fact proves that the water temperature is slightly change caused by thermal interference from the inlet tubes. A similar characteristic of the water temperature distribution is found in the four-tube GHE. In the multi-tube GHE, the water temperature decrease in the inlet tube is largest among those of other GHEs. The thermal interferences from the inlet tubes to the outlet tube are prevented by insulation of the outlet tube and then, the water temperature stays almost constant in the outlet.

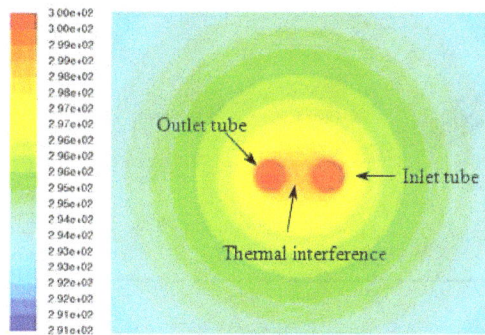

(a)U-tube

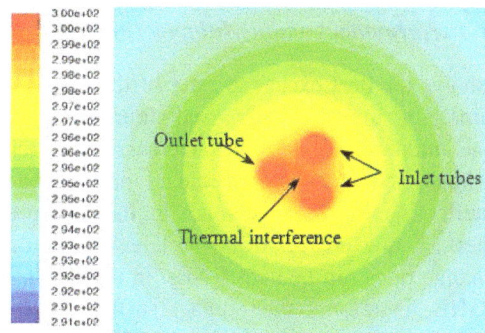

(b)Three-tube

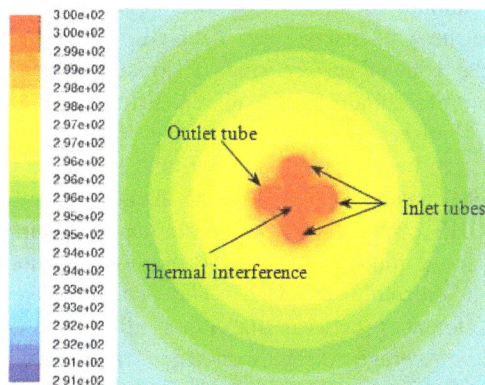

(c)Four-tube

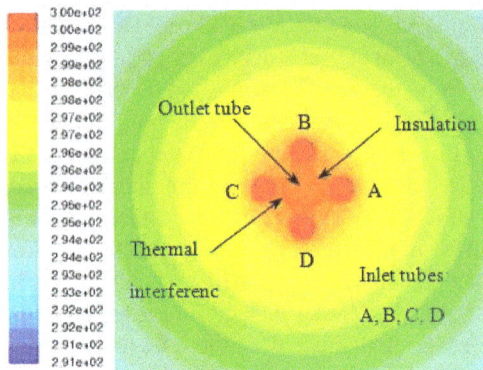

(d)Multi-tube

Figure 3. *The cross-sectional temperature distributions at 10 m depth after 72 h continuous operation.*

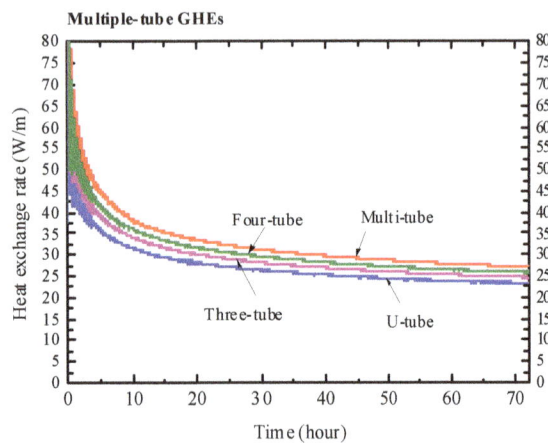

Figure 4. *Heat exchange rates of multiple-tube GHEs*

4.2. Thermal Performances of the Multiple-Tube GHEs

Figure 4 shows the heat exchange rates of the GHEs with operating time. In the three-tube and four-tube GHEs, increasing the number of inlet tubes increases the contact surface area and then leads to increase of heat exchange. However, ineffective of heat exchange in the outlet tube caused by thermal interferences from the inlet tubes reduces the heat exchange rates for the both GHEs. In the case of the multi-tube GHE, increasing the contact surface area and protecting thermal interference by insulation of the outlet tube contributes to the heat exchange rate. This result shows that installing a number of pipes inside the borehole does not increase the heat exchange rate of GHE linearly with increasing the contact surface area. Thermal interferences between the pipes should be considered. Effect of thermal interference between the tubes to the GHE thermal performance can be seen from the four-tube and multi-tube GHES which consist of four heat exchange pipes. The multi-tube GHE provides a better performance compared with that of in the four-tube GHE. It is due to insulation of outlet tube in the multi-tube GHE. This fact indicates that thermal interferences between the pipes affect the GHE performance. Average heat exchange rates of the multiple-tube GHEs after operating in 72 h are listed in Table 2. The GHE performances increase of 9.1 % for three-tube, of

13.6 % for four-tube, and of 20.1 % for multi-tube compared with that of the U-tube GHE.

Table 2. *Heat exchange rates of the multiple-tube GHEs after operating in 72 h.*

Multiple-tube GHEs	Average heat exchange rate in 72 h, $(\overline{Q}_{72h}/L)$ (W/m)
U-tube	27.09
Tri-tube	29.55
Four-tube	30.77
Multi-tube	32.53

5. Conclusions

The geothermal energy sources are locally available and environmentally friendly energy source in the case of reducing CO_2 emission. The well-known application is for space heating and cooling in residential and commercial buildings with using GSHP system. Recently, various types of GHEs are used in the GSHP system to exchange heat with the ground. The heat exchange performance of the GHE is an important subject of GSHP system design.

Multiple-tube GHEs are studied to investigate the thermal interference and performance of multiple-tube ground heat exchangers (GHEs). Heat exchange rate of the multiple-tube GHEs are affected significantly by thermal interferences between the inlet and outlet tubes. Increasing the number of inlet tubes increases the contact surface area and then leads to increase of heat exchange with the ground. However, ineffective of heat exchange in the outlet tube caused by thermal interferences from the inlet tubes reduces the heat exchange rates for the GHEs. The GHE performances increase of 9.1 % for three-tube, of 13.6 % for four-tube, and of 20.1 % for multi-tube compared with that of the U-tube. Thermal interferences between the pipes affect the GHE performance and should be considered in design of GHE.

References

[1]   ASHRAE: 2011 ASHRAE Handbook: HVAC Applications, SI Edition, American Society of Heating, Refrigerating and Air-Conditioning Engineers, Inc., 1791 Tullie Circle, N.E., Atlanta, GA 30329.

[2]   C. Yavuzturk, J.D. Spitler, , S.J. Rees, "A Transient Two-dimensional Finite Volume Model for the Simulation of Vertical U-tube Ground Heat Exchangers," ASHRAE Transactions, 1999, vol. 105(2), pp. 465-474.

[3]   C. Yavuzturk, J.D. Spitler, "Field Validation of a Short Time-Step Model for Vertical Ground Loop Heat Exchangers," ASHRAE Transactions, 2001, vol. 107(1), pp. 617-625.

[4]   V. Partenay, P. Riederer, T. Salquea, E. Wurtz, "The influence of the borehole short-time response on ground source heat pump system efficiency," Energy and Buildings, 2011, vol. 43, pp. 1280–1287.

[5]   L. Lamarche, B. Beauchamp, "New solutions for the short-time analysis of geothermal vertical boreholes," International Journal of Heat and Mass Transfer, 2007, vol. 50, pp. 1408–1419.

[6] G. Bandyopadhyay, W. Gosnold, M. Mann, "Analytical and semi-analytical solutions for short-time transient response of ground heat exchangers," Energy and Buildings, 2008, vol. 40, pp. 1816–1824.

[7] E.J. Kim, J.J. Roux, G. Rusaouen, F. Kuznik, "Numerical modelling of geothermal vertical heat exchangers for the short time analysis using the state model size reduction technique," Applied Thermal Engineering, 2010, vol. 30, pp. 706–714.

[8] A. Zarrella, M. Scarpa, M. De Carli, "Short time step analysis of vertical ground-coupled heat exchangers: The approach of CaRM," Renewable Energy, 2011, vol. 36, pp. 2357-2367.

[9] E. Zanchini, S. Lazzari, A. Priarone, "Long-term performance of large borehole heat exchanger fields with unbalanced seasonal loads and groundwater flow," Energy, 2012, vol. 38, pp. 66-77.

[10] P. Cui, H. Yang, Z. Fang, "Numerical analysis and experimental validation of heat transfer in ground heat exchangers in alternative operation modes," Energy and Buildings, 2008, vol. 40, pp. 1060–1066.

[11] H. Esen, M. Inalli, M. Esen, "Numerical and experimental analysis of a horizontal ground-coupled heat pump system," Building and Environment, 2007, vol. 42, pp. 1126–1134.

[12] H. Esen, M. Inalli, M. Esen, K. Pihtili, "Energy and exergy analysis of a ground-coupled heat pump system with two horizontal ground heat exchangers," Building and Environment, 2007, vol. 42, pp. 3606–3615.

[13] H. Esen, M. Inalli, "Thermal response of ground for different depths on vertical ground source heat pump system in Elazig, Turkey," Journal of the Energy Institute, 2009, vol. 82 (2), pp. 95-101.

[14] H. Esen, M. Inalli, Y. Esen, "Temperature distributions in boreholes of a vertical ground-coupled heat pump system," Renewable Energy, 2009, vol. 34, pp. 2672-2679.

[15] X. Li, Y. Chen, Z. Chen, J. Zhao, "Thermal performances of different types of underground heat exchangers," Energy and Buildings, vol. 38, pp. 543–547.

[16] Y. Hamada, H. Saitoh, M. Nakamura, H. Kubota, K. Ochifuji, "Field performance of an energy pile system for space heating," Energy and Buildings, 2007, vol. 39, pp. 517–524.

[17] J. Gao, X. Zhang, J. Liu, K.S. Li, J. Yang, "Thermal performance and ground temperature of vertical pile-foundation heat exchangers: A case study," Applied Thermal Engineering, 2008, vol. 28, pp. 2295–2304.

[18] C. Lee, M. Park, S. Min, S.H. Kang, B. Sohn, H. Choi, "Comparison of effective thermal conductivity in closed-loop vertical ground heat exchangers," Applied Thermal Engineering, 2011, vol. 31, pp. 3669-3676.

[19] L. Jun, Z. Xu, G. Jun, Y. Jie, "Evaluation of heat exchange rate of GHE in geothermal heat pump systems," Renewable Energy, 2009, vol. 34, pp. 2898–2904.

[20] Jalaluddin, A. Miyara, K. Tsubaki, S. Inoue, K. Yoshida, "Experimental study of several types of ground heat exchanger using a steel pile foundation," Renewable Energy, 2011, vol. 36, pp. 764-771.

[21] Jalaluddin, A. Miyara, "Thermal performance investigation of several types of vertical ground heat exchangers with different operation mode," Applied Thermal Engineering, 2012, vol. 33-34, pp. 167–174.

12

Fire Dynamics Simulation and Evacuation for a Large Shopping Center (Mall): Part I, Fire Simulation Scenarios

Khalid A. Albis[*], Muhammad N. Radhwi, Ahmed F. Abdel Gawad

Mechanical Engineering Department, College of Engineering & Islamic Architecture, Umm Al-Qura University, Makkah, Saudi Arabia

Email address:

kalbis31@gmail.com (K. A. Albis), mnradhwi@uqu.edu.sa (M. N. Radhwi), afaroukg@yahoo.com (A. F. A. Gawad)

Abstract: Malls like any retailing centers face exposure for a host of risks including fire, which is no stranger to shopping malls. Fires in closed malls, patronized by lots of people, can cause many fatalities among panicked people running and pushing to get out of these burning places and great damage to the property itself. This computational study covers the possibilities of smoke propagation and evacuation due to hazardous fires in a large shopping center (mall) in Makkah, Saudi Arabia. The mall occupies 50,753 m^2 and has two main floors. It contains 144 stores in the ground floor and 56 stores in the upper floor. It has five gates, one elevator, four escalators and five emergency exit stairs. The study is divided into two parts. Part I concerns four scenarios of fire simulation. Part II considers corresponding four scenarios of evacuation. The present results explain how fast the smoke may spread in such buildings and its mechanism to move from one floor to another. The smoke propagation/movement is highly affected by the architecture of the building and the type of activities inside it.

Keywords: Fire Dynamics Simulation, Smoke Propagation, Shopping Center (Mall)

1. Introduction

1.1. Fire Dynamics Simulator

Fire dynamics simulator (*FDS*) is a computational fluid dynamics (*CFD*) program of fire-driven fluid flow. The software solves numerically a large eddy simulation (*LES*) form of the Navier-Stokes equations appropriate for low-speed, thermally-driven flow, with an emphasis on smoke and heat transport from fires. *FDS* is free software developed by the National Institute of Standards and Technology (*NIST*) of the United States Department of Commerce, in cooperation with *VTT* Technical Research Centre of Finland. *Smokeview* is the companion visualization program that can be used to display the output of *FDS* [1]. Throughout its development, *FDS* has been aimed at solving practical fire problems in fire protection engineering, while at the same time providing a tool to study fundamental fire dynamics and combustion [1]. In the present investigation, PyroSim program [2] was used as interface software to construct the architecture of the investigated mall, and computational domain and mesh. It was also used for other programming aspects such as heat/smoke detectors, sprinklers, *etc*.

1.2. Objectives of the Present Study

This paper utilizes *FDS* to study the smoke propagation/movement in large shopping centers (malls). Usually, fires in closed malls that are attended daily by a big number of visitors can cause many fatalities among panicked people running and pushing to get out of the on-fire mall. Due to the large volume of malls and the complexity of their structure, it is difficult to predict the smoke movement in case of fire. Thus, evacuation plans cannot be set easily.

Therefore, the present research considers the smoke movement due to fire spots in different locations in a large shopping center (mall). This mall contains various activities such as clothing and fabric stores, boutiques, food markets, food court, entertainment facility, *etc*. Moreover, evacuation plans were proposed based on the predicted results of the smoke propagation to ensure safe and quick evacuation of the visitors as will appear in part II. In addition, *FDS* was used to quantify the performance of smoke/fire detectors in terms of their response and the effect of smoke and temperature for large open space applications.

2. Literature Review

2.1. Fire Dynamic Simulation

FDS was employed to model a fire case in which three fire fighters were killed in order to estimate the concentration of carbon monoxide present in the dwelling, which was the immediate cause of death of two of the fire fighters [3]. *FDS* was used to optimize the location of smoke detectors in a large building in Hong Kong for early fire detection. A set of computer simulations was carried out to calculate smoke movement, smoke detection performance, and the overall protection in accordance with performance-based fire safety system design methodologies [4].

Hot smoke tests (*HST*) were performed according to Australian Standard AS 4391-1999 in several buildings to compare with the predictions of *FDS*. It was sated that data from hot smoke tests, if gathered cost effectively, can be a valuable resource for computer model verification [5]. A newly developed integrated system-BFIRESAS was introduced to analyze the overall fire safety behavior of large space structures under real-fire conditions. The results showed that the structural respond and behavior of large space buildings under fire conditions have the systemic nature of a coherent whole, rather than the local effect [6].

To investigate how different configurations in a retail premises affect the smoke spread and temperatures during a fire, eleven tests were performed in scale 1:2 of a large room with small ventilation openings near the floor. Also, the fire tests were simulated using *FDS* to see how well *FDS* simulates under-ventilated fires. It was stated that there are cases where the temperatures from the simulations and measurements correspond relatively well with each other and yet other cases when the simulated temperature is higher than the measured temperature. This depends on the simulation case, the position in the set-up and the time period compared [7]. *FDS* was implemented to investigate smoke detector spacing for spaces (corridors) with deep beam pockets and level ceilings concerning varying beam depth, beam spacing, corridor width, and ceiling height. A subset of the modeled corridor configurations was conducted with full-scale experiments to validate the findings of the modeling study [8].

FDS was used to get the special characteristics of fire spread and smoke movement inside a supermarket as well as evacuation scenarios. The research provided some results for performance-based fire protection design in supermarket buildings [9]. In an attempt to investigate the accuracy of predictions of fire-induced flow into a compartment by *FDS*, finer grids and an inclusion of radiative heat in the combustion model were considered. Results revealed significant improvements to the prediction of mass flow rates for all the three positions of the fire source considered in the study [10].

A study on the fire dynamics within a large compartment with openings was carried out using *FDS,v5*. The study concerned a heat source placed within the basement that was ignited and the flames eventually reached the ceiling causing further ignition of the wood, continuing the spread of flame and heat [11]. Considering the characteristics of fire in the cabin of Civil Aviation, *FDS* and evacuation software *Evac* were utilized. An example of cabin fire was used to explain how to employ the software and apply the results to the practical evacuation and rescue of cabin fire [12].

Other studies concerned fire and smoke aspects in big multi-story buildings/structures can be found in [13-15].

2.2. Examples of Worldwide Fire Cases in Shopping Malls

In 2010, a fire completely destroyed the Garver Brothers Store, in Ohio, United States. Nineteen fire departments responded to the blaze, using so much water that they were forced to draw directly from the Tuscarawas River but they quickly saw that all their efforts were necessary simply to preserve neighboring buildings [16]. Thirteen children and six adults were killed due to sprinkler system's malfunction after a fire broke out in *Doha mall blaze* in the Qatari capital Doha in 2012. Firefighters were forced to break through roof to evacuate victims. Relative of one two-year-old victim said building did not appear to have fire alarms [17].

A fire accident occurred in a five-story marketplace in Kolkata, the capital city of West Bengal, India, in 2013. An estimated 19 people who were mostly laborers working in the market were killed in the accident. Initial reports indicated that the fire might have been initiated by a short circuit in the first floor of the market [18]. A stubborn fire lit up the night sky and sent debris and sparks flying high into the air as it destroyed a Route 9 mall; housing more than a dozen businesses. While it appeared to begin as a small fire, it spread rapidly as firefighters using aerial ladder and large hoses doused it from all angles, to no avail. The wood frame building assessed at $1,033,400, had several rooflines and ceilings and no sprinkles [19].

A fire sparked by a short circuit at a shop quickly engulfed a three-story shopping center in the eastern Pakistani city of Lahore killing at least 13 people in 2014.The fire broke out at the Anrkali Basaar. The building, which mostly houses shops that sells watches and clocks, had no firefighting equipment and no emergency exits. People inside were forced to flee through a single door, and most of the victims suffocated to death [20]. Fire ripped through stores at Main Street mall in Sayreville, in New Jersey, USA, in 2015. The fire consumed several stores, including a nail salon where the fire is believed to have started. The fire burned through at least three stores. But no injuries were reported [21].

Five people died and up to 25 people were missed due to a fire at a shopping centre (Admiral Centre) in the Russian city of Kazan in 2015. The fire started in a first-floor cafe and more than 600 people were cleared from the building. Part of the centre collapsed and forty people were injured in the blaze. It was reported that the building hadn't been equipped with fire safety and smoke control systems [22].

3. Governing Equations and Solution Procedure

This section presents the governing equations of *FDS* and an outline of the general solution procedure. The governing equations are presented as a set of partial differential equations, with appropriate simplifications and approximations. The numerical method essentially consists of a finite-difference approximation of the governing equations and a procedure for updating these equations in time. This section is based on [23].

3.1. Governing Equations

This section introduces the basic conservation equations for mass, momentum and energy for a Newtonian fluid. Note that this is a set of partial differential equations consisting of six equations for six unknowns, all functions of three spatial dimensions and time: density ρ, three components of velocity $u = [u,v,w]T$, temperature T, and pressure p.

3.1.1. Mass and Species Transport

Mass conservation can be expressed either in terms of the density, ρ,

$$\frac{\partial \rho}{\partial t} + \nabla \cdot \rho u = \dot{m}_b^m \tag{1}$$

or in terms of the individual gaseous species, $Y\alpha$:

$$\frac{\partial}{\partial t}(\rho Y\alpha) + \nabla \cdot \rho Y\alpha u = \nabla \cdot \rho D\alpha \nabla Y\alpha + \dot{m}_\alpha^m + \dot{m}_{b,\alpha}^m \tag{2}$$

Here, $\dot{m}_b^m = \sum\alpha\, \dot{m}_{b,\alpha}^m$ is the production rate of species by evaporating droplets or particles. Summing these equations over all species yields the original mass conservation equation because $\sum Y\alpha = 1$ and $\sum \dot{m}_\alpha^m = 0$ and $\sum \dot{m}_{b,\alpha}^m = \dot{m}_b^m$, by definition, and because it is assumed that $\sum \rho D\alpha \nabla Y\alpha = 0$. This last assertion is not true, in general. However, transport equations are solved for total mass and all but one of the species, implying that the diffusion coefficient of the implicit species is chosen so that the sum of all the diffusive fluxes is zero.

3.1.2. Momentum Transport

The momentum equation in conservative form is written as:

$$\frac{\partial}{\partial t}(\rho u) + \nabla \cdot \rho uu + \nabla p = \rho g + f_b + \nabla \cdot \tau_{ij} \tag{3}$$

The term uu is a diadic tensor. In matrix notation, with $u=[u,v,w]T$, the diadic is given by the tensor product of the vectors u and uT. The term $\nabla \cdot \rho uu$ is thus a vector formed by applying the vector operator $\nabla=(\partial\partial x,\ \partial\partial y,\ \partial\partial z)$ to the tensor. The force term f_b in the momentum equation represents external forces such as the drag exerted by liquid droplets. The stress tensor τ_{ij} is defined as:

$$\tau_{ij} = \mu\,(2\,S_{ij} - \tfrac{2}{3}\delta_{ij}\,(\nabla \cdot u);\ \delta_{ij} = \begin{cases} 1 & i = j \\ 0 & i \neq j \end{cases};\ S_{ij} = \tfrac{1}{2}\left(\frac{\partial ui}{\partial xj} + \frac{\partial uj}{\partial xi}\right)\, i,j = 1,2,3 \tag{4}$$

The term S_{ij} is the symmetric rate-of-strain tensor, written using conventional tensor notation. The symbol μ is the dynamic viscosity of the fluid. The overall computation can either be treated as a Direct Numerical Simulation (*DNS*), in which the dissipative terms are computed directly, or as a Large Eddy Simulation (*LES*), in which the large-scale eddies are computed directly and the subgrid-scale dissipative processes are modeled. The numerical algorithm is designed so that *LES* becomes *DNS* as the grid is refined. Most applications of *FDS* are *LES*. For example, in simulating the flow of smoke through a large, multi-room enclosure, it is not possible to resolve the combustion and transport processes directly. However, for small-scale combustion experiments, it is possible to compute the transport and combustion processes directly. For the purpose of outlining the solution procedure below, it is sufficient to consider the momentum equation written as:

$$\frac{\partial \rho}{\partial t} + F + \nabla H = 0 \tag{5}$$

and the pressure equation as

$$\nabla^2 H = -\frac{\partial}{\partial t}(\nabla \cdot u) - \nabla \cdot F \tag{6}$$

which is obtained by taking the divergence of the momentum equation.

3.1.3. Energy Transport

The energy conservation equation is written in terms of the sensible enthalpy, h_s:

$$\frac{\partial}{\partial t}(\rho hs) + \nabla \cdot \rho h_s u = \frac{Dp}{Dt} + \dot{q}''' - \dot{q}_b''' - \nabla \cdot \dot{q}'' + \varepsilon \tag{7}$$

The sensible enthalpy is a function of the temperature:

$$h_s = \sum_\alpha Y\alpha h_s,\alpha\ ;\ h_s,\alpha(T) = \int_{T_0}^T c_{p,\alpha}(T')dT' \tag{8}$$

Note the use of the material derivative, $D()/Dt = \partial()/\partial t + u \cdot \nabla()$. The term $\dot{q}'''$ is the heat release rate per unit volume from a chemical reaction. The term $\dot{q}_b''$ is the energy transferred to the evaporating droplets. The term $\dot{q}''$ represents the conductive and radiative heat fluxes:

$$\dot{q}'' = -k\nabla T - \sum_\alpha h_s,\alpha\rho D\alpha \nabla Y\alpha + \dot{q}''_r \tag{9}$$

Where, k is the thermal conductivity.

3.1.4. Equation of State

$$p = \frac{\rho RT}{W} \tag{10}$$

An approximate form of the Navier-Stokes equations appropriate for low Mach number applications is used in the model. The approximation involves the filtering out of acoustic waves while allowing for large variations in temperature and density. This gives the equations an elliptic character, consistent with low speed, thermal convective processes. In practice, this means that the spatially resolved pressure, $p(x,y,z)$, is replaced by an "average" or "background" pressure, $p_m(z,t)$, that is only a function of time and height above the ground.

$$\overline{P}m(z,t) = \rho TR \sum_\alpha Y\alpha / W\alpha \qquad (11)$$

Taking the material derivative of the background pressure and substituting the result into the energy conservation equation yields an expression for the velocity divergence, ∇u that is an important term in the numerical algorithm because it effectively eliminates the need to solve a transport equation for the specific enthalpy. The source terms from the energy conservation equation are incorporated into the divergence, which appears in the mass transport equations. The temperature is found from the density and background pressure via the equation of state.

3.2. Solution Procedure

FDS uses a second-order accurate finite-difference approximation to the governing equations on a series of connected recti-linear meshes. The flow variables are updated in time using an explicit second-order Runge-Kutta scheme. This section describes how this algorithm is used to advance in time the density, species mass fractions, velocity components, and background and perturbation pressure. Let ρ^n, Y_α^n, u^n, $\overline{p}_m^n$ and H^n denote these variables at the nth time step.

1. Compute the "patch-average" velocity field $\overline{u}^n$.

2. Estimate ρ, $Y\alpha$, and $\overline{P}m$ at the next time step with an explicit Euler step. For example, the density is estimated by

$$\frac{\rho^* - \rho n}{\delta t} + \nabla \cdot \rho^n \overline{u}^n = 0 \qquad (12)$$

3. Exchange values of ρ^* and Y_α^* at mesh boundaries.

4. Apply boundary conditions for ρ^* and Y_α^*.

5. Compute the divergence, $\nabla \cdot \overline{u}^*$, using the estimated thermodynamic quantities. Note that at this stage, the velocity field has not been estimated at the next time step, only its divergence.

6. Solve the Poisson equation for the pressure fluctuation with a direct solver on each individual mesh:

$$\nabla^2 H^n = -\left[\frac{\nabla \cdot u^* - \nabla \cdot \overline{u}^n}{\delta t}\right] - \nabla \cdot \overline{F}^n \qquad (13)$$

Note that the vector $\overline{F}^n = F(\rho^n, \overline{u}^n)$ is computed using patch-averaged velocities and that the divergence of the patch-averaged field is computed explicitly.

7. Estimate the velocity at the next time step

$$\frac{u^* - \overline{u}^n}{\delta t} + \overline{F}^n + \nabla H^n = 0 \qquad (14)$$

Note that the divergence of the estimated velocity field is identically equal to the estimated divergence, $\nabla \cdot u^*$, that was derived from the estimated thermodynamic quantities.

8. Check the time step at this point to ensure that

$$\delta t \max \left(\frac{|u|}{\delta x}, \frac{|v|}{\delta y}, \frac{|w|}{\delta z}\right) < 1 \; ; 2 \, \delta tv \left(\frac{1}{\delta x^2}, \frac{1}{\delta y^2}, \frac{1}{\delta z^2}\right) < 1 \qquad (15)$$

If the time step is too large, it is reduced so that it satisfies

both constraints and the procedure returns to the beginning of the time step. If the time step satisfies the stability criteria, the procedure continues to the corrector step.

This concludes the "Predictor" stage of the time step. At this point, values of H^n and the components of u^* are exchanged at mesh boundaries as follows:

1. Compute the "patch-average" velocity field $\overline{u}^*$.

2. Apply the second part of the Runge-Kutta update to the mass variables. For example, the density is corrected as

$$\frac{\rho^{n+1} - \frac{1}{2}(\rho^n + \rho^*)}{\delta t/2} + \nabla \cdot \rho^* \overline{u}^* = 0 \qquad (16)$$

3. Exchange values of ρ^n and Y_α^n at mesh boundaries.

4. Apply boundary conditions for ρ^n and Y_α^n.

5. Compute the divergence $\nabla \cdot u^{n+1}$ from the corrected thermodynamic quantities. Note again that the velocity field has not been corrected at the point.

6. Compute the pressure fluctuation using estimated quantities

$$\nabla^2 H^* = -\frac{\nabla \cdot u^{n+1} - \frac{1}{2}(\nabla \cdot \overline{u}^* + \nabla \cdot \overline{u}^n)}{\delta t/2} - \nabla \cdot \overline{F}^* \qquad (17)$$

7. Update the velocity via the second part of the Runge-Kutta scheme

$$\frac{u^{n+1} - \frac{1}{2}(\overline{u}^* + \overline{u}^n)}{\delta t/2} + \overline{F}^* + \nabla H^* = 0 \qquad (18)$$

Note again that the divergence of the corrected velocity field is identically equal to the divergence that was computed earlier.

8. At the conclusion of the time step, values of H^* and the components of u^{n+1} are exchanged at mesh boundaries.

3.3. Spatial Discretization

Spatial derivatives in the governing equations are written as second-order accurate finite-differences on a rectilinear grid. The overall domain is a rectangular box that is divided into rectangular grid cells. Each cell is assigned indices i, j and k representing the position of the cell in the x, y and z directions, respectively. Scalar quantities are assigned in the center of each grid cell; thus, ρ_{ijk}^n is the density at the nth time step in the center of the cell whose indices are i, j and k. Vector quantities like velocity are assigned at their appropriate cell faces. For example, u_{ijk}^n is the x-component of velocity at the positive-oriented face of the ijkth cell; $u_{i-1,jk}^n$ is defined at the negative-oriented face of the same cell.

4. Present Model (Mall) and Computational Aspects

4.1. Description of the Present Model (Mall)

The mall is actually a big shopping center that provides a complete and comprehensive shopping and recreation for all family members in Makkah city, Saudi Arabia, Fig. 1. The mall occupies 50,753 m^2 and has two main floors. It contains

144 stores in the ground floor and 56 stores in the upper floor. It has five main gates, one elevator, four escalators, and five emergency exit stairs, Fig. 2. Most of the mall's stores are for clothing and fabric.

The ground floor has five main entrances and three main corridors that are perpendicular to other three main corridors and includes mainly stores. The upper floor consists of stores as well; in addition to recreation facilities and a food court with five emergency stairs. The building is surrounded by parking lots especially in the two directions of the entrances; front and back.

Figure 1. *A picture of the front view of the mall.*

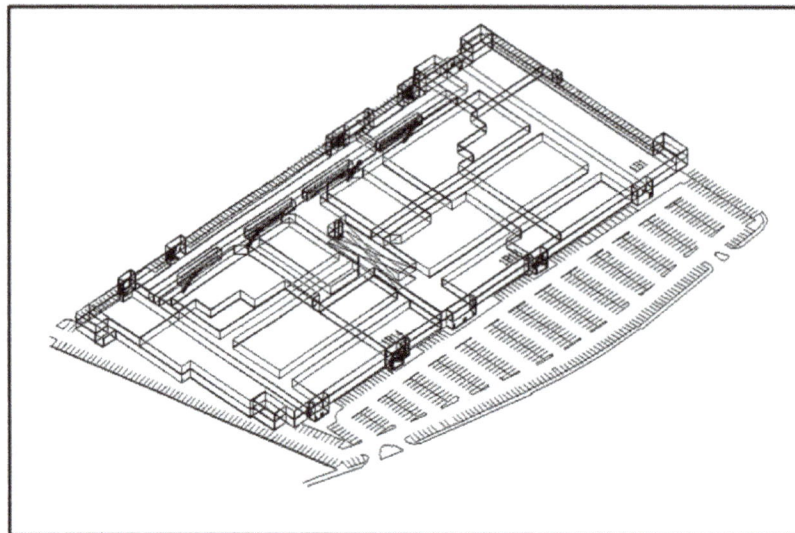

Figure 2. *Three-dimensional drawing of the Mall.*

4.2. Computational Grid (Mesh)

Usually, the computational grid of *FDS* is formed of uniform elements in the form of cuboids or cubes in the three dimensions of the building. The grid covers the whole space of the building. In the present study, cubic elements ($0.2 \times 0.2 \times 0.2\ m^3$) were used. The grid dimensions along the front of the mall (x-direction), along the width of the mall (y-direction), and along the height of the mall (z-direction) are 340 *m*, 179 *m*, and 20 *m*, respectively.

Figure 3(a) shows an overall view of the grid of the whole mall. The grid elements in two planes along the front of the mall (x-direction), and along the width of the mall (y-direction) are shown in Fig. 3(b). Figure 3(c) shows a zoomed *2D* view of the grid elements that take a square shape.

Figure 3(a). *verall view of the grid of the whole mall.*

Figure 3(b). rid elements in two planes along the front of the mall (x-direction), and along the width of the mall (y-direction).

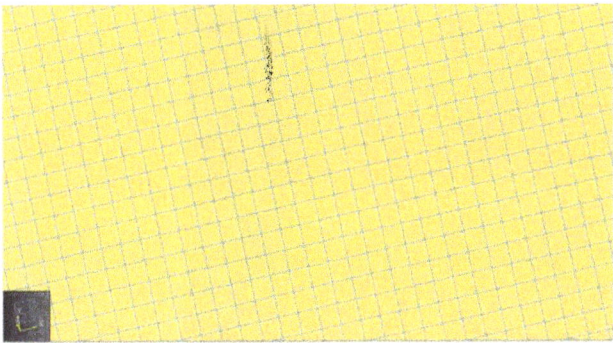

Figure 3(c). Zoomed 2D view of the grid elements.

Figure 3. Present computational grid.

4.3. Sprinklers and Heat/Smoke Detectors

Sprinkler is a device that spreads water on the fire location according to fire signal to suppress fire at its early stages [24-27]. Heat detector is a fire detector that detects either abnormally high temperature or rate of temperature rise or both. Smoke detector is a device that detects visible or invisible particles of combustion [28].

To simulate the actual situation of the mall, grids of sprinklers and heat/smoke detectors were constructed. The sprinklers and detectors were distributed in both floors according to their actual distribution in the mall.

Thus, the total number of sprinklers, smoke detectors, and heat detectors are 7848, 844, and 20, respectively in the whole mall.

Figure 4(a) shows a close view of the distribution of sprinklers (Blue), smoke detectors (Green), and heat detectors (Red). As heat detectors sense high temperature, they are placed in the kitchens of the food court. Figure 4(b) shows a plane of the ground floor that illustrates the sprinkles, and smoke detectors that cover the whole area of the ground floor. Figure 4(c) shows a plane of the upper floor that illustrates the sprinkles, and smoke and heat detectors that cover the whole area of the upper floor. As can be seen in Fig. 4(c), some spaces of the upper floor, to the far right and left, are still empty without stores. It is clear from Figs. 4(b) and 4(c) that huge numbers of sprinklers and smoke detectors were modeled in the present work, which needed big effort and consumed a lot of time.

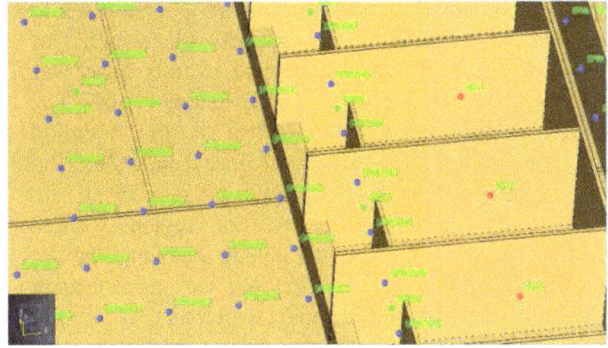

Figure 4(a). lose view of the distribution of sprinklers (Blue), smoke detectors (Green), and heat detectors (Red).

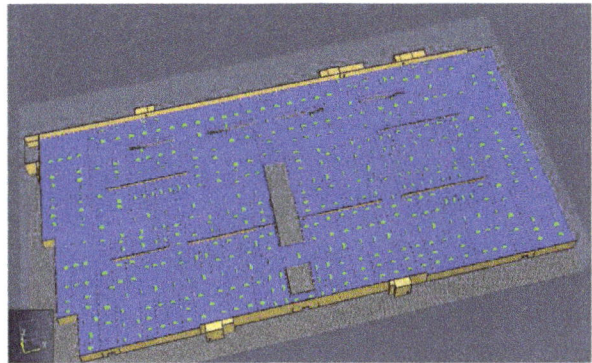

Figure 4(b). lane of the ground floor with sprinkles (Blue),and smoke detectors (Green) for the ground floor.

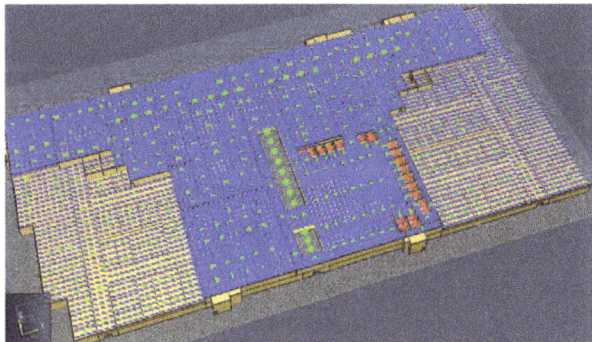

Figure 4(c). lane of the upper floor with sprinkles (Blue),smoke detectors (Green), and heat detectors (Red) for the upper floor.

Figure 4. istributions of sprinklers, and smoke and heat detectors.

5. Results and Discussions

5.1. Case Studies

To investigate the different possibilities of smoke propagation/movement in the mall, four case studies were considered. Two of these case studies concerned the fire ignition in two locations in the upper floor as occupants' evacuation would be more difficult. Two other case studies considered fire ignition in the ground floor. Based on these four fire simulations, four corresponding evacuation simulations were carried out, Part II.

In all cases, the Heat Release Rate per Unit Area (HRRPUA) was fixed at 250 kW/m^2 [29]. More estimates of HRRPUA for different applications can also be found in

[30,31]. The computations of fire simulation continued to 900 seconds (15 minutes) except for case (4), which was terminated at 420 seconds (7 minutes) as this time was sufficient to design the corresponding evacuation plan. The ambient temperature of the mall was kept at 20^oC. It should be noted that pressure fans of the emergency exits were not considered in the present study. As will be seen in all fire simulation cases, sprinkles in the fire region worked successfully. The sprinkles were not able to overcome the fire because the *HRRPUA* of the fire was set as constant. However, they slightly lowered the temperature in the fire region.

5.2. Fire Simulation Cases

5.2.1. Case Study (1)

The first case study concerned a fire ignition at one of the kitchens of the food court in the upper floor. To facilitate the following of the smoke propagation and movement with time, the important parts of the upper floor (main stores, corridors, emergency exits, *etc.*) are notified and numbered as can be seen in Fig. 5. The fire was set at kitchen no. 7.

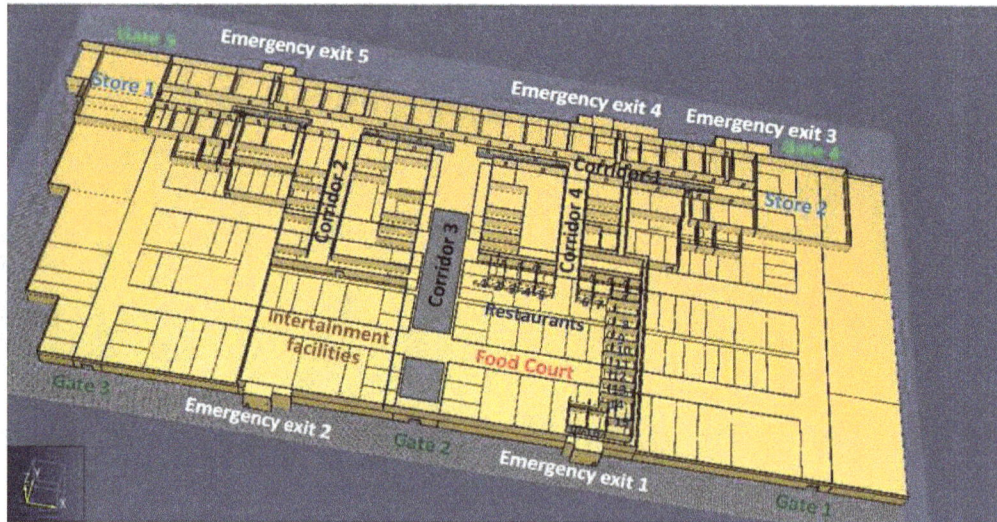

Figure 5. ain stores, corridors, and emergency exits in the upper floor.

The smoke propagation and movement with time is demonstrated in Fig. 6. The smoke movement with time can be monitored from Fig. 6 as follows:

1. After 5 *seconds* of fire start, smoke was still restricted to the inside of kitchen (7), Fig. 6(a).
2. After 10 *seconds*, the smoke spread from kitchen (7) to kitchen (8) and began to spread to the space of the food court, Fig. 6(b).
3. After 20 *seconds*, the smoke spread more to the space of the food court and kitchens (6,8,9,10,11). Also, smoke started to spread in corridor (4), Fig. 6(c).
4. After 40 *seconds*, smoke spread to kitchens(4,5,13,14) and completely filled kitchens (6,8,9,10,11,12). It also filled half the space of the food court and spread increasingly in corridor (4), Fig. 6(d).
5. After 60 *seconds*, smoke filled all kitchens except (1,2) and about three-fourth (3/4) of the food court and one-third (1/3) of corridor (4), Fig. 6(e).
6. After 80 *seconds*, smoke filled all kitchens, all the food court space, half (1/2) of corridor (4), and started to spread in corridor (3), Fig. 6(f).
7. After 100 *seconds*, smoke filled all kitchens, all the food court space, half (1/2) of corridor (3), and two-thirds (2/3) of corridor (4), Fig. 6(g).
8. After 120 *seconds*, smoke filled all kitchens, all the food court space, three-fourths (3/4) of corridor (4), more than half of corridor (3), and started to spread in the region of the entertainment facilities, Fig. 6(h).

9. After 200 *seconds*, smoke filled all kitchens, the food court space, corridor (4), three-fourths (3/4) of corridor (3), one-third (1/3) of the space of the entertainment facilities. Also smoke began to spread in corridors (1) and (2), and started to spread to the ground floor through escalators, Fig. 6(i).
10. After 300 *seconds* (5 *minutes*), smoke filled all kitchens, the food court space, corridors (3) and (4), half (1/2) of corridors (1) and (2), two-thirds (2/3) of the space of the entertainment facilities. Smoke also started to spread in the stores of corridors (1,3,4) and downwards in corridor (1)through the escalators, Fig. 6(j).
11. After 450 *seconds* (7.5 *minutes*), smoke entirely filled all kitchens, the food court space, corridors (2,3,4), more than three-fourths (3/4) of the space of the entertainment facilities, more than three-fourths (3/4) of corridor (1), the majority of the stores of corridors(1,3,4). Moreover, smoke started to spread to the stores of corridor (2) and the ground floor down corridors (1,3), Fig. 6(k).
12. After 600 *seconds* (10 *minutes*), smoke now completely filled all kitchens, the space of the food court, the space of the entertainment facilities, the majority of the stores of the corridors, in addition to one-fourth (1/4) of main store (2). Also, smoke continued to spread down the corridors (1,3) to the ground floor and get out of the emergency exit (3),

Fig. 6(l). It should be noted that pressure fans of the emergency exits were not considered in the present study.

13. After 750 *seconds* (12.5 *minutes*), smoke filled all of the spaces in the upper floor except main stores (1,2). It filled just half of store (2) and one-third (1/3) of store (1). In the ground floor, the smoke entirely filled corridor (4), and started to spread to corridors (1,2,3) and some stores of corridor (4). Also, smoke get to emergency exits (1,2,3), Fig. 6(m).

14. After 900 *seconds* (15 *minutes*), smoke filled all of the spaces in the upper floor except main stores (1,2). Smoke filled just three-fourths (3/4) of store (2) and two-thirds (2/3) of store (1). In the ground floor, smoke entirely filled corridor (4), and continued to spread in corridors (1,2,3) and some stores of corridor (4). Also, smoke continued to get out of emergency exits (1,2,3), Fig. 6(n). Figure 6(o) shows the smoke getting out of the emergency-exit stairs (3) and (5) as can be seen from the back of the mall.

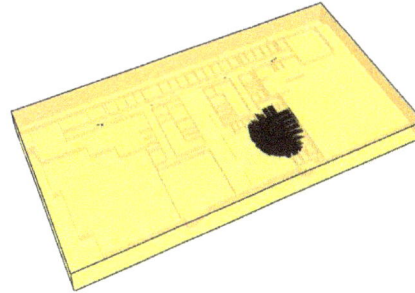

Figure 6(d). 40 Sec.

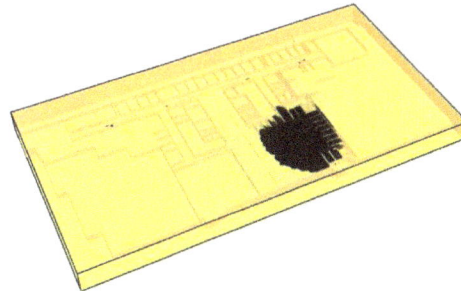

Figure 6(e). 60 Sec. (1 Min.).

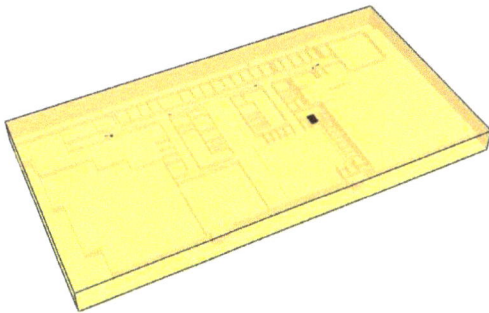

Figure 6(a). 5 Sec.

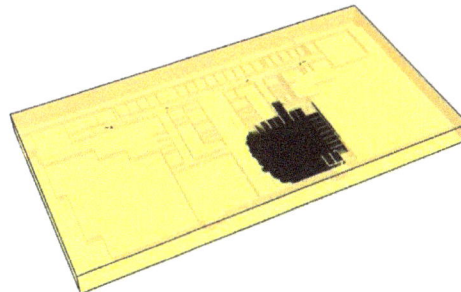

Figure 6(f). 80 Sec.

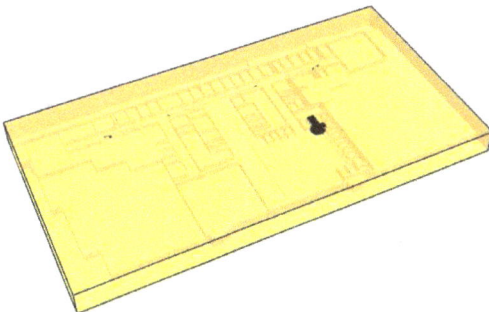

Figure 6(b). 10 Sec.

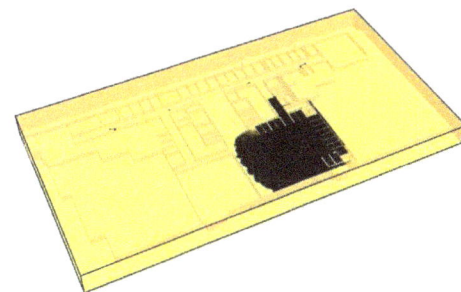

Figure 6(g). 100 Sec.

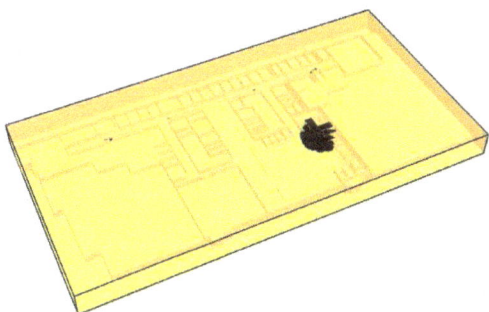

Figure 6(c). 20 Sec.

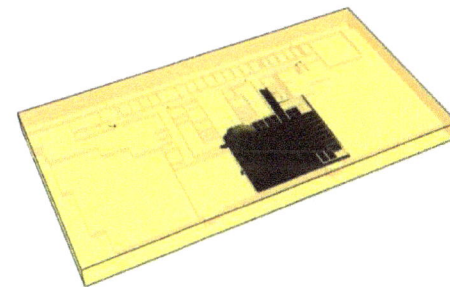

Figure 6(h). 120 Sec. (2 Min.).

Figure 6(i). 200 Sec.

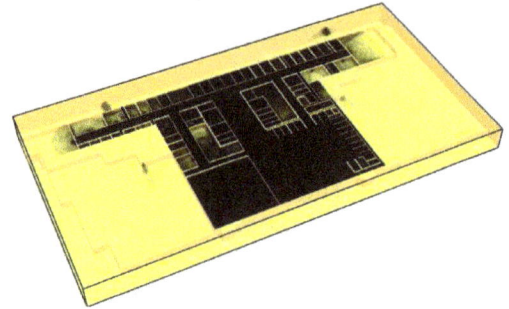

Figure 6(j). 300 Sec. (5 Min.).

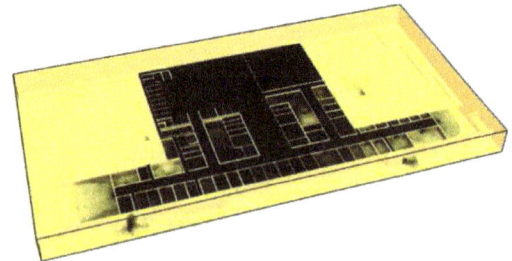

Figure 6(k). 450 Sec. (7.5 Min.).

Figure 6(l). 600 Sec. (10 Min.).

Figure 6(m). 750 Sec. (12.5Min.).

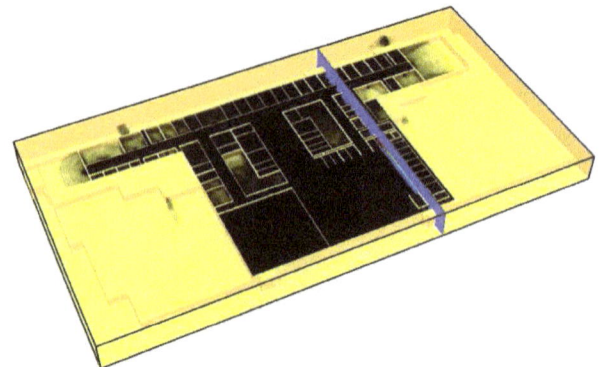

Figure 6(n). 900 Sec. (15 Min.).

Figure 6(o). Smoke getting out of the emergency-exit stairs 3 and 5 (back view of the mall).

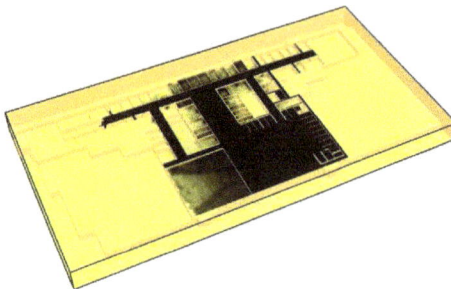

Figure 6. Smoke propagation with time, Case (1).

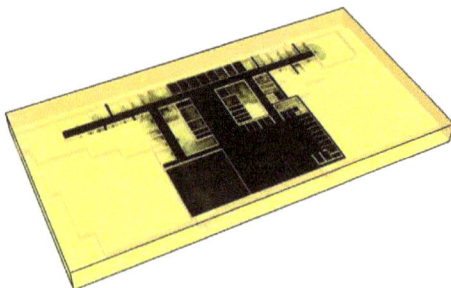

Figure 7 shows the temperature distribution near the area of fire. Figure 7(a) shows the temperature distribution in a vertical plane passing through the fire spot and in the normal direction to the mall front. It is clear that the maximum temperature of 105^oC is reached in the fire spot. Figure 7(b) shows the temperature distribution in another vertical plate, which is parallel to the plane of Fig. 7(a) but 10 *m* away from the fire spot. The maximum temperature is 45^oC. Figure 7(c) shows the temperature distribution in a vertical plane parallel to the mall front and 15 *m* away from the fire spot. The maximum temperature is about 30^oC. Figure 7(d) indicates that the sprinkles in the fire region worked successfully.

Figure 7(a). Temperature distribution in a vertical plane passing through the fire spot and in the normal direction to the mall front.

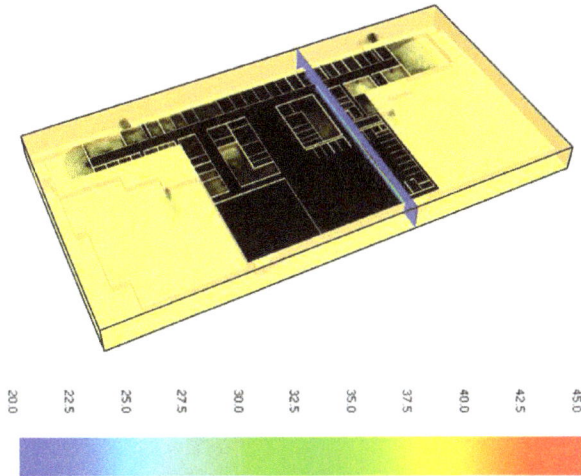

Figure 7(b). Temperature distribution in a vertical plane 10 m away from the fire spot and in the normal direction to the mall front.

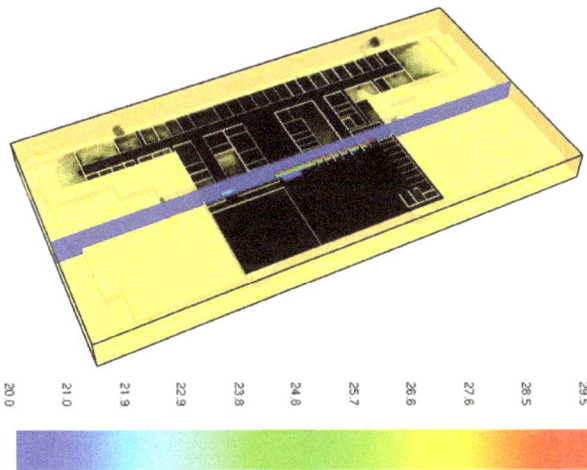

Figure 7(c). Temperature distribution in a vertical plane parallel to the mall front and 15 m away from the fire spot.

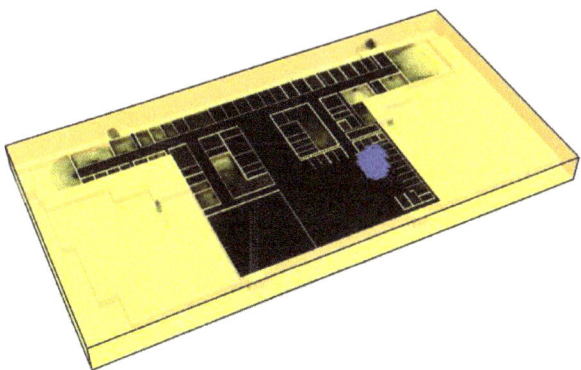

Figure 7(d). The working sprinkles in the fire region.

Figure 7. Temperature distribution near the area of fire, Case (1).

5.2.2. Case Study (2)

The second case study concerned a fire ignition at main store (2)in the upper floor, Fig. 5. The smoke propagation/movement with time is demonstrated in Fig. 8. The smoke movement with time can be monitored from Fig.

8 as follows:

1. After 5 seconds of fire start, smoke just began to spread in store (2), Fig. 8(a).
2. After 10 seconds, smoke filled one-third (1/3) of store (2), Fig. 8(b).
3. After 20 seconds, smoke filled more than three-fourths (3/4) of store (2). Smoke was still trapped in store (2), Fig. 8(c).
4. After 40 seconds, smoke completely filled store (2) and started to spread to corridor (1), Fig. 8(d).
5. After 60 seconds, smoke completely filled store (2), continued spreading in corridor (1), and started to spread to the first escalator in front of store (2), Fig. 8(e).
6. After 80 seconds, smoke completely filled store (2), one-fourth (1/4) of corridor (1), and most of the space of the first escalator in front of store (2), Fig. 8(f).
7. After 100 seconds, smoke completely filled store (2), the space of the first escalator in front of store (2), and more than one-fourth (1/4) of corridor (1), Fig. 8(g).
8. After 120 seconds, smoke completely filled store (2), the space of the first escalator in front of store (2), and more than one-third (1/3) of corridor (1). Also, smoke approached the next escalator, Fig. 8(h).
9. After 200 seconds, smoke completely filled store (2), the space of the two escalators in front of store (2), half (1/2) of corridor (1). Also, smoke started to spread in corridor (4) and in the corridor that leads to the mosque and the rest (WC) rooms, Fig. 8(i).
10. After 300 seconds (5 minutes), smoke completely filled store (2), the space of the three escalators in front of store (2), corridor (4), the corridor that leads to the mosque and rest (WC) rooms. Also, smoke filled two-thirds (2/3) of corridor (1) and approached the food court zone, Fig. 8(j).
11. After 450 seconds (7.5 minutes), smoke completely filled store (2), the three escalators in front of store (2), corridors 3 and 4, the corridor that leads to the mosque and rest rooms. Also, smoke filled three-fourths (3/4) of corridor (1) and one-third of corridor (2), and started to spread in the space of the food court, Fig. 8(k).
12. After 600 seconds (10 minutes), smoke completely filled store (2), corridor (1), the four escalators that exist in corridor (1), corridors(3) and (4). Also, smoke filled three-fourths (3/4) of corridor (2) and one-fourth(1/4) of the space of the food court. Moreover, smoke started to spread in the mosque and rest (WC) rooms adjacent to store (2), Fig. 8(l).
13. After 750 seconds (12.5 minutes), smoke completely filled store (2), corridors (1,2,3,4), the corridors that lead to the mosque and rest (WC) rooms on that floor, the four escalators that exist in corridor (1). Also, smoke filled three-fourths(3/4) of the space of the food court and spread to the majority of the kitchens and the stores of all corridors, Fig. 8(m).

14. After 900 seconds (15 minutes), smoke completely filled store (2), all corridors, all kitchens, the four escalators that exist in corridor (1), the entire space of the food court, and the majority of stores of all corridors. Moreover, smoke started to spread to the space of the entertainment facilities, Fig. 8(n). Figure 8(o) shows the smoke getting out of the emergency-exit stairs (3) and (4) as can be seen from the back of the mall.

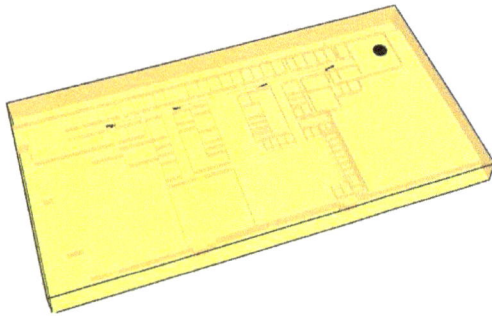

Figure 8(e). 60 Sec. (1 Min.)

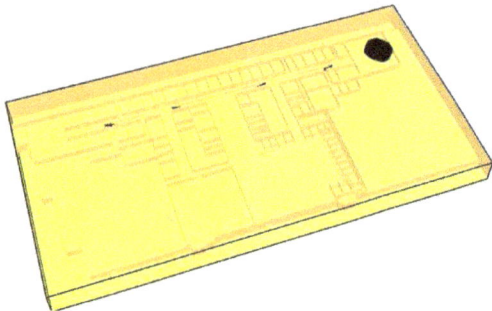

Figure 8(a). 5 Sec.

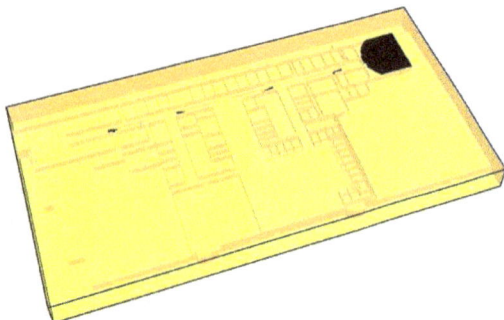

Figure 8(f). 80 Sec.

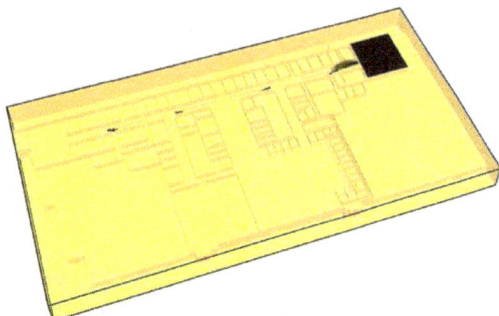

Figure 8(b). 10 Sec.

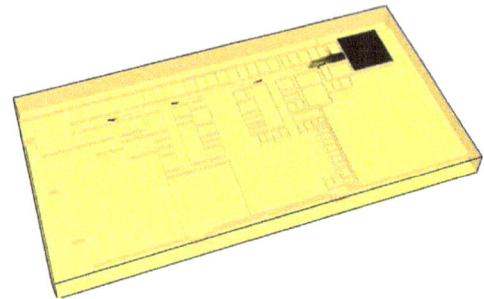

Figure 8(g). 100 Sec.

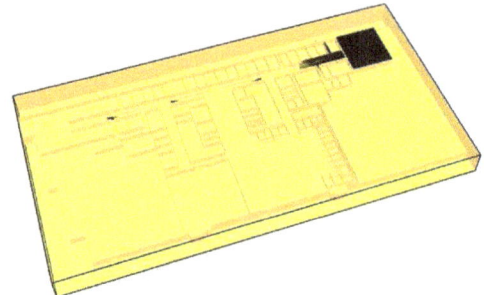

Figure 8(c). 20 Sec.

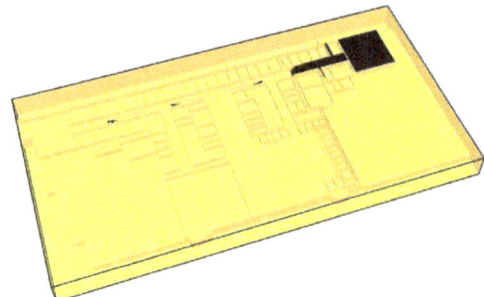

Figure 8(h). 120 Sec. (2 Min.).

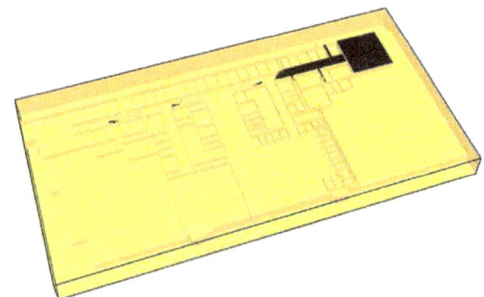

Figure 8(d). 40 Sec.

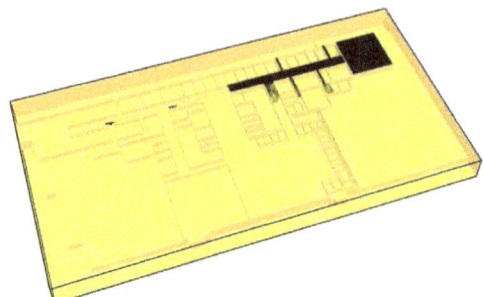

Figure 8(i). 200 Sec.

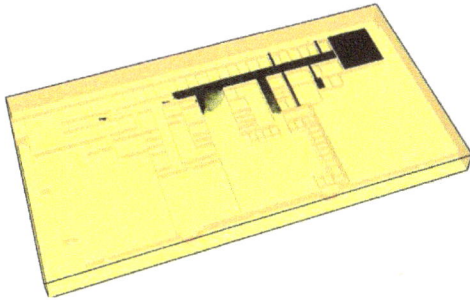

Figure 8(j). 300 Sec. (5 Min.)

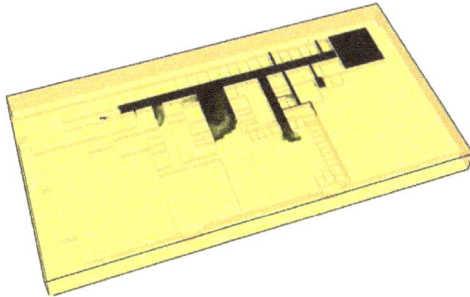

Figure 8(k). 450 Sec. (7.5 Min.).

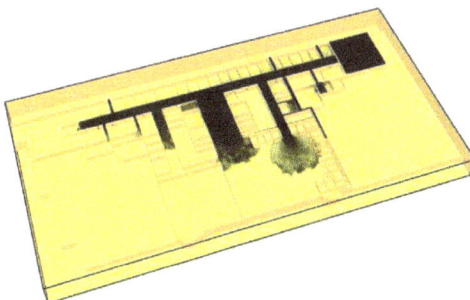

Figure 8(l). 600 Sec. (10 Min.)

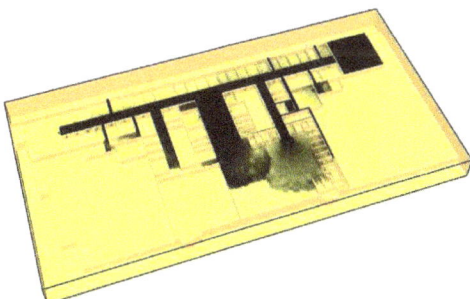

Figure 8(m). 750 Sec. (12.5 Min.).

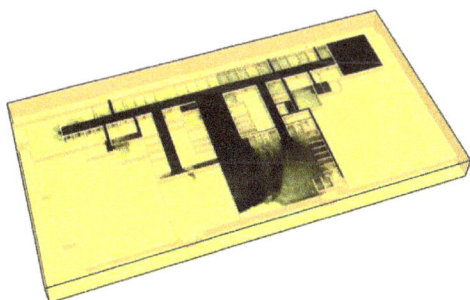

Figure 8(n). 900 Sec. (15 Min.).

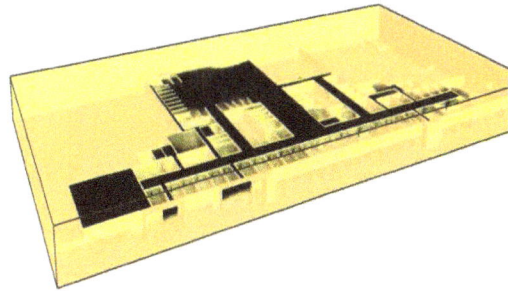

Figure 8(o). Smoke getting out of the emergency-exit stairs 3 and 4 (Back view of the mall).

Figure 8. Smoke propagation with time, Case 2.

Figure 9 shows the temperature distribution near the area of fire. Figure 9(a) shows the temperature distribution in a vertical plane passing through the fire spot and in the normal direction to the mall front. It is clear that the maximum temperature of $105^{o}C$ is reached in the fire spot. Figure 9(b) shows the temperature distribution in another vertical plate, which is parallel to the plane of Fig. 9(a) but 10 *m* away from the fire spot. The maximum temperature is $90^{o}C$. Figure 9(c) shows the temperature distribution in a vertical plane parallel to the mall front and 15 *m* away from the fire spot. The maximum temperature is about $80^{o}C$. Figure 9(d) indicates that the sprinkles in the fire region worked successfully.

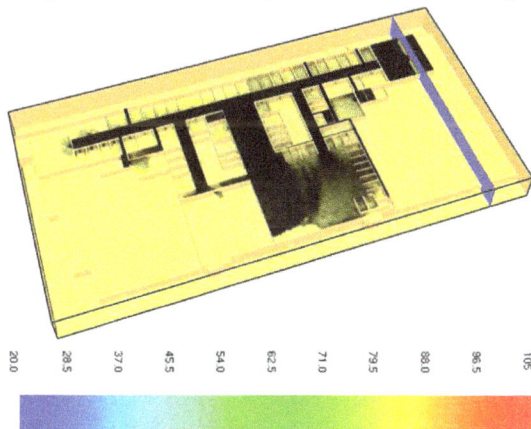

Figure 9(a). Temperature distribution in a vertical plane passing through the fire spot and in the normal direction to the mall front.

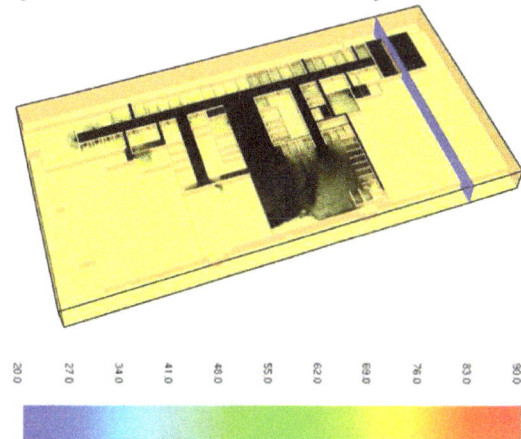

Figure 9(b). Temperature distribution in a vertical plane 10 m away from the fire spot and in the normal direction to the mall front.

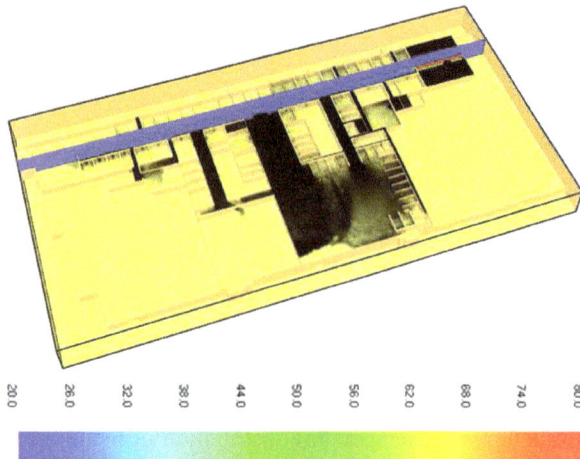

20.0 26.0 32.0 38.0 44.0 50.0 56.0 62.0 68.0 74.0 80.0

Figure 9(c). *Temperature distribution in a vertical plane parallel to the mall front and 15 m away from the fire spot.*

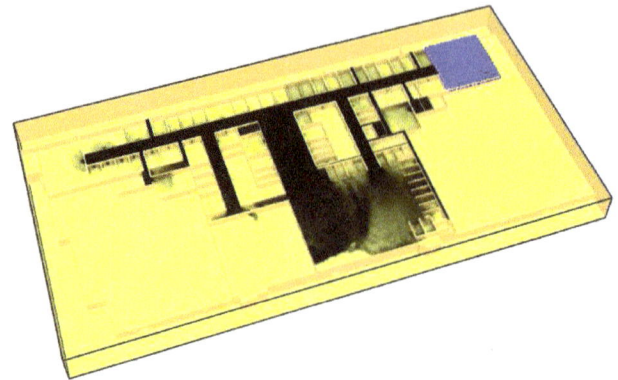

Figure 9(d). *The working sprinkles in the fire region.*

Figure 9. *Temperature distribution near the area of fire, Case(2).*

5.2.3. Case Study (3)

The third case study concerned a fire ignition at main store (3)in the ground floor. Again, to facilitate the following of the smoke propagation and movement with time, the important parts of the ground floor (main stores, gates, corridors, emergency exits, *etc.*) are notified and numbered as can be seen in Fig. 10.

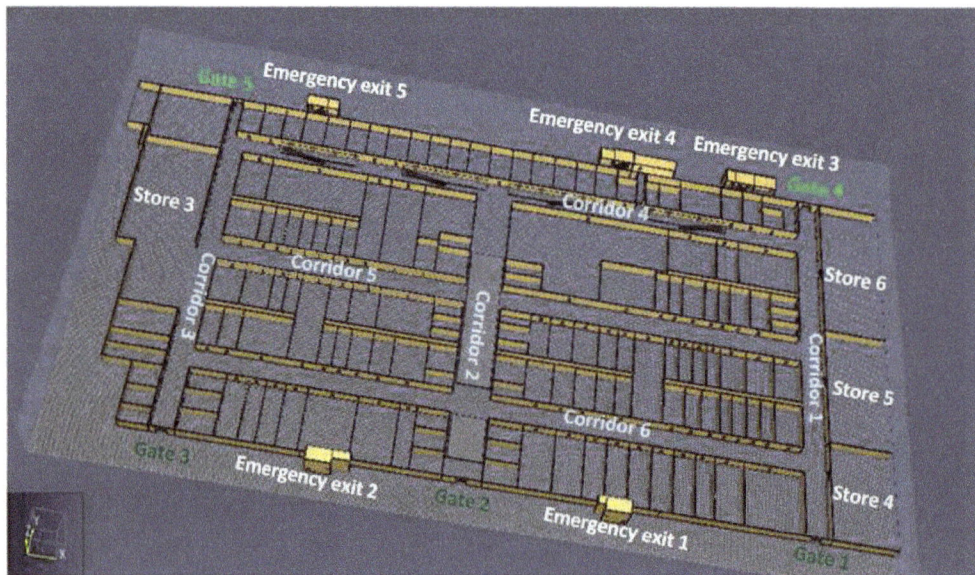

Figure 10. *Main stores, corridors, and gates in the ground floor.*

The smoke propagation/movement with time is demonstrated in Fig. 11. The smoke movement with time can be monitored from Fig. 11 as follows:

1. After 5 *seconds* of fire start, smoke started to spread inside store (3), Fig. 11(a).
2. After 10 *seconds*, smoke propagated in store (3) and was still trapped inside it, Fig. 11(b).
3. After 20 *seconds*, smoke filled more than half of store (3) and still trapped inside it, Fig. 11(c).
4. After 40 *seconds*, smoke filled more than three-fourths (3/4) of store (3) and started to spread in corridor (3), Fig. 11(d).
5. After 60 *seconds*, smoke completely filled store (3), filled one-third (1/3) of corridor (3), and started to

spread in corridor (5), Fig. 11(e).
6. After 80 *seconds*, smoke completely filled store (3), more than half of corridor (3), and continued to spread in corridor (5), Fig. 11(f).
7. After 100 *seconds*, smoke filled completely store (3), three-fourths (3/4) of corridor (3), continued to spread in corridor (5), and started to spread in corridor (6) and to the stores adjacent to store (3) in corridor (3), Fig. 11(g).
8. After 120 *seconds*, smoke filled completely store (3), more than three-fourths (3/4) of corridor (3), and continued to spread in corridors (5) and (6) and the stores of corridor (3), Fig. 11(h).
9. After 200 *seconds*, smoke completely filled store (3),

corridor (3) and the stores in corridor (3), about one-third (1/3) of corridor (5), about one-fourth (1/4) of corridor (6), and started to spread to the stores of corridors(5) and (6) and get out through gates(3) and (5), Fig. 11(i).

10. After 300 *second*s (5 *minutes*), smoke completely filled store (3), corridor (3), all stores of corridor (3), half (1/2) of corridors (5) and (6), some stores in the first half of corridors (5) and (6), and about one-fourth (1/4) of corridor (4). Also, it started to spread in corridor (2) and continued to get out of gates (3) and (5), Fig. 11(j).

11. After 450 *seconds* (7.5 *minutes*), smoke completely filled store (3), corridors(2) and (3), all stores in corridor (3), two-thirds (2/3) of corridor (6), half (1/2) of corridor (5), the majority of stores in the first half of corridors (5) and (6), one-third (1/3) of corridor (4), and continued to get out of gates (3) and (5), Fig. 11(k).

12. After 600 *seconds* (10 *minutes*), smoke completely filled store (3), corridors (2) and (3), all stores in corridors (2) and (3), three-fourths (3/4) of corridors (5) and (6), the majority of stores in the first half of corridors (4), (5) and (6), two-thirds (2/3) of corridor (4). Also, smoke continued to get out of gates (3) and (5) and started to get out of gate (2), Fig. 11(l).

13. After 750 *seconds* (12.5 *minutes*), smoke completely filled store (3), corridors (2) and (3), all stores in corridors (2) and (3), the majority of other corridors except corridor (1), the majority of stores in corridors (5) and (6), some stores in corridor (4). Also, smoke continued to get out through gates (2),(3) and (5), Fig. 11(m).

14. After 900 *seconds* (15 *minutes*), smoke completely filled store (3), corridors (2), (3) and (6), and all stores in corridors (2), and (3). Also, smoke filled the majority of stores in corridor (6), three-fourths (3/4) of corridors (4) and (5), and the majority of stores in them, Fig. 11(n). Figure 11(o) shows the smoke getting out through gates (2) and (3). Figure 11p shows the smoke getting out through gate (5) as seen from the back of the mall.

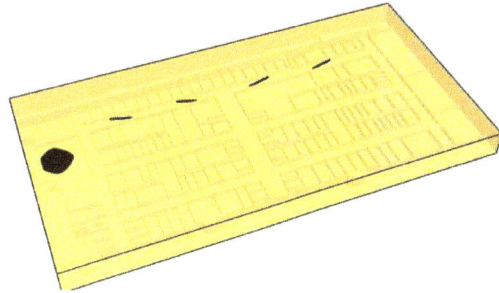

Figure 11(a). 5 Sec.

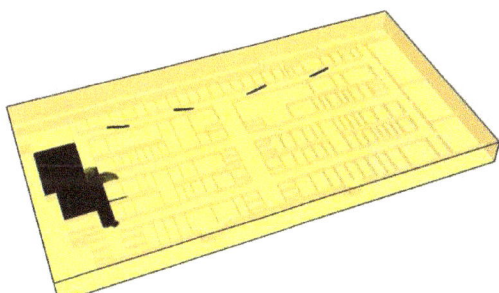

Figure 11(b). 10 Sec.

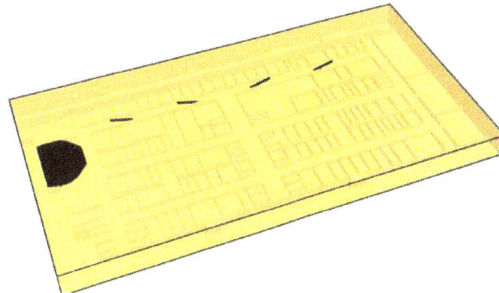

Figure 11(c). 20 Sec.

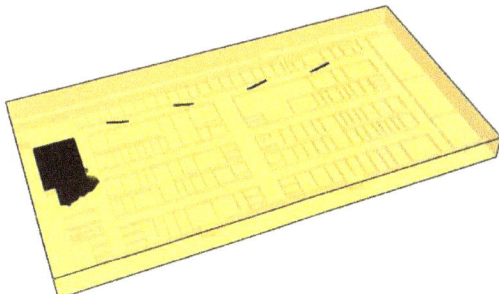

Figure 11(d). 40 Sec.

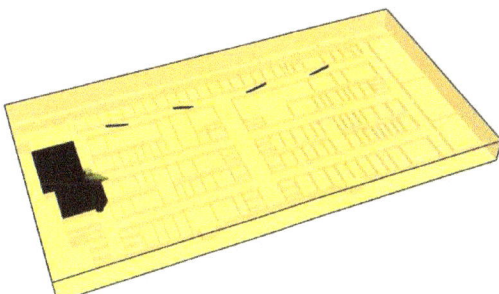

Figure 11(e). 60 Sec. (1 Min.).

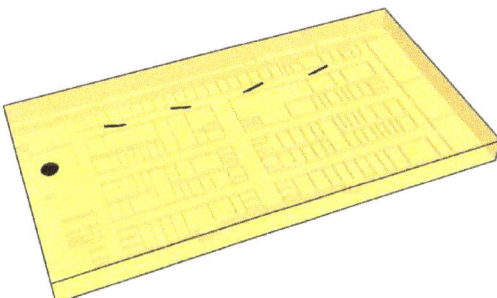

Figure 11(f). 80 Sec.

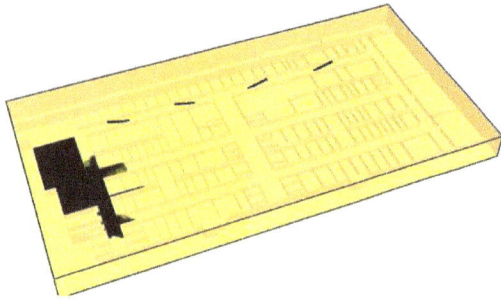

Figure 11(g). 100 Sec.

Figure 11(l). 600 Sec. (10 Min.)

Figure 11(h). 120 Sec. (2 Min.).

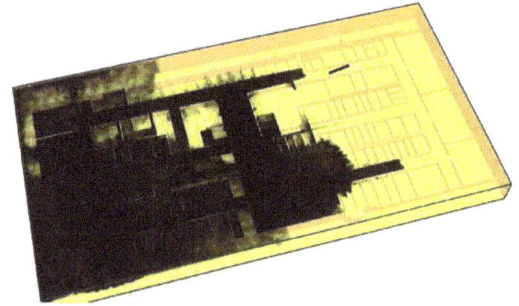

Figure 11(m). 750 Sec. (12.5 Min.).

Figure 11(i). 200 Sec.

Figure 11(n). 900 Sec. (15 Min.).

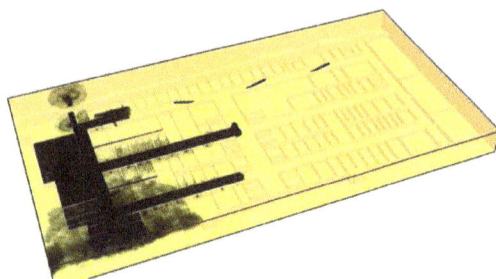

Figure 11(j). 300 Sec. (5 Min.).

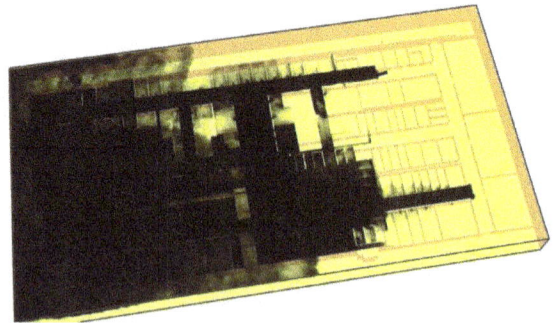

Figure 11(o). Smoke getting out through gates (2) and (3).

Figure 11(k). 450 Sec. (7.5 Min.).

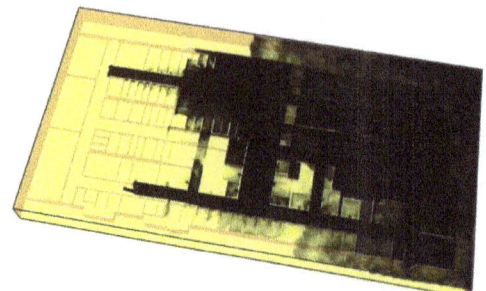

Figure 11(p). Smoke getting out through gate (5) (Back view of the mall).

Figure 11. Smoke propagation with time, Case (3).

Figure 12 shows the temperature distribution near the area of fire. Figure 12(a) shows the temperature distribution in a vertical plane passing through the fire spot and in the normal direction to the mall front. It is clear that the maximum temperature of $85^{o}C$ is reached in the fire spot. Figure 12(b) shows the temperature distribution in another vertical plate, which is parallel to the plane of Fig. 12(a) but 10 m away from the fire spot. The maximum temperature is $65^{o}C$. Figure 12(c) shows the temperature distribution in a vertical plane parallel to the mall front and 15 m away from the fire spot. The maximum temperature is about $40^{o}C$. Figure 12(d) indicates that the sprinkles in the fire region worked successfully.

Figure 12(d). *The working sprinkles in the fire region.*

Figure 12. *Temperature distribution near the area of fire, Case (3).*

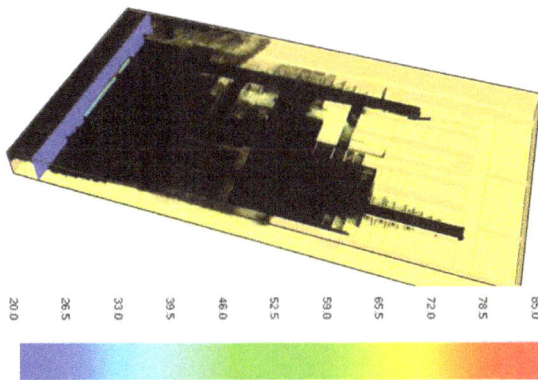

Figure 12(a). *Temperature distribution in a vertical plane passing through the fire spot and in the normal direction to the mall front.*

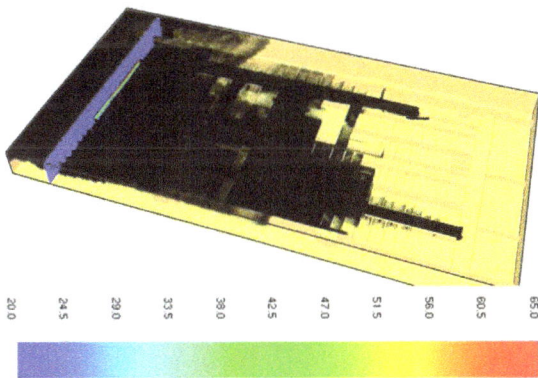

Figure 12(b). *Temperature distribution in a vertical plane 10 m away from the fire spot and in the normal direction to the mall front.*

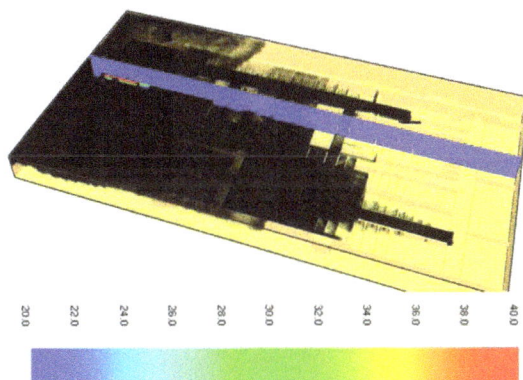

Figure 12(c). *Temperature distribution in a vertical plane parallel to the mall front and 15 m away from the fire spot.*

5.2.4. Case Study (4)

The fourth case study concerned a fire ignition at main store (5) in the ground floor, Fig. 10. The smoke propagation and movement with time is demonstrated in Fig. 13. The smoke movement with time can be monitored from Fig. 13 as follows:

1. After 5 *seconds* of fire start, smoke started to spread inside store (5), Fig. 13(a).
2. After 10 *seconds*, smoke propagated in store (5) and was still trapped inside it, Fig. 13(b).
3. After 20 *seconds*, smoke approximately filled three-fourths (3/4) of store (5) and started to spread outside the store through its main door, Fig. 13(c).
4. After 40 *seconds*, smoke completely filled store (5) and moved outside the store through its main door in corridor (1), Fig. 13(d).
5. After 60 *seconds*, smoke completely filled store 5 and continued to spread in corridor (1), Fig. 13(e).
6. After 80 *seconds*, smoke completely filled store (5), approximately filled one-fourth (1/4) of corridor (1), and started to spread in corridor 5, Fig. 13(f).
7. After 100 *seconds*, smoke completely filled store (5), approximately filled one-third (1/3) of corridor (1), and continued to spread in corridor (5), Fig. 13(g).
8. After 120 *seconds*, smoke completely filled store (5), approximately filled half (1/2) of corridor (1), continued to spread in corridor (5), and started to spread in corridor (6), Fig. 13(h).
9. After 200 *seconds* (5 *minutes*), smoke completely filled store (5), approximately filled 80% of corridor (1), continued to spread in corridors (5) and (6), Fig. 13(i).
10. After 300 *seconds*, smoke completely filled store (5) and corridor (1), continued to spread in corridors(5) and (6), started to enter stores (4) and (6), and went outside the mall through main gates (1) and (4), Fig. 13(j).
11. After 420 *seconds* (7 *minutes*), smoke completely filled store (5) and corridor (1), continued to spread in corridors (4), (5) and (6), spread more in stores (4) and (6), and continued to move outside the mall through main gates (1) and (4), Fig. 13(k).

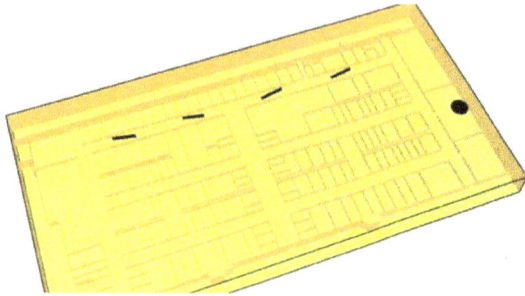

Figure 13(a). 5 Sec.

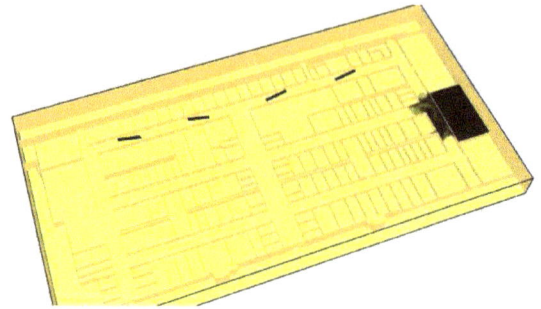

Figure 13(b). 10 Sec.

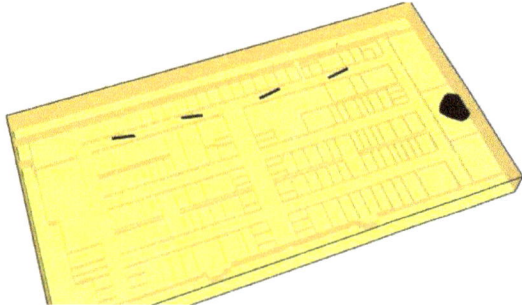

Figure 13(c). 20 Sec.

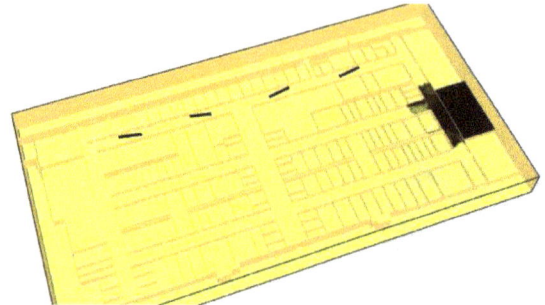

Figure 13(d). 40 Sec.

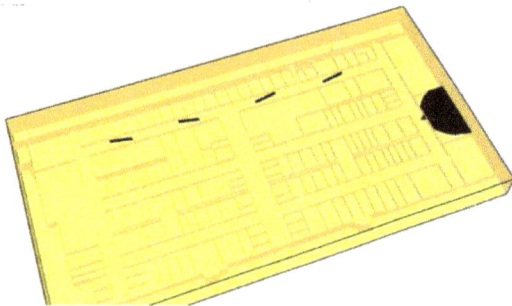

Figure 13(e). 60 Sec. (1 Min.).

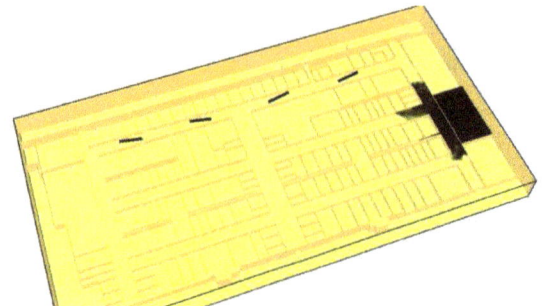

Figure 13(f). 80 Sec.

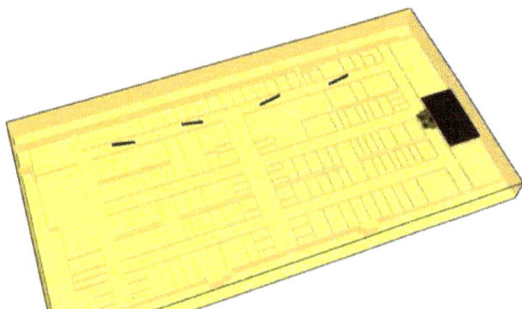

Figure 13(g). 100 Sec.

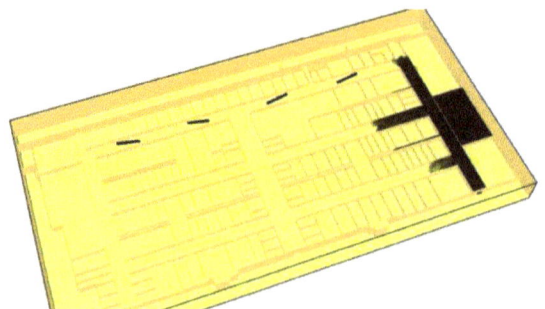

Figure 13(h). 120 Sec. (2 Min.).

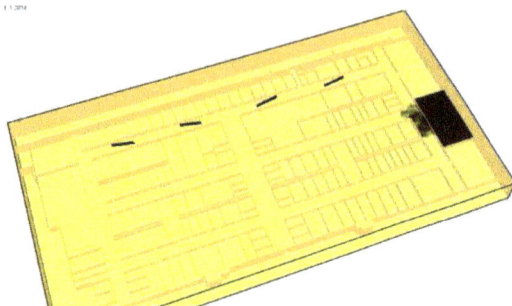

Figure 13(i). 200 Sec.

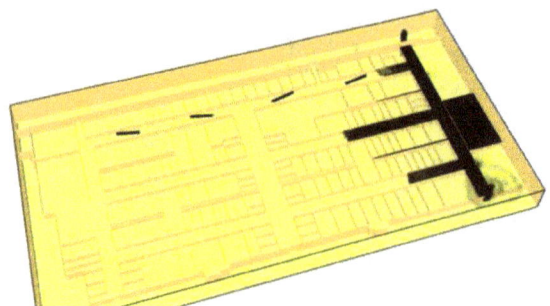

Figure 13(j). 300 Sec. (5 Min.)

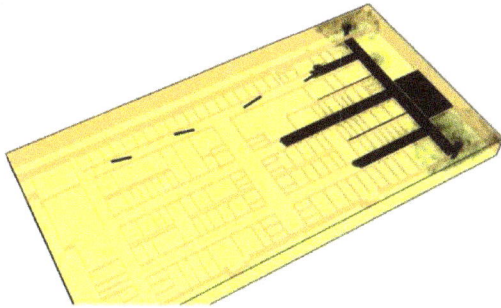

Figure 13(k). *420 Sec. (7 Min.).*

Figure 13. *Smoke propagation with time, Case (4).*

Figure 14 shows the temperature distribution near the area of fire. Figure 14(a) shows the temperature distribution in a vertical plane passing through the fire spot and in the normal direction to the mall front. It is clear that the maximum temperature of $100^o C$ is reached in the fire spot. Figure 14(b) shows the temperature distribution in another vertical plate, which is parallel to the plane of Fig. 14(a) but 10 *m* away from the fire spot. The maximum temperature is $80^o C$. Figure 14(c) shows the temperature distribution in a vertical plane parallel to the mall front and 15 *m* away from the fire spot. The maximum temperature is about $50^o C$. Figure 14(d) indicates that the sprinkles in the fire region worked successfully.

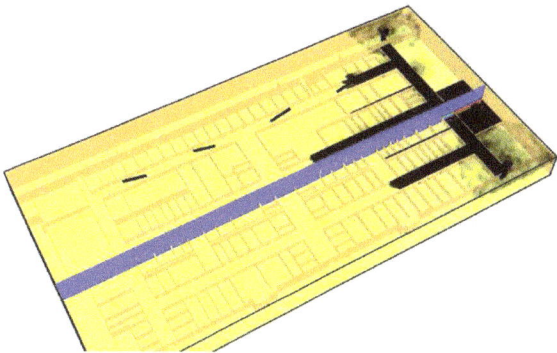

Figure 14(a). *Temperature distribution in a vertical plane passing through the fire spot and in the normal direction to the mall front.*

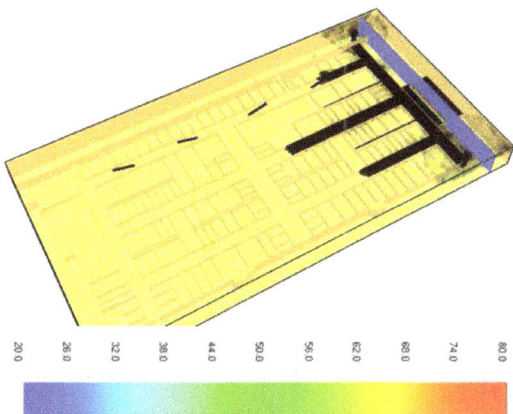

Figure 14(b). *Temperature distribution in a vertical plane 10 m away from the fire spot and in the normal direction to the mall front.*

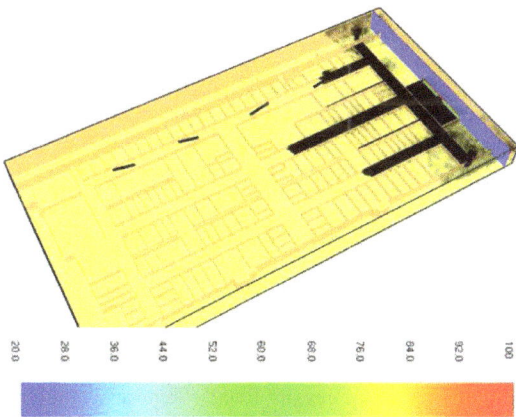

Figure 14(c). *Temperature distribution in a vertical plane parallel to the mall front and 15 m away from the fire spot.*

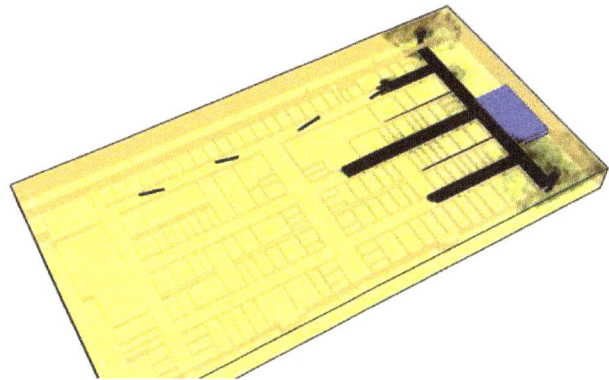

Figure 14(d). *The working sprinkles in the fire region.*

Figure 14. *Temperature distribution near the area of fire, Case(4).*

6. Conclusions

The present investigation concerns smoke propagation and evacuation due to fires in a large shopping center (mall) in Makkah, Saudi Arabia. The mall is a two-story building. Four fires cases were considered. Half of the cases were implemented for the ground floor and the other half were implemented for the upper floor.

Based on the present results, the following concluding points can be stated:

1- The large volume of such malls and the complexity of their structure make it very difficult to predict the smoke movement in case of fire without actual computational studies. Thus, the location of fire has a noticeable effect on the smoke propagation due to the interference of the stores, corridors, facility, *etc.* of the mall.

2- Smoke may spread rapidly in such malls. It also moves from one floor to another through architecture openings (after about 3 minutes in the present study).

3- As the used area of the upper floor is less than its total area, about 900 *Sec.* (15 *Min.*) were sufficient to approximately fill the occupied area of the floor with smoke. However, only about two-thirds of the ground *floor was full with smoke after 900 Sec. (15 Min.).*

4- For all the present four fire cases, the sprinkles in the fire region operated successfully. The sprinkles were not able to overcome the fire because the *HRRPUA* of the fire was set as constant. However, they slightly lowered the temperature in the fire region.

5- The maximum computed temperature was 105^oC at the fire spot, which is sufficient to cause severe injuries to occupants.

Nomenclature

$\dot{m}_b^m$	: Production rate of species by evaporating droplets or particles
$\bar{p}_m^n$	: Background pressure at the nth time step
$\dot{q}''$	: Conductive and radiative heat fluxes
$\dot{q}'''$	: Heat release rate per unit volume from a chemical reaction
$\dot{q}_b'''$	: Energy transferred to the evaporating droplets
$\bar{u}^n$	: Patch-average velocity field
H^n	: Perturbation pressure at the nth time step
Y_α^n	: Species mass fractions at the nth time step
u^n	: Velocity components at the nth time step
$u_{i-1,jk}^n$	: x-component of velocity at the negative-oriented face of the ijk^{th} cell
u_{ijk}^n	: x-component of velocity at the positive-oriented face of the ijk^{th} cell
D()/Dt	: Material derivative
f_b	: External forces such as the drag exerted by liquid droplets
h_s	: Sensible enthalpy
i, j, k	: Indices representing the position of the cell in the x, y,z directions, respectively
K	: Thermal conductivity
P	: Pressure
p(x,y,z)	: Spatially resolved pressure
$p_m(z,t)$	: Average or background pressure
S_{ij}	: Symmetric rate-of-strain tensor
T	: Temperature
u =[u,v,w]T	: Three components of velocity
uu	: Diadic tensor
$Y\alpha$	: Individual gaseous species

Greek

ρ_{ijk}^n	: Density at the nth time step in the center of the cell whose indices are i, j,k
ρ^n	: Density at the nth time step
$\nabla\dot{u}$	: Velocity divergence
μ	: Dynamic viscosity
ρ	: Density
τ_{ij}	: Stress tensor

Abbreviations

2D	: Two-dimensional
CFD	: Computational Fluid Dynamics
DNS	: Direct Numerical Simulation
Evac	: Evacuation software
FDS	: Fire Dynamics Simulator
HRRPUA	: Heat Release Rate Per Unit Area
HST	: Hot Smoke Tests
LES	: Large Eddy Simulation
NIST	: National Institute of Standards and Technology

References

[1] P. Smardz, Validation of Fire Dynamics Simulator (FDS) for Forced and Natural Convection Flows, Master of Science in Fire Safety Engineering, University of Ulster, United Kingdom, October 2006.

[2] PyroSim User Manual, Thunderhead Engineering, 2014.

[3] A. M. Christensen, and D. J. Icove, "The Application of NIST's Fire Dynamics Simulator to the Investigation of Carbon Monoxide Exposure in the Deaths of Three Pittsburgh Fire Fighters", J. Forensic Sci., Jan. 2004, Vol. 49, No. 1, pp.104-107.

[4] Use FDS to Assess Effectiveness of Air Sampling Smoke Detection in Large Open Spaces, Xtralis VESDA White Paper, Doc. 16998_00.

[5] A. Webb, FDS Modelling of Hot Smoke Testing, Cinema and Airport Concourse, Worcester Polytechnic Institute, USA, M.Sc. degree in Fire Protection Engineering, November 29, 2006.

[6] S. Jianyong, and C. Longzhu, "Analysis and Assessment of Whole Structural Fire Safety for Public Buildings", ISGSR2007 First International Symposium on Geotechnical Safety & Risk, Shanghai Tongji University, China, Oct. 18-19, 2007.

[7] A. Lönnermark, and A. Björklund, "Smoke Spread and Gas Temperatures during Fires in Retail Premises-Experiments and CFD Simulations", SP Technical Research Institute of Sweden, Fire Technology SP Report 2008:55, ISBN: 978-91-86319-16-8, ISSN: 0284-5172.

[8] D. Gottuk, C. Mealy, and J. Floyd, "Smoke Transport and FDS Validation", Fire Safety Science, Vol. 9, pp. 129-140, 2008. Doi: 10.3801/IAFSS.FSS.9-129.

[9] D. Ling, and K. Kan, "Numerical Simulations on Fire and Analysis of the Spread Characteristics of Smoke in Supermarket", International Conference CESM 2011, Wuhan, China, June 18-19, 2011, pp. 7-13, doi: 10.1007/978-3-642-21802-6_2.

[10] L. Wang, J. Lim, and J. G. Quintiere, "Validation of FDS Predictions on Fire-Induced Flow: A Follow-Up to Previous Study", 2011 Fire and Evacuation Modeling Technical Conference, Baltimore, Maryland, USA, Aug. 15-16, 2011.

[11] M. Tabaddor, Fire Modeling of Basement with Wood Ceiling, Underwriters Laboratories Inc., USA, December 2011.

[12] Z.-j. Yu, O. Xu, and J.-w. Han, "The Application of FDS Used in the Cabin Fire Simulation and Human Evacuation of Civil Aviation", 2012 International Conference on Mechanical Engineering and Material Science (MEMS 2012), Nov. 2012.

[13] A. F. Abdel-Gawad, and H. A. Ghulman, "Fire Dynamics Simulation of Large Multi-story Buildings, Case Study: Umm Al-Qura University Campus", International Conference on Energy and Environment 2013 (ICEE2013), Universiti Tenaga Nasional, Putrajaya Campus, Selangor, Malaysia, 5-6 March 2013. [Institute of Physics (IOP) Conference Series: Earth and Environmental Science, Vol. 16, No. 1, 2013, doi:10.1088/1755-1315/16/1/012040].

[14] A. F.AbdelGawad, "Multidisciplinary Engineering for the Utilization of Traditional Automated Storage and Retrieval System (ASRS) for Firefighting in Warehouses", American Journal of Energy Engineering (AJEE), Special Issue: Fire, Energy and Thermal Real-life Challenges, Vol. 3, No. 4-1, , pp. 1-22, July 2015. doi: 10.11648/j.ajee.s.2015030401.11.

[15] A. F. Abdel Gawad, and H. A.Ghulman,"Prediction of Smoke Propagation in a Big Multi-Story Building Using Fire Dynamics Simulator (FDS)", American Journal of Energy Engineering (AJEE), Special Issue: Fire, Energy and Thermal Real-life Challenges, Vol. 3, No. 4-1, pp. 23-41, July 2015. doi: 10.11648/j.ajee.s.2015030401.12.

[16] B. Duer, "Arson Fire Destroys Strasburg Landmark", Canton Repository, 15 October 2010.

[17] Daily mail reporter, 28 May 2012.

[18] S. Malik,"17 Killed in Kolkata Market Fire", NDTV,27 February 2013.

[19] gazettenet.com, October 27, 2013.

[20] A. Raja, CNN, December 30, 2014.

[21] A. Attrino and S. Epstein, NJ Advance Media for NJ.com, February 26, 2015.

[22] BBC news, 12 March 2015.

[23] K. McGrattan, S. Hostikka, J. Floyd, H. Baum, R. Rehm, W. Mell, and R. McDermott, Fire Dynamics Simulator (Version 5), Technical Reference Guide, Volume 1: Mathematical Model, National Institute of Standards and Technology, NIST Special Publication 1018-5, 2010.

[24] Standard for the Installation of Sprinkler Systems, National Fire Protection Association (NFPA), 2013 Edition, ISBN: 978-145590455-6.

[25] Standard for the Installation of Sprinkler Systems in One- and Two-Family Dwellings and Manufactured Homes, National Fire Protection Association (NFPA), 2013 Edition, ISBN: 978-145590456-3.

[26] Standard for the Installation of Sprinkler Systems in Low-Rise Residential Occupancies, National Fire Protection Association (NFPA), 2013 Edition, ISBN: 978-145590457-0.

[27] Recommended Practice for Fire Department Operations in Properties Protected by Sprinkler and Standpipe Systems, National Fire Protection Association (NFPA), 2010 Edition, ISBN: 978-087765967-9.

[28] National Fire Alarm and Signaling Code, National Fire Protection Association (NFPA), 2013 Edition, ISBN: 978-145590464-8.

[29] L. Staffansson, Selecting Design Fires , Department of Fire Safety Engineering and Systems Safety, Lund University, Sweden, Report 7032, 2010, ISSN: 1402-3504.

[30] J. Hietaniemi, and E. Mikkola, Design Fires for Fire Safety Engineering, VTT Working Papers 139, VTT Technical Research Centre of Finland, 2010, ISBN 978-951-38-7479-7.

[31] Z.-C. Grigoraş, and D. Diaconu-Şotropa, "Establishing the Design Fire Parameters for Buildings", Bul. Inst. Polit. Iaşi, t. LIX (LXIII), f. 5, pp. 133-141, 2013.

13

Fire Dynamics Simulation and Evacuation for a Large Shopping Center (Mall), Part II, Evacuation Scenarios

Khalid A. Albis*, Ahmed F. Abdel Gawad, Muhammad N. Radhwi

Mechanical Engineering Department, College of Engineering & Islamic Architecture, Umm Al-Qura University, Makkah, Saudi Arabia

Email address:

kalbis31@gmail.com (K. A. Albis), afaroukg@yahoo.com (A. F. A. Gawad), mnradhwi@uqu.edu.sa (M. N. Radhwi)

Abstract: Evacuation plans in large malls are very essential in cases of emergency especially fire. Many fatalities may occur among panicked people running and pushing to get out of these burning places. The present study considers evacuation cases corresponding to the fire cases that were considered in part I. It is established that evacuation planning is vital to help people get out of the mall without being exposed to suffocation due to smoke inhaling or having accidents due to panic rush in corridors and over the stairs/escalators.

Keywords: Fire Dynamics Simulation, Smoke Propagation, Evacuation, Shopping Center (Mall)

1. Literature Review

A simplified elevator service model was developed to evaluate effectiveness of evacuation by elevators and conducted some case studies to examine the feasibility and problems of elevator use for evacuation of aged people in Japan. It was found that the diverging point of the advantage of evacuation by elevator in comparison to evacuation by stairs appears roughly on the 14th floor to the 16th floor [1]. The factors that affect the evacuation process were considered to foresee the human behavior in case of fire. These factors are related to the occupants' characteristics, the building characteristics and the fire characteristics. It was concluded that keen consideration should be given to the interplay of these factors to gain a better understanding of human behavior in fire and to improve the design and implementation of fire safety systems in buildings [2].

An analysis of staff behavior in five unannounced evacuations of Marks and Spencer retail stores was carried out. The retail stores participating in the study comprised two three-story city-center stores and one single-story out-of-town store in different locations in the United Kingdom. It was found that staff responses to the alarm, both in terms of time and nature of their responses, varied depending on the setting they were in and their associated responsibilities. Contrary to their training, they did not always respond immediately by evacuating customers but, in the majority of cases, first sought confirmation of the need to evacuate. They did, however, have a significant impact on customer response, not only in overcoming customers' initial evacuation inertia but also in directing them towards suitable exits [3]. *EXODUS* software was used to study the emergency evacuation in supermarkets. Thus, many actual factors were considered such as the number of customers, the ratio of personnel's age, the walking speed. The safety measures of supermarket management for evacuation were proposed [4].

The definition of "panic" was reviewed based on actual fire incidents that may have been misreported or misinterpreted as "panic". It was stated that despite the numerous evidence that panic is a very rare occurrence in fires, the idea of panic and the term continue to be used by the public as well as fire experts. Thus, it is necessary to demystify the misconception that panic is an essential element of a fire and identify any scientific justification for continuing using this concept [5]. A comprehensive review was carried out to provide guidance to engineers involved in the design of underground transportation systems for evacuation planning in case of fire. The review revealed many potential solutions to commonly observed evacuation problems [6]. Code *FDS+Evac* was used for *3D* simulations of fire dynamics and evacuation for a building of 20 floors in China with an area of 2060 m^2 (Technological Department of Wuhan University of Technology). The purpose of the work was to improve the speed of occupants' evacuation [7]. Recently, other evacuation studies were addressed [8-12].

2. Evacuation Cases

Based on the results of the four fire cases, part I, four evacuation cases were studied. The four evacuation cases suggest the scenarios of people evacuation in the four cases of fire. To consider the differences of the evacuated persons (gender, age, *etc.*), their moving speed was set between 2.1 and 2.5 *m/s*. As will be seen in all cases, it is obvious that the flow of persons was faster than the smoke movement. Hence, all persons left the floor safely through the emergency exit

and the smoke did not reach anyone of them. Thus, the evacuation process was successful.

2.1. Case Study (1)

The first case study concerned evacuation in case of fire ignition at kitchen (7) of the food court in the upper floor, Fig. 1. The evacuation process is based on one hundred persons that are located in the area of the food court.

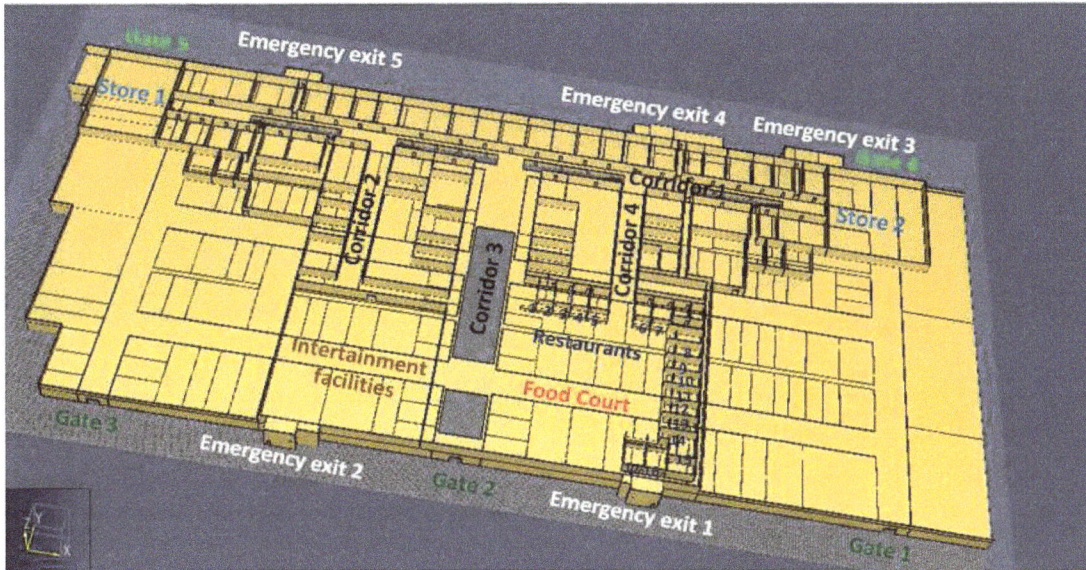

Figure 1. Main stores, corridors, and emergency exits in the upper floor.

The evacuation process and the persons' movement with time are demonstrated in Fig. 2. The persons' movement with time can be monitored from Fig. 2 as follows:

1- After 5 *seconds* of fire start, persons started to move in the area of the food court, Fig. 2(a).

2- After 10 *seconds*, persons in the food court area moved to exit (5) through corridor (3), Fig. 2(b).

3- After 20 *seconds*, three-fourths (3/4) of persons moved from the food court area to corridor (3), Fig. 2(c).

4- After 40 *seconds*, most persons moved through corridor (1) that leads to exit (5). Some others entered a store in corridor (3) through one entrance and exited from another entrance in corridor (4) that leads also to exit (5), Fig. 2(d).

5- After 60 *seconds*, most persons moved in corridor (1) that leads to exit (5). Only eight persons moved in corridor (4), Fig. 2(e).

6- After 70 *seconds*, the first person entered exit (5), then the rest of persons started to follow him, Fig. 2(f).

7- After 90 *seconds*, most persons entered the small corridor that leads to exit (5), Fig. 2(g).

8- After 120 *seconds*, the majority of persons had left exit (5), Fig. 2(h).

9- After 130 *seconds*, the last person entered exit (5) and left the upper floor, Fig. 2(i).

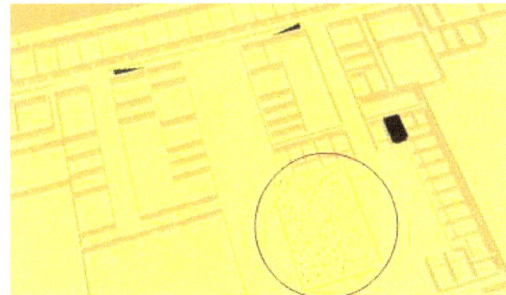

Figure 2(a). 5 Sec.

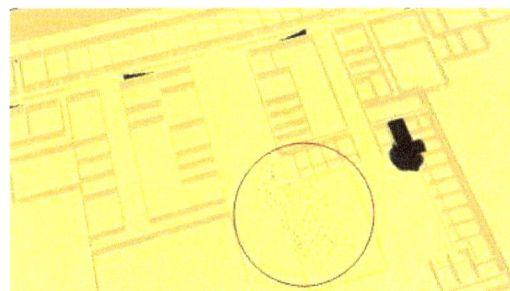

Figure 2(b). 10 Sec.

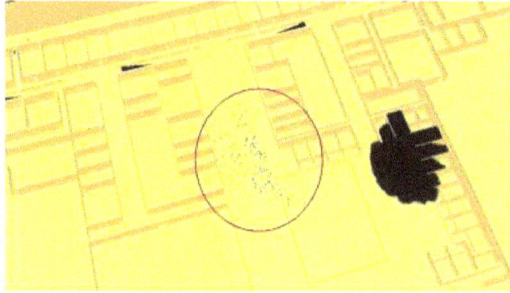

Figure 2(c). 20 Sec.

Figure 2(d). 40 Sec.

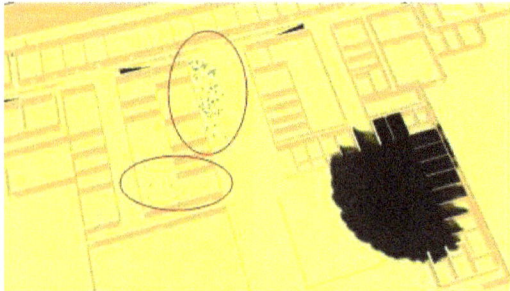

Figure 2(e). 60 Sec. (1 Min.).

Figure 2(f). 70 Sec.

Figure 2(g). 90 Sec.

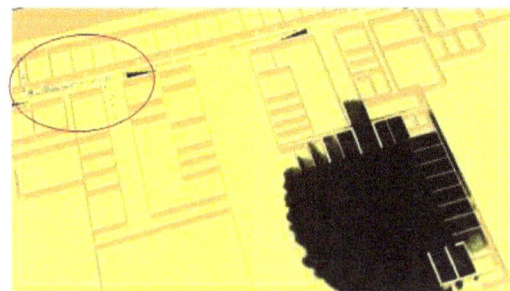

Figure 2(h). 120 Sec. (2 Min.).

Figure 2(i). 130 Sec.

Figure 2. *Development of evacuation process with time, Case (1).*

2.2. Case Study (2)

The second case study concerned evacuation in case of fire ignition at main store (2) in the upper floor, Fig. 1. The evacuation process is based on eighty persons that are located in corridor (1) in front of main store (2).

The evacuation process and the persons' movement with time are demonstrated in Fig. 3. The persons' movement with time can be monitored from Fig. 3 as follows:

1- After 5 *seconds* of fire start, persons started to move in corridor (1) in front of store (2), Fig. 3(a).

2- After 10 *seconds*, persons moved on corridor (1) towards emergency exit (4), Fig. 3(b).

3- After 20 *seconds*, persons approached emergency exit (4), Fig. 3(c).

4- After 30 *seconds*, almost 50% of persons moved into the small corridor that leads to emergency exit (4), Fig. 3(d).

5- After 40 *seconds*, the majority of persons entered in the small corridor that leads to emergency exit (4), Fig. 3(e).

6- After 50 *seconds*, most persons escaped through emergency exit (4), Fig. 3(f).

7- After 60 *seconds*, the last person escaped through emergency exit (4), Fig. 3(g).

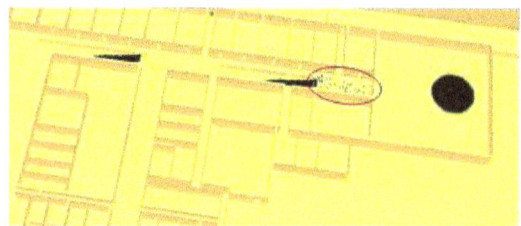

Figure 3(a). 5 Sec.

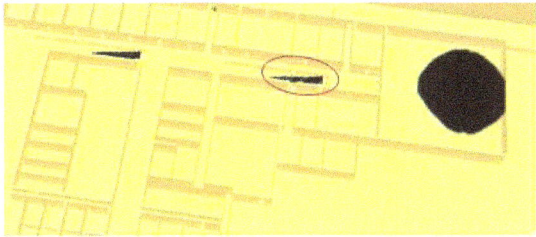

Figure 3(b). 10 Sec.

Figure 3(c). 20 Sec.

Figure 3(d). 30 Sec.

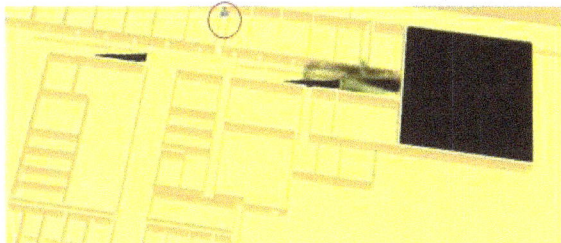

Figure 3(e). 40 Sec.

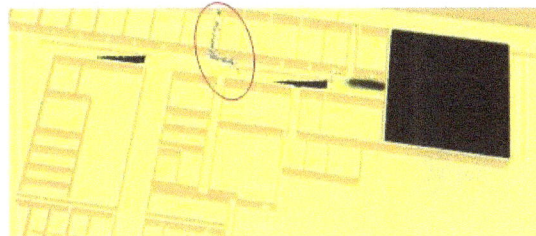

Figure 3(f). 50 Sec.

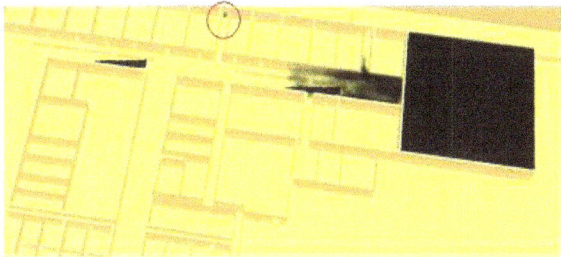

Figure 3(g). 60 Sec. (1 Min.).

Figure 3. Development of evacuation process with time, Case (2).

2.3. Case Study (3)

The third case study concerned evacuation in case of fire ignition at main store (3) in the ground floor, Fig. 4. The evacuation process is based on one hundred persons that are located in the main store (3).

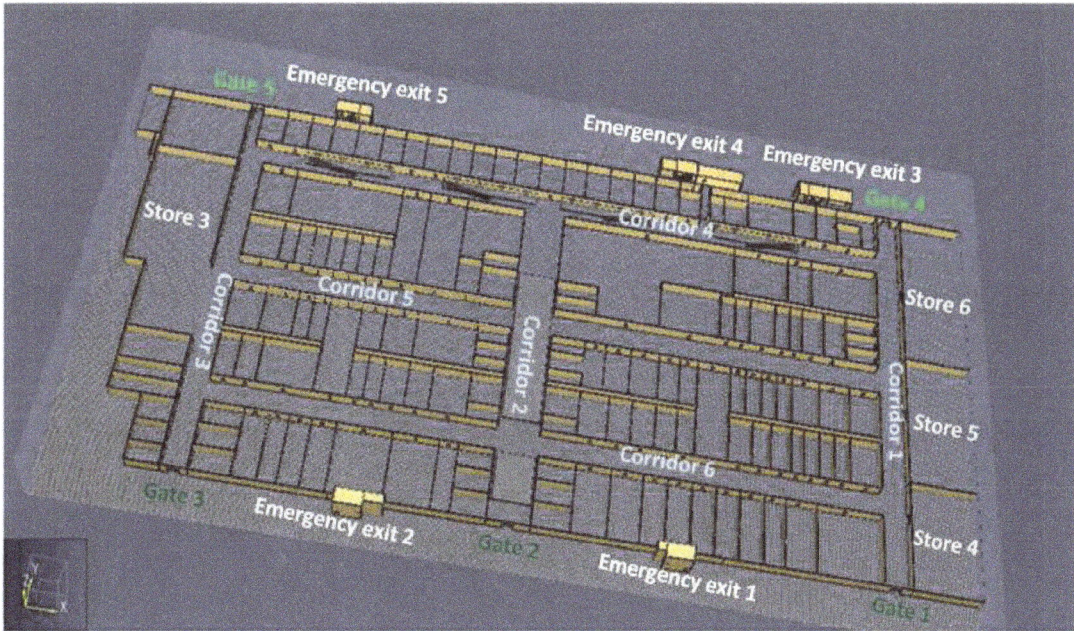

Figure 4. Main stores, corridors, and emergency exits in the ground floor.

The evacuation process and the persons' movement with time are demonstrated in Fig. 5. The persons' movement with time can be monitored from Fig. 5 as follows:

1- After 5 seconds of fire start, persons started to move in store (3), Fig. 5(a).
2- After 10 seconds, most persons moved to corridor (3) on their way to main gate (3), Fig. 5(b).
3- After 20 seconds, all persons moved to corridor (3) and the first person exited main gate (3), Fig. 5(c).
4- After 30 seconds, many persons exited from main gate (3), Fig. 5(d).
5- After 40 seconds, the last person exited from main gate (3), Fig. 5(e).

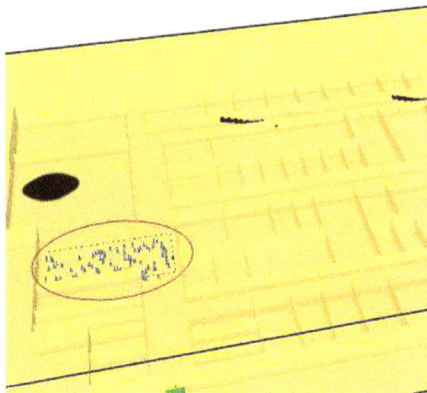

Figure 5(a). 5 Sec.

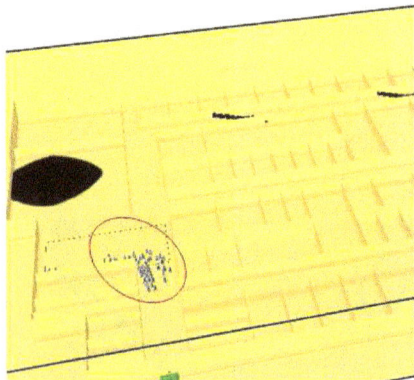

Figure 5(b). 10 Sec.

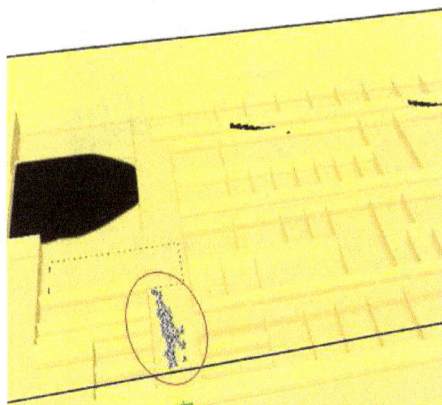

Figure 5(c). 20 Sec.

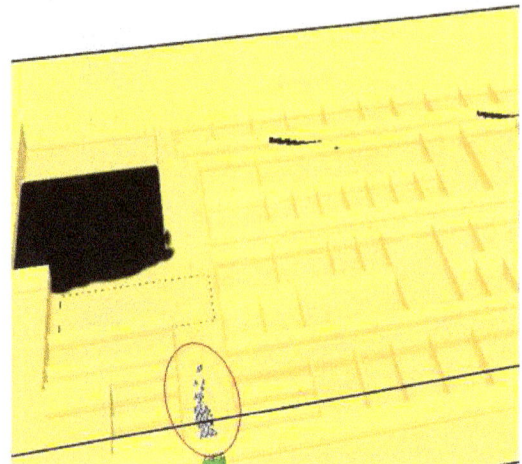

Figure 5(d). 30 Sec.

Figure 5(e). 40 Sec.

Figure 5. Development of evacuation process with time, Case (3).

2.4. Case Study (4)

The fourth case study concerned evacuation in case of fire ignition at main store (5) in the ground floor, Fig. 4. The evacuation process is based on one hundred persons that are located in corridor (1) in front of main store (5).

The evacuation process and the persons' movement with time are demonstrated in Fig. 6. The persons' movement with time can be monitored from Fig. 6 as follows:

1- After 5 *seconds* of fire start, persons started to move in corridor (1) in front of store 5, Fig. 6(a).
2- After 10 *seconds*, persons moved in corridor (1) towards main gate (1), Fig. 6(b).
3- After 20 *seconds*, persons approached main gate (1), Fig. 6(c).
4- After 30 *seconds*, persons started to get out the mall through main gate (1), Fig. 6(d).
5- After 40 *seconds*, all persons had left the mall through main gate (1), Fig. 6(e).

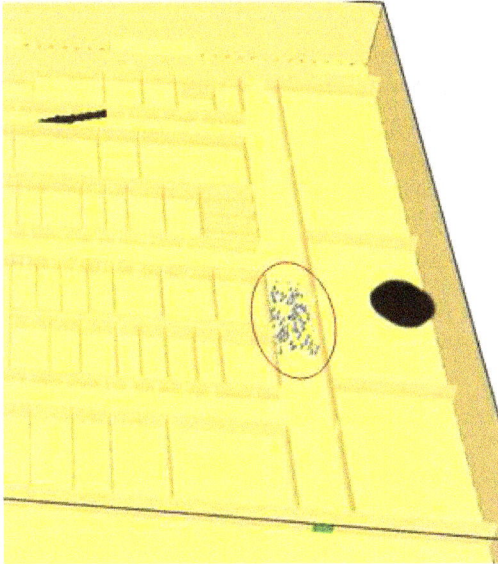

Figure 6(a). 5 Sec.

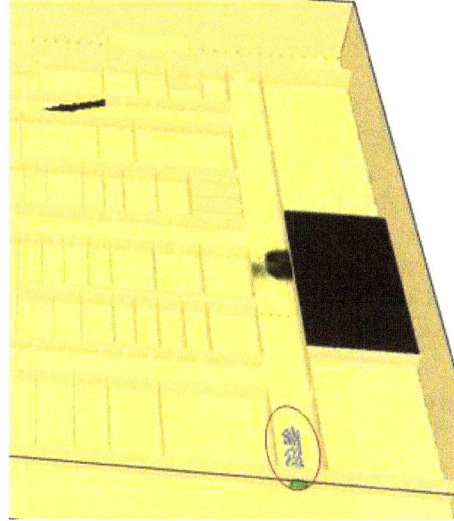

Figure 6(b). 10 Sec.

Figure 6(c). 20 Sec.

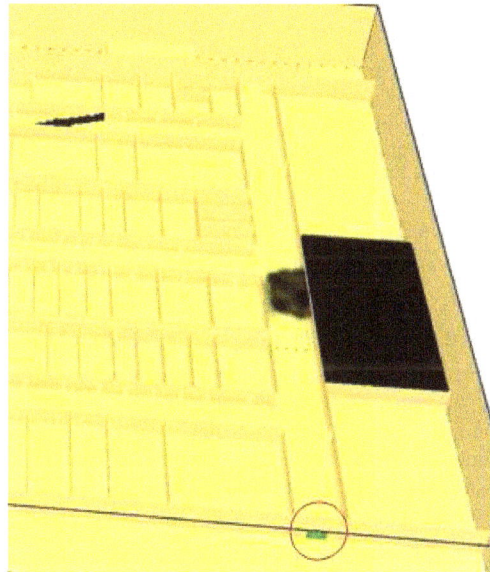

Figure 6(d). 30 Sec.

Figure 6(e). 40 Sec.

Figure 6. Development of evacuation process with time, Case (4).

3. Conclusions

The present investigation concerns smoke propagation and evacuation due to fires in a large shopping center (mall) in Makkah, Saudi Arabia. The mall is a two-story building. Four evacuation cases were considered. Half of the cases were implemented for the ground floor and the other half were implemented for the upper floor.

Based on the present results, the following concluding points can be stated:

1- It is well-established that evacuation planning is vital to help people get out of the mall without being exposed to suffocation due to smoke inhaling or having accidents due to panic rush in corridors and over the stairs/escalators.

2- For all the present four cases of evacuation, all persons left the corresponding floor safely through the nearest emergency exit and the smoke did not reach anyone of them. Thus, the evacuation processes were totally successful.

3- For the present cases, it was found that 60 seconds were enough to evacuate all persons from the upper floor. However, 40 seconds were enough to evacuate all persons from the ground floor.

4- It is clear that the assigned velocity range of persons during evacuation (2.1 and 2.5 *m/s*) was suitable for the four evacuation cases considering the differences of persons (gender, age, *etc.*).

Nomenclature

3D	: Three-dimensional
Evac	: Evacuation software
EXODUS	: Evacuation software
FDS	: Fire Dynamics Simulator

References

[1] A. Sekizawa, S. Nakahama, Y. Ikehata, M. Ebihara, and H. Notake, "Study on Feasibility of Evacuation by Elevators in a High-Rise Building-A Case Study for the Evacuation in the Hiroshima Motornachi High-Rise Apartments", AOFST 4, International Association for Fire Safety Science, pp. 217-225, 2000.

[2] G. Proulx, "Occupant Behaviour and Evacuation", 9th International Fire Protection Symposium, Munich, 25-26 May 2001.

[3] D. A. Samochine, K. Boyce, and T. J. Shields, "An Investigation into Staff Behaviour in Unannounced Evacuations of Retail Stores-Implications for Training and Fire Safety Engineering", Fire Safety Science 8, pp. 519-530, 2005. Doi: 10.3801/IAFSS.FSS.8-519.

[4] L. Qiang, and J. Hong-yu, "The Study on Safety Evaluation of Evacuation in a Large Supermarket", 5th Conference on Performance-based Fire and Fire Protection Engineering, Procedia Engineering, Vol. 11, pp. 273-279, 2011. Doi:10.1016/j.proeng.2011.04.657.

[5] R. F. Fahy, G. Proulx, and L. Aiman, "Panic or not in Fire: Clarifying the Misconception", Fire and Materials, Special Issue on Human Behaviour in Fire, Vol. 36, No. 5-6, pp. 328-338, Oct. 2012.

[6] K. Fridolf, D. Nilsson, and H. Frantzich, "Fire Evacuation in Underground Transportation Systems: A Review of Accidents and Empirical Research", Fire Technology, Vol. 49, No. 2, pp. 451-475, April 2013.

[7] A. L. da Silva, "Study of Building Fire Evacuation and Geometric Modeling Based on Continuous Model FDS+AVEC", European Scientific Journal, May 2014, SPECIAL/ edition ISSN: 1857 – 7881 (Print) e - ISSN 1857-7431.

[8] M. Xi, and S. P. Smith, "Exploring the Reuse of Fire Evacuation Behaviour in Virtual Environments", Proceedings of the 11th Australasian Conference on Interactive Entertainment (IE 2015), Vol. 27, p. 30, 2015.

[9] Y. Chen, Y. Cai, P. Li, and G. Zhang, "Study on Evacuation Evaluation in Subway Fire Based on Pedestrian Simulation Technology", Mathematical Problems in Engineering, Vol. 2015, 2015.

[10] X. Che, Y. Niu, B. Shui, J. Fu, G. Fei, P. Goswami, and Y. Zhang, "A Novel Simulation Framework Based on Information Asymmetry to Evaluate Evacuation Plan", The Visual Computer, May 2015.

[11] J. Kang, I.-J. Jeong, and J.-B. Kwun, "Optimal Facility-Final Exit Assignment Algorithm for Building Complex Evacuation", Computers & Industrial Engineering, Vol. 85, pp. 169-176, July 2015.

[12] M. T. Kinateder, E. D. Kuligowski, P. A. Reneke, and R. D. Peacock, "Risk Perception in Fire Evacuation Behavior Revisited: Definitions, Related Concepts, and Empirical Evidence", Fire Science Reviews, Vol. 4, No. 1, pp. 1-26, 2015.

14

Numerical analysis of fluid flow properties in a partially perforated horizontal wellbore

Mohammed Abdulwahhab Abdulwahid[1, *]**, Sadoun Fahad Dakhil**[2]**, I. N. Niranjan Kumar**[3]

[1]Marine Engineering Department, Andhra University, Visakhapatnam, India
[2]Fuel & Energy Department, Basrah Technical College, Basrah, Iraq
[3]Marine Engineering Department, Andhra University, Visakhapatnam, India

Email address:

Mohw2002@yahoo.com (M. A. Abdulwahid), drsadoun2@gmail.com (S. F. Dakhil), neeru9@yahoo.com (I. N. N. Kumar)

Abstract: The pressure drops in horizontal wellbores, acceleration, wall friction, perforation roughness, and fluid mixing are analyzed in a partially perforated wellbore. It was demonstrated that the perforation inflow actually reduced the total pressure drop. The pressure drop due to perforation roughness was eliminated by the perforation inflow when the ratio of radial perforation flow to axial pipe flow rate reached a certain limit. Three dimensional numerical simulations on a partially perforated pipe with 150 perforations, geometrically similar with wellbore casing (12 SPF, and 60 phasing) were presented and analyzed. Numerical simulations by commercial code CFX were also conducted with Reynolds numbers ranging from 28,773 to 90,153 and influx flow rate ranging from 0 to 899 lit/hr to observe the flow through perforated pipe, measure pressure drops, friction factors and pressure loss coefficients. The acceleration pressure drop might be important compared with the frictional pressure drop. The numerically calculated results using k-ε model were compared with the experimental results. The numerical solutions agreed well with the experimental data.

Keywords: Pressure Drop, Perforation, Numerical, Radial Flow, Wellbore

1. Introduction

Over the last decade, flows through horizontal wells have become a well-established technology for the recovery of oil and gas. A considerable amount of analytical and experimental work has been published on various aspects of horizontal-well production, including transient flow, stabilized inflow models, productivity indices, coning and cresting behavior. Although these methods provide insight into the behavior of horizontal wells, only a few of them are considered the pressure drop along the wellbore assuming the infinite conductivity essentially. Horizontal well productivity is limited by the pressure drop within the wellbore, especially when the pressure drop is compared with the reservoir drawdown. A better understanding of the factors affecting the total pressure drop within the wellbore is essential. Knowledge of different pressure drops that effects the horizontal wellbores is crucial in designing successful horizontal wells and optimizing well performance.

In 1990, [1] proposed the first semi analytical model to evaluate the production performance of a horizontal well with the consideration of the wellbore-pressure drop resulting from turbulent flow. Later the study continued by others [2-9] who have presented different coupling models for wellbore flow and reservoir inflow through perforations. However, in certain case studies the pressure drop along a wellbore was studied just by considering only the frictional component. In most circumstances; the pressure drop is studied taking the acceleration into consideration by neglecting the other effects like inflow, mixing etc. Reference [10] in their study revealed that, because of the existence of perforation inflow, acceleration pressure drop is an important factor relative to the frictional component. This significantly will influence the well-flow rates under certain conditions.

With the increase in flow velocity, the momentum influences the pressure drop in addition to the friction pressure drop. This part of the pressure drop has been addressed by several authors in recent years [5, 8, 11, and 12]. Apart from that the perforation holes act like roughness elements which increase the friction factor of the wellbore [13].

Reference [14] performed experimental studies of turbulent

air flow in a porous circular pipe with uniform air injection through the pipe wall. The fully developed turbulent air flow, at Reynolds numbers of 28,000 to 82,000, entered the pipe while air was injected uniformly through the wall at different ratios ranging from 0.00246 to 0.0584 of injection velocity to the average velocity at the entrance.

It is quite interesting to note that the characteristics of pipe flow with wall mass transfer are different from those of channel flow or flow past a flat plate. For example, considering the laminar flow case, the local friction factor increases with an increase of wall Reynolds number for pipe flow but decreases for the channel flow [15].

Reference [16] studied flow resistance in a perforated pipe, both with and without flow injection through the pipe wall, by conducting experiments on a pipe of $6^{5/6}$ inch outside diameter and 17 ft in length. Reference [11] stated that the total pressure drop along a perforated pipe is made up of wall friction and inflow acceleration and computed the wall friction factor in the same way for a regular, unperforated pipe. Reference [17] studied channel flow with continuous influx into the horizontal channel from an oil-reservoir model. They stated that the pressure gradients increased almost uniformly in the test channel because of the confluence of influx and axial flow, and the resulting pressure drop increases linearly with influx velocity.

A careful set of single-phase experiments in a perforated pipe with radial inflow has been conducted by [18 and 19]. In these experiments, water is used as the working fluid. References [18 and 19] attempt to account for the effects of radial inflow by assuming that the pressure drop in a perforated pipe is the sum of three contributions: the frictional pressure drop, the pressure drop associated with the acceleration of the fluid in the pipe, and finally, a mixing pressure drop. References [18 and 19] showed that most of the pressure drops in the perforated pipe is due to frictional and accelerational effects. However, the mixing pressure drop is significant and its contribution to the pressure drop is often negative. It is, therefore, suggested that the radial inflow lubricates the pipe flow. Although this seems reasonable in the case where the velocity entering through the radial perforations is small compared with the axial velocity in the pipe, they would not expect lubrication to occur when the radial velocity is of the same order of magnitude as the axial velocity. Namely, when radial and axial velocities are of the same magnitude, a jet will penetrate the axial pipe flow resulting in a certain degree of blockage of the pipe. This would lead to an increase of the pressure drop in the pipe.

In this paper, the theoretical that study of the pressure drop in a partially perforated wellbore is presented. That incorporates not only frictional, accelerational pressure drops but also the pressure caused by inflow. The main differences between the theoretical study in this paper with the experiments carried out by [18 and 19] are the diameter of perforations and the perforation density of the pipe. SG has used a perforation diameter of 3mm and 158 perforations, where as in this present study, a perforation of 4mm diameter and 150 perforations have been used. The objective of this paper is to determine theoretically the various factors that contribute to the total pressure drop in a perforated pipe. In addition to the pressure profiles along a blank section downstream of a perforated section were measured, and new wall-friction-factor correlations for pipe flow with perforation influx were calculated. In line with [18 and 19], it was noticed that the lubrication of the pipe flow occurred when the ratio of the total perforation flow rate to the total flow rate at the pipe outlet was small.

2. Theoretical Model

Theoretical analysis was carried out to determine the total pressure drops, frictional, acceleration and additional pressure drops. Fluid flow in a wellbore is considered as shown in Fig. 1 and assumed single-phase flow of an incompressible Newtonian fluid under the isothermal conditions with no heat transfer to and from the fluid to the environment. The test pipe is a partly perforated one and the rest is a plain pipe without perforation. Pipe and perforation geometry for experimental and theoretical study is listed in Table 1. The computational domain taken up in this study is same as that of the dimensions considered in the experimental rigs [18 and 19]. The geometry has been analyzed using three dimensional Computational Fluid Dynamics (CFD). Fig. 2 is the structured computational grids; the mesh consists of 146221 nodes and 542121 elements with five boundary layers. The calculations were carried out with commercial finite volume code ANSYS FLUENT 14 CFX5 using a first scheme and turbulent with k epsilon model.

Table 1. Geometry of the test pipe.

Item	Experimental	Theoretical
Outer Diameter	30 mm	-
Inner Diameter	21.94 mm	22 mm
Perfo. Diameter	3.0 mm	4.0 mm
Total perfo. number	158	150
Perfo. phasing	60 °	60 °
Perfo. density	12 SPF	12 SPF

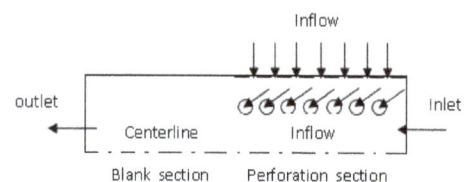

Figure 1. Configuration of partly perforated test pipe (not to scale).

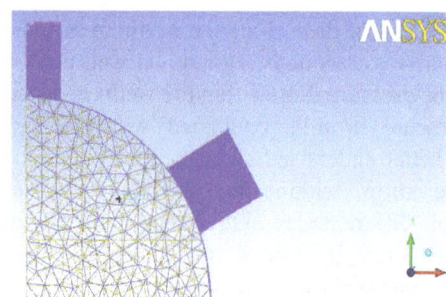

Figure 2. The unstructured mesh for partly perforated pipe.

3. Simulation Parameters

The fluid considered for the simulations is water with constant density of 998.2 kg/m^3 and dynamic viscosity of 0.001 kg/m s. Three tests were carried out with Reynolds number of the inlet flow ranging from 28,773 to 90,153. In each of the tests, flow rate through the perforations was increased from zero to maximum value. The roughness of the test pipe wall was 0.03 mm; the type of the test pipe was PVC.

Test details are summarized in Table 2. Uniform water mass flow is introduced at the inlet of a partially perforated pipe. Two boundary conditions are considered. At the inlet mass flow is taken into consideration both axially and radially where as at the exit outlet pressure is considered as the boundary condition. It is assumed that no-slip boundary conditions occur along the walls. Water enters at a uniform temperature (T) of 25°C. For the symmetry lines both velocity and pressure is kept constant.

Table 2. *Parameters of partly perforated pipe tests.*

Test	Inlet Flow Rate (lit/hr)	Perforation inlet Flow Rate (lit/hr)	Inlet Flow(Re)
Test 1	5,157 to 5,618	0-841	82,756 to 90,153
Test 2	3,361 to 3,836	0-854	53,935 to 61,557
Test 3	1,793 to 2,318	0-899	28,773 to 37,198

4. Pressure Drop in a Pipe with Radial Inflow

Over a long period of time the pressure drop in a fully developed turbulent pipe flow is being studied by several researchers and investigators. The pressure drop in a straight pipe has been determined in numerous experiments. The total pressure drop in a perforated horizontal wellbore can be divided into a reversible pressure drop and an irreversible pressure drop. The reversible pressure drop is due to the momentum change (flow acceleration) as where more fluid enters the wellbore through perforations. While the irreversible pressure drop is that due to the pipe wall friction, perforation roughness and mixing effects. The following relationship gives the four pressure drop terms that make up the total pressure drop in a perforated horizontal well

$$\Delta p = \Delta p_{acc.} + \Delta p_{wall} + \Delta p_{perfo.} + \Delta p_{mix.} \quad (1)$$

The last two terms of Equation (1) combine into one term $\Delta p_{add.}$, which is the pressure drop due to the combined effects of fluid mixing and perforation roughness. Equation (1) can then be rewritten as

$$\Delta p = \Delta p_{acc.} + \Delta p_{wall} + \Delta p_{add.} \quad (2)$$

Applying the conservation of linear momentum to the control volume in the axial direction for each perforation unit has equal length ΔL, results in the sum of the forces acting on the control volume surfaces towards the downstream direction of the pipe axis

$$\sum F = m_{out}' u_{out} - m_{in}' u_{in} \quad (3)$$

where the mass flow rate is

$$m' = \rho A u \quad (4)$$

When radial inflow occurs, the static pressure in the pipe is not uniform, and the velocity profile is not fully developed. In addition to the force contributed by the pressure difference across the control volume and wall shear force, the sum of the forces acting on the control volume surface includes a force due to the combined effects of the irreversible process of fluid mixing and the presence of the perforation hole, including the effect of non-uniformly distribution of static pressure and non-fully developed velocity profile,

$$\sum F = (p_{in} A - p_{out} A) - \tau_w (\pi D \Delta L) - F_{add.} \quad (5)$$

From the above equations, this can be rearranged to get the following total pressure drop,

$$p_{in} - p_{out} = \rho(u_{out}^2 - u_{in}^2) + \Delta p_{wall} + \Delta p_{add.} \quad (6)$$

Equation (6) indicates that the total pressure drop consists of three different components:
- The pressure drop due to kinematic energy change (acceleration effects). This demonstrates the first term on the right side of Equation (6).
- The frictional pressure drop due to wall friction in a perforation unit, Δp_{wall}, is based on the average velocity u_{out} downstream of the perforation, and can be calculated from the Darcy-Wesibach equation White [20],

$$\Delta p_{wall} = \frac{f}{2} \frac{\Delta L}{D} \rho u_{out}^2 \quad (7)$$

When the relative roughness of the pipe is known, an accurate and convenient relationship for the friction factor in the turbulent pipe flow is the Haaland equation

$$f = \left\{ -1.8 \log_{10} \left[\frac{6.9}{\text{Re}} + \left(\frac{\varepsilon}{3.7D} \right)^{1.11} \right] \right\}^{-2} \quad (8)$$

For a hydraulically smooth pipe, surface roughness ε should be set to zero.

This equation applies to both laminar and turbulent flow.

The pressure drop due to perforation roughness, $\Delta p_{perfo.}$, is the extra pressure drop due to the presence of the perforations. It represents the extra friction caused by the perforations acting as roughness elements in the pipe wall. The pressure drop due to perforation roughness is most important when there is no flow through the perforations. It has been shown that the magnitude of the pressure drop due to perforation roughness depends on the pipe-perforation geometry and the perforation density [21].

The pressure drop due to mixing effects, Δp_{mix}, is an irreversible pressure drop which cannot be further classified. This pressure effect arises from the complex interaction between perforation flow and wellbore flow, which causes disturbances in the boundary layer and hence affects the pressure drop. The irreversible pressure drop due to mixing needs to be determined by experiments [21].

5. New Wall-Friction-Factor Correlations [10]

Mass transfer through the pipe wall affects the wall-friction shear. The influence of either inflow or outflow depends on the regime present on the wellbore. The inflow (production well) increases the wall friction for laminar flow while decreasing it for turbulent flow. In contrast, outflow (injection well) decreases the wall friction for laminar flow while increasing it for turbulent flow. In other words, the wall friction is different from that of pipe flow with no inflow or outflow. Therefore, friction –factor correlations for pipe flow without inflow or outflow cannot be used for wellbore flow with both axial flow in the pipe and inflow or outflow through perforations [10].

A new correlation for the local wall friction factor for turbulent flow has been developed in [10] using Olson and Eckert's experimental data[14] for turbulent air flow in a porous pipe with uniform air injection through the pipe wall. The new correlation is of the form [10]

$$f = f_o \left[1 - 29.03 \left(\frac{\text{Re}_w}{\text{Re}} \right)^{0.8003} \right] \qquad (9)$$

It was found that the ratio between the local friction factor and the no-wall flow friction factor does not depend on the wall Reynolds number to axial Reynolds number ratio; instead, it depends only on the wall Reynolds number. Therefore, a new correlation for the local friction factor was developed [10]

$$f = f_o \left[1 - 0.0153 \, \text{Re}_w^{0.3978} \right] \qquad (10)$$

Eq. 10 is a satisfactory correlation for local wall friction factor for single-phase turbulent wellbore flow [10].

6. Results and Discussions

6.1. Pressure Drops in Perforated Section

In this paper, theoretically were carried out on the pipe that was simulated with the experimental pipe. Three tests with different pipe flow rate were carried out for the perforated pipe.

The analytical results were examined in terms of the total pressure drop, as shown in Fig. 3. The individual tests had an average outlet flow Reynolds number in the range of 37,460 to 108,940. The total flow rate ratio (q) is the total perforation flow rate divided by the total flow rate at pipe outlet.

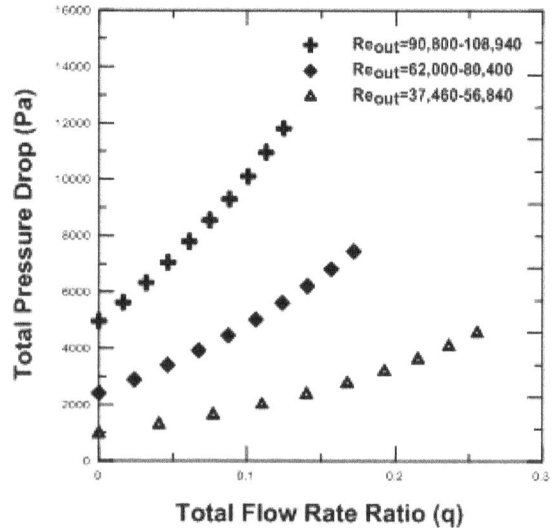

Figure 3. *Total pressure drop across perforated section.*

The total pressure drop was found to be higher for higher Reynolds numbers. This effect was caused by the larger wall frictional pressure drop under higher flow velocity. As the rate of flow through the perforations increases, the flow rate ratio increases and the total pressure drop increases. The main reason is that a higher flow rate through the perforations giving a larger acceleration pressure drop. In addition, it was found that greater wall friction was due to larger average flow velocity in the pipe, which was caused by inflow through the perforations and increased mixing effect.

The pressure drop due to momentum change (acceleration pressure drop) was calculated from Equation (6) (the first term of the right side). It was noticed that the values of acceleration pressure drop for partly perforated wellbore were higher than the values of the frictional pressure drop.

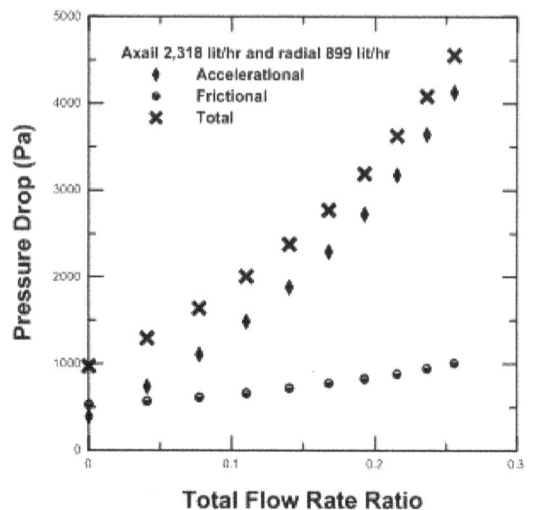

Figure 4. *Pressure drops across perforated section for Test3.*

The acceleration pressure drop is also very important and cannot be neglected from the inference of Fig. 4. Reference [10] explained that for partly perforated wellbore, the ratio of the acceleration pressure drop to the frictional pressure drop R_{af} is higher and changes from 1.27 to 0.71. The cumulative acceleration pressure drop from starting point of the perforations is more or less the same as the cumulative frictional pressure drop. It is shown that the acceleration pressure drop may or may not be important compared to the frictional component depending on the specific pipe geometry, fluid properties and flow conditions.

The increase in total flow rate ratio q leads to an increase in the pressure drop and further this leads to increase in the axial pressure drop.

Fig. 5 represents the total pressure drop and acceleration pressure drop for three tests. For tests 1 and 2 the acceleration pressure drop is approximated to zero value but test 3 has values greater than values of test 1 and test 2 at zero value of total flow rate ratio. The acceleration pressure drop contributes to the important part of the total pressure drop for all the three tests. The acceleration component contributes in the range of 36.6% to 99.9% for test 1, from 13% to 99.9% for test 2, and for test 3 from 9.3% to 59.8%.

From Equation (6), subtracting the acceleration pressure drop from the total pressure drop $\Delta p - \rho(u_{out}^2 - u_{in}^2)$, the results are equal to the summation of pressure drop due to wall friction and additional pressure drop. The additional pressure includes pressure drop due to perforation roughness and mixing flow $\Delta p_{add.}$.

Fig. 6 depicts the behavior of the remaining pressure drop. The remaining pressure drop decreases with increase in total flow rate ratio. As the rate of flow through the perforations increases, the flow rate ratio increases correspondingly and the total pressure drop increases. It is due to the larger acceleration pressure drop for higher flow rate through the perforations. The remaining pressure drop decreases with the increasing value of the total flow rate ratio. Due to this, the acceleration pressure drop increases with the increase in total flow rate ratio.

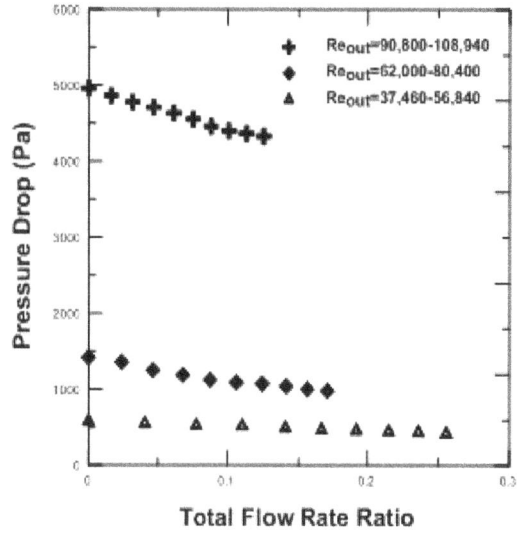

Figure 6. *Pressure drop due to wall friction, perforation roughness, and flow mixing.*

6.2. Additional Pressure Drop in Perforated Section

The additional pressure drop is the resultant of the mixing effect and the perforation roughness. The additional pressure drop is obtained by subtracting the frictional pressure drop from the remaining pressure drop. The wall friction pressure drop of a perforated section was calculated using Equation (7) for a unit perforation. The friction factors were calculated from Equation (8) for perforated pipe using 0.03 mm. The roughness of the pipe was taken into consideration because the pipe was made of PVC.

The wall friction pressure drop for uniformly perforated section was calculated Equation (11). Using Equation (8), the friction factor was calculated. The friction factor can also be found from Equation (9) because it depends on the ratio of $\left(\dfrac{Re_w}{Re} \right)$ or from Equation (10) because it depends only on the wall Reynolds number [10]

$$\Delta p_{wall} = \sum_{i=1}^{N-1} \left(f_i \frac{\Delta L_i}{D} \frac{\rho u_i^2}{2} \right) \qquad (11)$$

where N is the total number of perforations and D is the pipe diameter. The value of ΔL_i is the distance in the axial direction of the pipe between two adjacent perforations minus the equivalent length that is occupied by a perforation. For a uniformly perforated pipe, ΔL_i is a constant, calculated as

$$\Delta L_i = \frac{L}{N-1} - \frac{d^2}{4D} \qquad (12)$$

where d is the perforation diameter and L the total length between the first and the last perforation of the perforated section.

Fig. 7 represents the additional pressure drop with total flow rate ratio for all the tests. The additional pressure drop

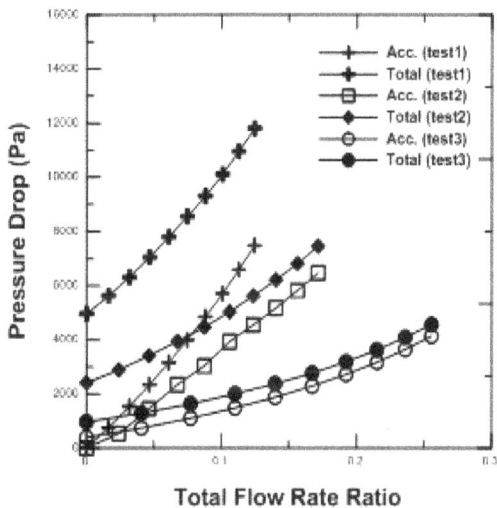

Figure 5. *Total and acceleration pressure drops for the three tests.*

decreases as the flow rate ratio increases. When the flow rate ratio is zero, the additional pressure drop is caused by perforation roughness only. When the total flow rate ratio reaches a certain limit, which is about 0.045 for test 1 and 2, the pressure drop caused by the perforation is eliminated by the smoothing effect. Beyond this limit of the flow rate ratio, the additional pressure drop has a negative value, which means that the pressure drop due to wall friction is reduced by the smoothing effect. For test 3, we show that all the additional pressure drop values are negative because of increases in radial flow, except for the value at zero flow rate ratio.

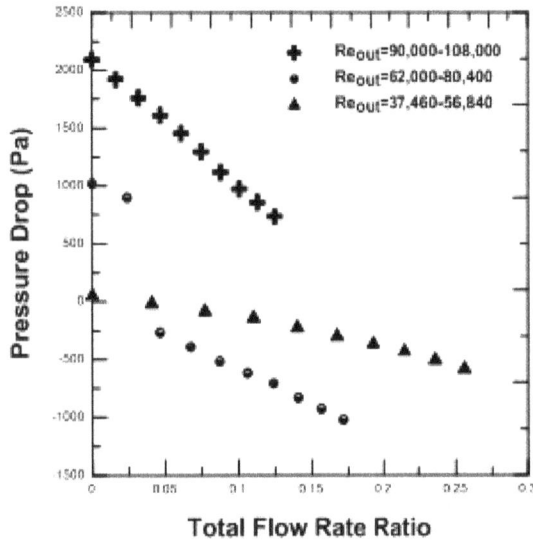

Figure 7. Additional pressure drop.

6.3. Total Pressure Drop in Blank Section

Total pressure drop was calculated for plain pipe section without perforations with flow rate ratio as shown in **Fig. 8**. The perforated section is followed by the plain section. The values of the total pressure drop in the plain section are lower than the values of total pressure drop for the perforated section because the pressure drop in the plain section of the pipe is mainly due to the pipe wall friction.

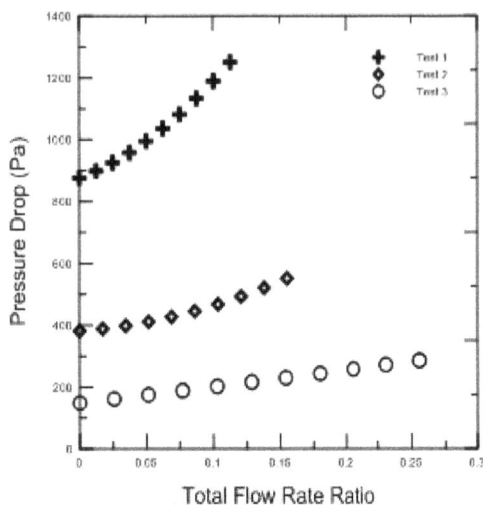

Figure 8. Total pressure drop for blank section.

6.4. Pressure Drop Coefficients

Pressure drop in a perforated pipe is the function of the flow rate in the pipe. A pressure drop coefficient is defined as the pressure drop across the perforated section divided by the kinetic energy at the outlet of the pipe.

$$K = \frac{\Delta p}{0.5 * \rho * u_{out}^2} \tag{13}$$

where u_{out} is the average flow velocity at the outlet of the test pipe.

The pressure drop coefficients were calculated for total and additional pressure drops for perforated section for all the three tests. The data points for each test follow a straight line as shown in Fig. 9. Data points of the three tests for total pressure drop followed parally and closely except for some points of those tests conducted with Reynolds number range from 62,000 to 80,400 and from 37,460 to 56,840. The pressure drop coefficients increase linearly with increasing total flow rate ratio.

Data points of different tests for additional pressure are shown in Fig. 10. Here also the graphical lines follow closely and parallel at low Reynolds number except for the tests conducted with Reynolds number ranges from 90,800 to 108,940 (due to high effect of pressure drop with high Re). So increase Re range increase the pressure drop, increase of mass flow with high Re and low kinetic energy at the perforated section. The data points decrease with increasing total flow rate ratio as shown in Fig 6.

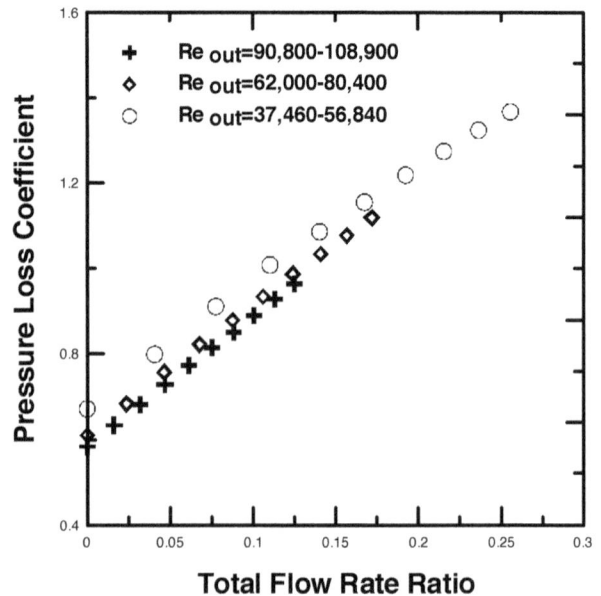

Figure 9. Pressure drop coefficient for total pressure drop.

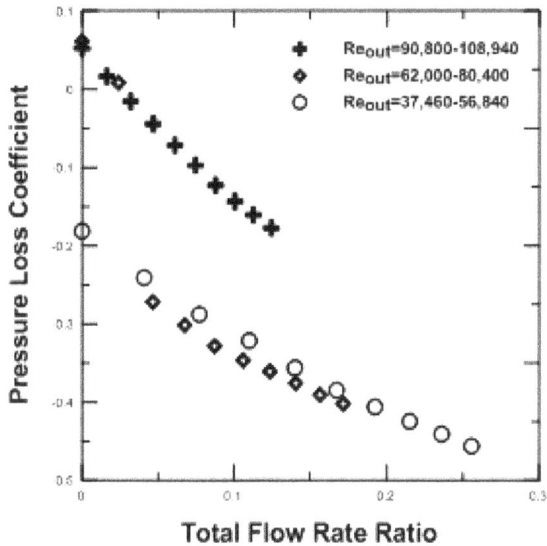

Figure 10. Pressure drop coefficient for additional pressure drop.

Fig 11 depicts the Numerical Simulation results in this paper and the experimental results conducted in tests[21] were used for the comparison of the total pressure drop in the perforated section between the first perforation and the last perforation.

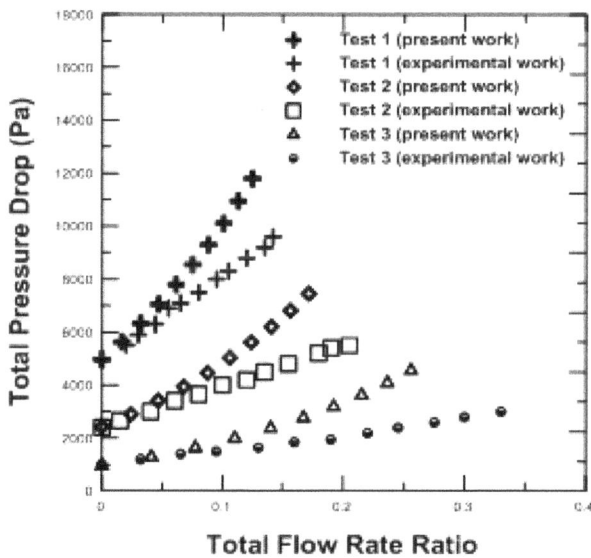

Figure 11. Comparison of numerical and experimental data[21].

It further shows that there is a difference for all the tests especially when the total flow rate increases and the curves are diverging from the experimental tests. The reason for this is at the range of values of the Reynolds numbers is different and higher than from the experimental tests.

6.5. Friction Factor

Friction factor is a dimensionless parameter extensively used in pipe flows to express the pressure drop due to frictional effect. Fig.12 represents the new correlation prediction by Equation (9) for local friction factor ratio f / f_o with injection Reynolds number Re_w. It was observed that the

local friction factor ratio for a perforated pipe with fluid injection for all the tests conducted above decreases when the radial flow increases. For test 1, the values of the local friction factor ratio is higher than tests 2 and 3. From Equation (9), it was observed that with the increase in the value of the radial flow, the local friction factor ratio decreased.

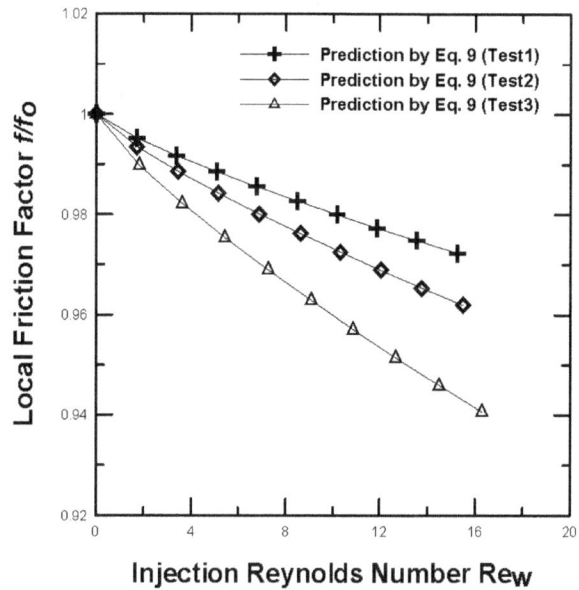

Figure 12. Wall friction factor correlation for turbulent pipe flow with inflow.

Fig. 13 represents the new correlation prediction by Equation (10) for local friction factor ratio f / f_o with injection Reynolds number Re_w. It was observed that all the data points of the three tests pertaining to the friction factor for a perforated pipe with fluid injection were close to each other because the results depended on Re_w only.

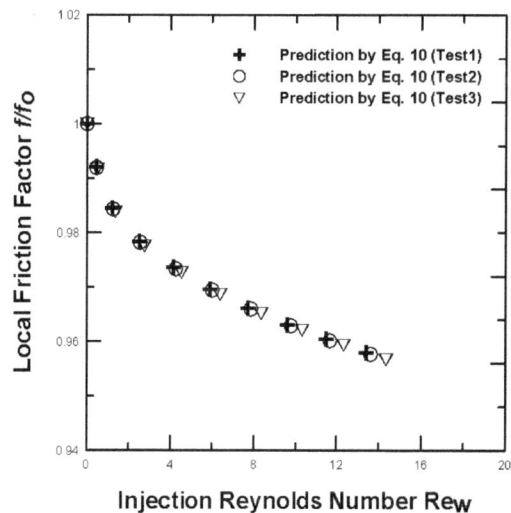

Figure 13. Wall friction factor correlation for turbulent pipe flow with inflow.

7. Conclusions

Numerical simulations have been carried out on the flow in a partly perforated pipe with inflow through perforations. The geometry of the pipe used was similar to the pipe used in the experimental tests [18, 19 and 21]. The total pressure drop in a horizontal wellbore is the sum of the pressure drops due to momentum change (acceleration), wall friction, perforation roughness, and fluid mixing. The acceleration pressure drop cannot be ignored compared with the frictional pressure drop. The total pressure drop for perforated section was larger than the total pressure drop in the plain section without perforations. The additional pressure drop caused by the perforation roughness was eliminated by the smoothing effect once the flow rate ratio reached a certain limit. It was observed that the local friction factor ratio for a perforated pipe with fluid injection decreased with increase of the radial flow.

References

[1]   Dikken, B.J., "Pressure Drop in Horizontal Wells and its Effect on Production Performance," JPT (November 1990) 1426.

[2]   Islam, M.R. and Chakma, A., "Comprehensive Physical and Numerical Modeling of a Horizontal Well," paper SPE 20627 presented at the 1990 SPE Annual Technical Conference and Exhibition, New Orleans, 23-26 September.

[3]   Folefac, A.N. et al. "Effect of pressure Along Horizontal Wellbore on Well Performance, Aberdeen, 3-6 September.

[4]   Ozkan, E., Sarica, C., and Haciislamoglu, M.: "Effect of Conductivity on Horizontal Well Pressure Behavior," paper SPE 24683 presented at the 1992 SPE Annual Technical Conference and Exhibition, Washington, Dc, 4-7 October.

[5]   Ihara, M. and Shimizu, N., "Effect of Acceleration Pressure Drop in a Horizontal Wellbore," paper SPE 26519 presented at the 1993 SPE Annual Technical Conference and Exhibition, Houston, 3-6 October.

[6]   Seines, K. et al., "Considering Wellbore Friction Effects in Planning Horizontal Wells," JPT (October 1993) 994.

[7]   Landman, M.J., "Analytical Modeling of Selectivity Perforated Horizontal Wells," J. Petroleum Science and Engineering (1994) 10, 179.

[8]   Sarica, C. et al., "Influence of Wellbore Hydraulics on Pressure Behavior and Productivity of Horizontal Wells," paper SPE 28486 presented at the 1994 SPE Annual Technical Conference and Exhibition, New Orleans, 25-28 September.

[9]   Novy, R.A., "Pressure Drops in Horizontal Wells: When Can They be Ignored?" SPERE (1995) 29.

[10]   Ouyang, L.B. et al., "A Single-Phase Wellbore-Flow Model for Horizontal, Vertical, and Slanted Wells," SPE Journal 3(2), 1998, pp. 124-133.

[11]   Asheim, H. et al., "A Flow Resistance Correlation for Completed Wellbore," J. Petrol. Sci. Eng., 1992, 8 (2), pp. 97-104.

[12]   Marett, B.P., Landman, M.J., "Optimal Perforation Design for Horizontal Wells in Reservoir with Boundaries," paper SPE 25366 presented at the 1993 SPE Asia Pacific Oil and Gas Conference and Exhibition, Singapore, February 8-10.

[13]   Su, Z., Gudmundsson, J.S., "Friction Factor of Perforation Roughness in Pipes," SPE 26521 presented at the 1993 SPE 68th Annual Technical Conference and Exhibition, Houston, TX, USA, October 3-6.

[14]   Olson, R.M. and Eckert, E.R.G., "Experimental Studies of Turbulent Flow in a Porous Circular Tube with Uniform Fluid Injection through the Tube Wall," J. Applied Mechanics (1966) 33, No. 1, 7.

[15]   Rabithby, G., "Laminar Heat Transfer in the Thermal Entrance Region of Circular Tubes and Two-Dimensional Rectangular Ducts with Wall Suction and Injection," Intl. J. Heat and Mass Transfer (1971) 14, No. 2, 223.

[16]   Kloster, J., "Experimental Research on Flow Resistance in Perforated Pipe," Master thesis, Norwegian Int. of Technology, Trondheim, Norway (1990).

[17]   Ihara, M. et al., "Flow in Horizontal Wellbores with Influx through Porous Walls," paper SPE 28485 presented at the 1994 SPE Annual Technical Conference and Exhibition, New Orleans, 25-28 September.

[18]   Su, Z. and Gudmundsson, J.S., "Pressure Drop in Perforated Pipes," PROFIT Projected Summary Reports, Norwegian Petroleum Directorate, Stavanger (1995).

[19]   Su, Z. and Gudmundsson, J.S., "Pressure Drop in Perforated Pipes," report, Department of Petroleum Engineering and Applied Geophysics, U. Trondheim, Norway (1995).

[20]   White, F.M., "Fluid Mechanics," McGraw-Hill, Inc. 1986.

[21]   Su, Z. and Gudmundsson, J.S.: "Perforation Inflow Reduces Frictional Pressure Loss in Horizontal Wellbores," J. Petrol. Sci. Eng., 1998, 19, pp. 223-232.

15

Scale-up of flat bladed mixer in orange juice concentrate process

S. R. Mostafa[1], M. A. Sorour[2], S. M. Bo Samri[3]

[1]Chemical Engineering Dept., Faculty of Eng., Cairo Univ., Giza. Egypt
[2]Food Eng. and Packaging Dept., Food Tech. Research Institute, Agric. Research center, Giza, Egypt
[3]Public Authority of Education Applied & Training Healthy Science, Food Processing Nutrition, Kuwait, Kuwait

Email address:

Salwa_raafat@hotmail.com (S. R. Mostafa) manal.sorour@yahoo.com (M. A. Sorour), fsnd_1@yahoo.com (S. M. B. Samri)

Abstract: Mixing of orange concentrates to be homogenized was investigated using flat – bladed impeller. The rheological properties of orange juice concentrate were studied over the range 10-70°C, solid concentration 66 wt% and speed of spindle 50-250 rpm. Shear stress-shear rate data indicate that the concentrate behaves as non-Newtonian pesudoplastic fluid. Geometry was studied by varying the impeller to a column diameter. An impeller mixer was connected to an ammeter in order to predict the power of the mixer. The relation between a power number, blend number, pumping number and Reynolds's number were calculated at different D/T. Scale-up of the mixing process from the laboratory to the production plant scale was carried out utilizing the aforementioned correlations.

Keywords: Scale-Up, Mixing, Flow Behavior of Orange Juice, Power Number, Flat Bladed Mixer, Mixing of Shear Thinning Fluids

1. Introduction

Mixing processes are complex, multi-faceted in nature and require an understanding of the fluid flow behavior along with an understanding of the mechanical and power requirement aspects of the equipment. [1]

Mixing plays an important role in the chemical, biochemical and food industries. To obtain high-quality products and high-efficiency processes, mixing must satisfy not only the needs of heat and mass transfer, but also the required homogeneity in the vessel in the shortest possible time. It has been widely reported throughout the literature that mixing conditions are related to product quality in a variety of processes. [2-4]

Most of the chemical and allied process industries frequently use a mixing operation to increase the degree of homogeneity of a property such as viscosity, color, concentration, and temperature [5]. These industries often involve non-Newtonian fluids with a yield stress; namely, pulp suspension, food substances such as ketchup and mayonnaise, paint, cement, pigment slurries, certain polymer and biopolymer solutions and wastewater sludge. [6].

The mixing of carrot concentrate (66%) to be homogenized were studied using a flat- bladed impeller. The relation between Power number, Blend number, Pumping number and Reynold's number were plotted at different D/T ratio. Scale-up of the mixing process from the laboratory to the production plant scale was carried out utilizing the aforementioned correlations. [7]

The rheological properties of apricot jam puree over the range 30-80°C, solid concentration 45, 55, 65 wt% of apricot jam puree, and speed of spindle 50-250 rpm were studied. Shear stress-Shear rate data indicated that the puree behaves as non-Newtonian pseudoplastic fluid which fitted well to power law. An impeller mixer was connected with Ammeter in order to predict the power of the mixer, then to predict the power number and Reynolds number at different temperatures and revolution per second. [8]

Experimental measurements on the influence of geometry of the pendulum agitators with clapping blades and of the physical parameters of mixed fluid on the homogenization time, the power consumption and the energy of mixing were analyzed and original formulas were proposed for the determination of the above mentioned mixing variables by

Masiuk and Kawecka [9].

Scaling- up for product quality requires simultaneous consideration of both mixing and heat transfer and accordingly industrial laminar flow mixers and heat exchangers need to be specially designed [10-13].

The objectives of this paper are to develop the effect of impeller to tank diameter ratio D/T on the plot of the dimensionless groups; power number, blend number, flow number, versus Reynold's number, then to use the obtained relationships to scale up the mixer process in non-Newtonian fluids. Since fluid rheology strongly impacts the design of an agitation system, a flow behavior investigation of the fluid (orange juice concentrate) is carried out to serve for the objectives of the paper.

2. Material and Methods

2.1. Material

Orange juice and its concentrate were prepared in El-Marwa company (6 October City) at 12% and 66% respectively.

2.2. Method

2.2.1. Flow Behavior of Orange Juice Concentrates

Rheological parameters (shear stress, shear rate, viscosity) of orange juice were measured at different temperatures using Brookfield Engineering labs DV-III Rheometer. The rheological properties of orange juice concentrate were studied over the range 4-70°C, solid concentration 66 wt% of orange concentrates, and shear rate 2.2-22 sec^{-1}.

2.2.2. Mixing of Orange Juice Concentrates

The mixing of orange juice concentrates to be homogenized using flat–bladed impeller. The relation between Power number, Blend number, Pumping number and Reynold's number were calculated at different D/T. Scale-up of the mixing process from the laboratory to the production plant scale was carried out utilizing the aforementioned correlations.

There are several important dimensionless numbers that are required to design mixers. All must be determined experimentally for a given impeller configuration. These numbers can be used to quantify the performance characteristics of an impeller. Dimensionless numbers are affected by geometric factors, such as the ratio of impeller diameter (D) to tank diameter (T), D/T and the ratio of clearance from the tank bottom to tank diameter,C/T. [14-15]

The impeller power number, N_P, is used to predict the power of the mixer, P, directly and torque, t, indirectly:

$$N_P = \frac{P}{\rho N^3 D^5} \tag{1}$$

Where, N_p is the power number; P is the power of the mixer, watt; D is the diameter of the impeller, m; N is revolutions per second and p is the density of the concentrate, kg/m^3.

The impeller blend number; NB is used to predict the blend time, θ, in a a mixed process. number, N_B, attempts to predict the effect of impeller D/T on the results:

$$N_B = N\theta \left(\frac{D}{T}\right)^{2.3} \tag{2}$$

Where, N_B is the blend number; N is revolutions per second; D is the diameter of the impeller, m; T is the tank diameter; m and θ is the time in second.

The impeller pumping number, N_Q, is used to predict the impeller pumping rate, q, directly and bulk fluid velocity, V_{bf}, indirectly.

$$N_Q = \frac{q}{ND^{2.3}} \tag{3}$$

$$v_{bf} = 4q / \pi T^2 \tag{4}$$

Where, q is volumetric flow rate of fluid leaving the impeller blades, m^3, N is revolutions per second; D is impeller diameter ,v_{bf} bulk fluid velocity, m/s and T is tank diameter, m.

Finally, Reynold's number (Re), for non-Newtonian fluids measures the ratio of inertial to viscous forces within the mixing environment. The generalized Reynolds number is calculated from the following equation. [16]

3. Results and Discussion

3.1. Flow Behavior of Orange Juice Concentrates

Shear rate-Shear stress relations are plotted in Fig.1. for orange concentrate (66 wt %) at different temperatures (4, 10, 30, 40, 50, 60, 70°C)

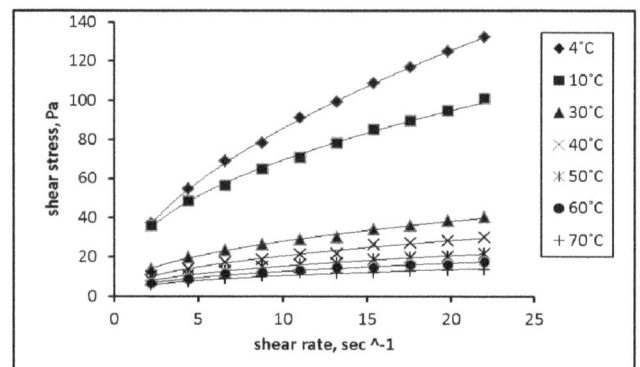

Figure 1. Relation between shear stress and shear rate at 66% solid concentration of orange juice and different temperature.

The results show that all the samples exhibited non-Newtonian pseudoplastic behavior at all the studied temperatures and concentrations.

Shear stress-shear rate data obtained fitted well to the constitutive equation:

$$\tau = k\gamma^n \tag{5}$$

Where, τ is the shear stress, Pa

k is the plastic viscosity (consistency index)

γ is the shear rate, sec^{-1}

n is the flow behavior index.

The values of flow behavior index (n) and consistency index (k) at different temperatures are shown in Table 1.

Table 1. Relation between k, n with temperature

Temperature, °C	k	n
4	24.02	0.531
10	24.85	0.446
30	9.986	0.447
40	7.042	0.465
50	5.651	0.433
60	4.853	0.416
70	4.349	0.38

3.2. Mixing of Orange Juice Concentrates

3.2.1. Power Calculation

All the dimensionless numbers discussed in previous section are correlated with Reynold's number.

For calculating the power of the mixer, an impeller was connected to an Ammeter to measure the current at different revolutions per minute of the mixer.

$$P = I\,V \qquad (6)$$

Where P is the power of the mixer, watt, I is the current, Ampere, and V is the voltage, volt

3.3. Effect of Impeller to Tank Diameter Ratio on Dimensionless Groups

3.3.1. Power Number – Reynolds Number Plot

The Power number is analogous to a friction factor; it is proportional to the ratio of the drag force acting on a unit area of the impeller and the inertial stress; that is, the flow of momentum associated with the bulk motion of the fluid (McCabe, Smith, 2001) [17].

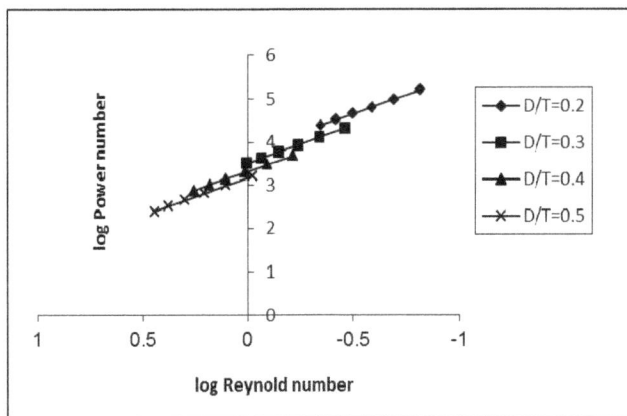

Figure 2. Relation between power number (Np) and Reynolds number of (Re) at different D/T ratio.

The relation between Reynold's number and power number was fairly fitted to the following equation.

$$\text{Log } N_P = \log A + B \log Re \qquad (7)$$

Where, P_0 is the power number; Re is Reynold's number; A and B are constants.

Figure (2) shows the relation between power number and Reynold's number at different impeller to a column diameter (D/T) ratio.

3.3.2. Blend Number – Reynolds Number Plot

This correlation depicts the trends observed for flat-blade impeller type for different values of D/T. Fig. 3 shows that as D/T increases, Reynolds number decrease (i.e., revolutions per second decrease) Blend number increases, (i.e., same time of mixing)

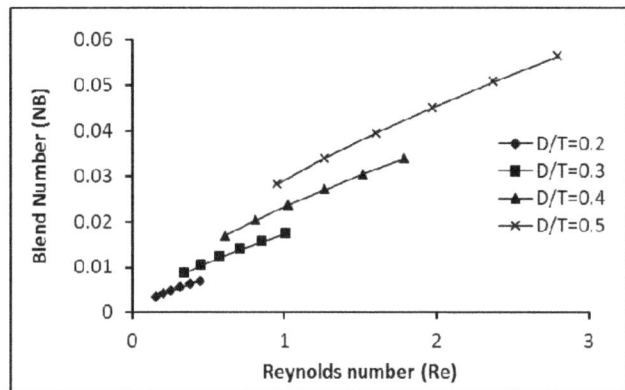

Figure 3. Relation between blend number (N_B) and Reynolds number (Re) at different D/T ratio.

3.3.3. Pumping Number- Reynolds Number Plot

Figure (4) presents the effect of D/T ratio on pumping number. Clearly the increase in D/T ratio causes a decrease in pumping number at the same Reynolds number. This may be explained due to the fact that constant volumetric flow rate of fluid leaving impeller, q, requires the decrease of pumping number with an increase in impeller diameter.

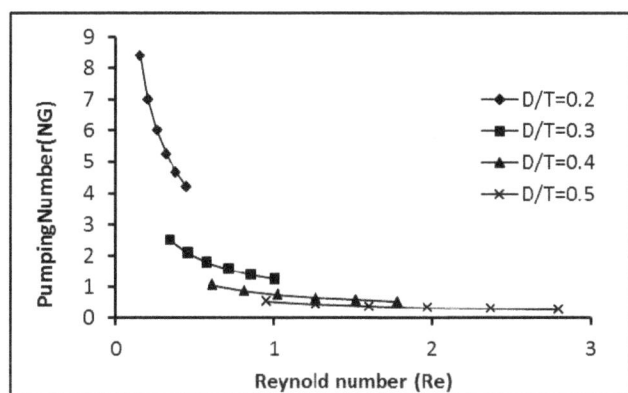

Figure 4. Pumping number (N_Q) as function of Re at different D/T ratio

3.4. Mixer Scale-Up

The scale-up procedure applied throughout this paper is based upon the assumption that lab-scale and plant-scale are geometrically similar and holding power per volume, P/V, fixed that is probably the most commonly used criterion in mixing scale-up because it is easily understandable and practical. Furthermore, fixed P/V correlates well with mass transfer characteristics in the mixer, and it is conservative enough to provide adequate performance in production scale equipment.

To calculate P/V, rearrange Eq. 1 to solve for P and V is calculated by:

$$V' = \frac{\pi T^2 Z}{4} \qquad (8)$$

Where, V' is the tank volume, m^3, T is the tank diameter, m, Z is the tank height, m.

Table 2. contains the scale-up calculation summary of a plant-scale mixer. It is noticed that the successive increase in revolutions per second, n, results in higher value of motor power, PM. Although the values of motor power (PM) range between 0.86 - 1 hp for the investigated rotational speeds 34 – 87 rpm respectively. The scale-up rule (based on fixed P/V) that must be applied to determine the plant-scale mixer speed (n_{ps}) to be used to duplicate the lab-scale results using (n_{ls}) is:

$$n_{ls} D_{ls}^{2/3} = n_{ps} D_{ps}^{2/3} \qquad (9)$$

Where, n_{ls} and n_{ps} are revolutions per second in lab and plant scale, respectively; Dls and Dps are impeller diameter in lab and plant scale, respectively.

$$D_{ls} = 0.08 \text{ m}, \ T_{ls} = Z_{ls} = 0.34 \text{ m}$$

$$D_{ps} = 0.12 \text{ m}, \ T_{ps} = Z_{ps} = 0.6 \text{ m}$$

$$P_M = P / 0.85$$

Table 2. Summary of scale-up analysis for D/T=0.2

Run	N (rps)	Re	N$_p$	P, watt	P/V'	P$_M$, hp
Lab(1)	0.833	0.271	37395.71	99	4697.613	0.156
Plant(1)	0.437	0.686	11465.02	549	4697.59	0.865
Lab(2)	1	0.359	23083.77	105.6	5109.307	0.166
Plant(2)	0.525	0.911	6998.837	597.123	5109.351	0.942
Lab(3)	1.166	0.456	15142.42	110	5184.903	0.173
Plant(3)	0.613	1.158	4611.012	605.87	5184.205	0.955
Lab(4)	1.333	0.562	10550.01	114.4	5375.499	0.18
Plant(4)	0.701	1.425	3212.242	628.2117	5375.375	0.991
Lab(5)	1.5	0.675	7694.591	118.8	5605.096	0.187
Plant(5)	0.788	1.711	2335.259	652.16	5605.292	1.028
Lab(6)	1.666	0.795	5713.234	121	5755.894	0.191
Plant(6)	0.876	2.016	1755.765	672.6	5755.189	1.061

4. Conclusion

Orange concentrate exhibited non-Newtonian pseudoplastic fluid at temperatures (4-70°C) and solid concentration 66% wt. An impeller mixer was used to predict the power number, blend number and pumping number as a function of Reynold's number at different impeller to tank

diameter ratios at 10°C and 66% concentration. The effect of D/T on N$_P$, N$_B$ and N$_Q$ was explainable. The design logic described throughout this paper depends on having reliable values of N$_P$, N$_B$ and N$_Q$ over the laminar flow regime for the impeller system being analyzed. These data allow selection of the appropriate scale-up criterion, ultimately leading to economic scale-up of pseudoplastic fluid mixing.

Notation

A and B	are constants in equation (12), dimensionless.
D	impeller diameter, m.
I	current intensity, Ampere.
k	consistency index, Pa.sec
n	revolutions per second
N$_P$	power number
N$_B$	blend number.
N$_Q$	pumping number
P	power of the mixer, watt
P$_M$	motor power, hp
q	volumetric flow rate of fluid leaving the impeller blades, m^3
Re	Reynolds number
T	tank diameter, m
V	voltage, volt
V'	volume, m^3
Z	liquid height in the tank, m
γ	shear rate, sec^{-1}
μ	Viscosity, Pa.sec
ρ	Density, kg/m^3
τ	shear stress, Pa

References

[1] K. Rajeev, Ch. Thakur, Vial, G. Djelveh, M. Labbafi, Mixing of complex fluids with flat-bladed impellers: effect of impeller geometry and highly shear-thinning behavior" Chemical Engineering and Processing, 43, pp. 1211–1222, 2004.

[2] Z. Maache-Rezzoug, J. M. Bouvier, K. Allaf, and C. Patras, Study of mixing in connection with the rheological properties of biscuit dough and dimensional characteristics of biscuits, J. Food Eng., 35, pp. 43–56, 1998.

[3] C. A. Kim, J. T. Kim, K. Lee, H. J. Choi, and M. S. Jhon, Mechanical degradation of dilute polymer solutions under turbulent flow, Polymer, 41, pp. 7611–7615, 2000.

[4] W. S. Kim, I. Hirasawa, and W. S. Kim, Effects of experimental conditions on the mechanism of particle aggregation in protein precipitation by polyelectrolytes with a high molecular weight, Chem. Eng. Sci, vol. 56, pp. 6525–6534, 2001.

[5] R. P. Chhabra, J.F. Richardson, Non-Newtonian Flow and Applied Rheology, Engineering Application, second ed. Elsevier, Butterworth-Heinemann, Amsterdam, 2008.

[6] A.W. Etchells, W.N. Ford, D.G.R. Short, Mixing of Bingham plastics on an industrial scale, Inst. Chem. Eng. Prog. Symp. Ser., 108, pp. 1–10, 1987.

[7] S.R.Mostafa, and M.A. Sorour, Effect of impeller geometry on mixing of carrot concentrate. TESCE, 32, pp. 1-13, 2006.

[8] M.A. Sorour, Prediction of power number in mixing of apricot jam puree, Journal of Engineering and Applied Science, 53, pp. 133-144, 2006.

[9] S., Masiuk, T. J. Kawecka, Mixing energy measurements in liquid vessel with pendulum agitators, Chemical Engineering and Processing, 43, 91-99, 2004.

[10] C.B. Elias, J. B. Joshi, Role of hydrodynamic shear on activity and structure of proteins", In: Scheper, T. (Ed.), Advances in Biochemical Engineering/Biotechnology, Springer, 59, pp. 47–71, 1998.

[11] B. McNeil, L.M. Harvey, Viscous fermentation products. Critical Reviews in Biotechnology 13, pp. 275–304, 1993.

[12] S. Saito, K. Arai, K. Takahashi, M. Kuriyama, Mixing and agitation of viscous fluids., In: Cheremisinoff, N.P. (Ed.), Encyclopedia of Fluid Mixing, vol. 2. Gulf Publishing, pp. 901–948, 1986.

[13] D.B. Todd, Mixing of highly viscous fluids, polymers, and pastes, In: Paul, E.L., Atiemo-Obeng, V.A., Kresta, S.M. (Eds.), Handbook of Industrial Mixing: Science and Practice. Wiley, pp. 987–1025, 2004.

[14] C. J. Geankoplis, 1983, Transport processes and unit operations, Allyn and Bacon, Inc., 2nd Ed., 1983

[15] R. J. Wilkens, C. Henry, L.E. Gates, Chemical Engineering Progress, pp. 44-52, 2003.

[16] Z. Xueming, H. Zondong, A. W. Nienowd, C.A. Kent, Rheological characteristics power consumption, mass and heat transfer during xanathan gum fermentation, Chinese Journal of Chemical Engineering, 2, pp. 198- 210, 1994.

[17] McCabe, W.L., and Smith, J.C., 2001, Unit Operations of Chemical Engineering, 6th Ed., McGraw-Hill, New York, NY.

16

Management of thermal quantity of hydrogen and sulphur during combustion of Kosova's lignite

Ahmet Haxhiaj, Nyrtene Deva, Mursel Rama

Faculty of Geosciences, University of Mitrovica "Isa Boletini", Mitrovica, Republic of Kosovo

Email address

ahaxhiaj52@yahoo.com (A. Haxhiaj), ahmet.haxhijaj@uni-pr.edu (A. Haxhijaj)

Abstract: The aim of paper is analyzing the issue related to the management of effective thermal quantity during lignite combustion and emissions of gases that are product of complete and non-complete combustion of lignite in the boilers. The real thermal amount of total burning lignite is 7524 kJ kg^{-1}. Thus, the losses of thermal quantity at ignition and during combustion of lignite directly depend on the diameter of lignite pieces intended for combustion. Paper contains the analyses of the thermal value, the composition of combustible and non-combustible matters of Kosova's lignite. The research on theoretical and practical field of management of thermal quantity of lignite is based and verified by DIN 1942, 1952, 1956 standards which describe effective thermal quantity. The paper reflects positively on management of the effective thermal quantity.

Keywords: Boilers, Combustion, Hydrogen, Sulphur, Thermal

1. Introduction

Boilers, combustion of lignite coal and its components in "Kosovo Energy Corporation" are basic processes for economic and environmental sustainability of Energetic Corporation. The paper discusses the combustion process of hydrogen and sulphur in oxidizing zone while is minimized the carbon combustion associated with hydrogen. The main subject of this paper is related to the chemical composition, power and thermal value of lignite coal which is 7524 kJ kg^{-1} and depends on the percentage of carbon, hydrogen, sulphur, moisture and sterile parts. [1,2]

In particular will be analyzed the balance of thermal quantity for hydrogen and sulphur, of Kosova lignite. During complete combustion of hydrogen is acquired the good utilization of thermal quantity and as product we have gases with minimal thermal quantity. Furthermore is analyzed the amount of thermal balance of sulphur combustion, whereas as products are benefited sulphur oxides and sulphur, also appears that the thermal amount is lost during the combustion process in boilers. With incomplete combustion we will have gases richer with carbon monoxide, hydrogen and sulphur, which carrying away the thermal quantity in environment, pollute the environment and reduce economic sustainability in the

technological process in energetic corporations.[3,5,10]

2. The Composition of Fuel

The main components of the fuel are carbon, hydrogen, sulphur, moisture and heterogeneous composition of oxygen, nitrogen and ash. The high content of moisture and ash in fuel, reduce its quality and is high impact factors for the quality determination [4].

2.1. Analytical Composition the Lignite

Lignite as fuel is composed of non-combustible and combustible matters which determine its thermal value. Carbon in the fuel is a free and as such is determining the thermal value of the fuel, is also associated with hydrogen as methane, ethane, propane, etc. (2.1)

$$C+H+O+N+S+W+A=100\% \qquad (2.1)$$

Based on the analysis such as experimental, analytical and XR,[2,5] which are realized at the Laboratory of Energetic Corporation of Kosovo, results that the Kosova's lignite has the average percentage of elements that are described in bellow table (2.1):

Table 2.1. Average percentage of elements in lignite

Element	C	H	O	N	S	W	A
Percentage %	27	2.20	13,63	0,0100	0,63	40,7	14,8

3. Combustion of Lignite

Combustion of the lignite coal is a complex process and includes 4 zones: heating zone, reductionzone, oxidation zone and ash zone.

Oxidation – combustion zone is subject of studying in this paper.

During the complete combustion of lignite as products are obtained CO_2 and H_2O, whereas during the incomplete combustion of lignite products are CO and H_2.[6,7]

3.1. Combustion of Hydrogen

Hydrogen is matter that burns in fuels, is located as free and associated with other fuel elements. [6,8]

Complete combustion of hydrogen is accomplished by reaction (3.1).

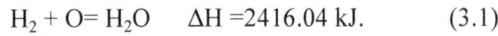

$$H_2 + O = H_2O \quad \Delta H = 2416.04 \text{ kJ.} \quad (3.1)$$

To find the enthalpy for absolute combustion of one kilogram of hydrogen below expression is used (3.2):

$$\Delta H_H = 2416.04 \times 100\%/MWH \text{ kJ/kg} \quad (3.2)$$

100%- is percentage of hydrogen associated with the oxygen (complete combustion of hydrogen)

MWH- is molecular weight of hydrogen

Hydrogen enthalpy for complete combustion is described through the expression (3.3).

$$\Delta H_H = 2416.04 \times 100\%/PMH = 2416.04 \times 1//2$$
$$= 1208.02 \text{ kJ/kg} \quad (3.3)$$

Alternative I with relative combustion, where 80% of hydrogen is related to oxygen, then enthalpy is described through the expression (3.4):

$$\Delta H_H' = 2416.04 \times 80\%/PMH = 966.41 \text{ kJ/kg} \quad (3.4)$$

Alternative II with low combustion, where 50% of hydrogen is related to oxygen, then enthalpy is described through the expression (3.5):

$$\Delta H_H'' = 2416.04 \times 50\%/PMH = 604.01 \text{ kJ/kg} \quad (3.5)$$

To find the thermal quantity earned with combustion of hydrogen we use the expression (3.6):

$$Q = m \times \Delta H \text{ kJ/h} \quad (3.6)$$

m-weight of hydrogen
ΔH-enthalpy
By 100% hydrogen combustion, the thermal amount

earned is based in expression (3.7):

$$Q = 100\% \times \Delta H_H = 1208.02 \text{ kJ/h} \quad (3.7)$$

Alternative I with 80% of hydrogen combustion, the thermal amount earned is realized according the expression

$$Q' = 80\% \times \Delta H'_H = 773.13 \text{ kJ/h} \quad (3.8)$$

Alternative II with 50% of hydrogen combustion, the thermal amount earned is realized according the expression (3.9) :

$$Q'' = 50\% \times \Delta H''_C = 302.005 \text{ kJ/h} \quad (3.9)$$

3.2. Combustion of Sulphur

Sulphur as an element in the composition of lignite is matter that is burned and release thermal quantity, which should be well used in the process of lignite combustion in boilers. Sulphur in fuels is located as free and as sulphate, which is considered as mineral matter with negative effects on the quality of fuel and in the environment. [6,8]

Combustion of sulphur becomes by the reaction (3.10):

$$S + O_2 = SO_2 \quad \Delta H = 279 \text{ kJ} \quad (3.10)$$

To find the enthalpy for the complete combustion of one kilogram of sulphur is used the expression (3.11):

$$\Delta H_S = 279 \times 100\%/MWS \text{ kJ/kg} \quad (3.11)$$

100% - is sulphur percentage associated with oxygen (complete combustion of sulphur).

MWS- is molecular weight of sulphur

To find the enthalpy for the complete combustion of one kilogram of sulphur is used the expression (3.12):

$$\Delta H_S = 279 \times 100\%/PMS = 279 \times 1/32 = 87.187 \text{ kJ/kg} \quad (3.12)$$

Alternative I with relative combustion, where 80% of sulphur is associated with oxygen, then enthalpy is calculated through the expression (3.13):

$$\Delta H'_S = 279 \times 80\%/PMS = 69.75 \text{ kJ/kg} \quad (3.13)$$

Alternative II with poor combustion, where 50% of sulphur is associated with oxygen, then enthalpy is is calculated through the expression (3.14):

$$\Delta H''_S = 279 \times 50\%/PMS = 43.53 \text{ kJ/kg} \quad (3.14)$$

To find the thermal amount earned during sulphur combustion the below expression is used (3.15):

$$Q = m \times \Delta H_S \text{ kJ/h} \quad (3.15)$$

m-weight of sulphur ,
ΔH_S - enthalpy of Sulphur
By 100% of sulphur combustion, the thermal amount earned is based in expression (3.16) :

$$Q = 100\% \times \Delta H_S = 87.187 \text{ kJ/h} \quad (3.16)$$

Alternative I with 80% of sulphur combustion the thermal amount earned is based in expression (3.17) :

$$Q'=80\% \text{x } \Delta H'_S = 55.8 \text{ kJ/h} \qquad (3.17)$$

Alternative II with 50% of sulphur combustion the thermal amount earned is based in expression (3.18) :

$$Q''=50\% \text{x } \Delta H''_S = 21.76 \text{ kJ/h} \qquad (3.18)$$

Table 3.1 and figure 3.1 presents the data related the thermal quantity gained by combustion of hydrogen and sulphur.

Table 3.1. *Thermal quantity of hydrogen and sulphur combustion*

Percentage of hydrogen and sulphur combustion (%)	Weight of hydrogen and sulphur (kg)	Enthalpy of hydrogen (kJ/kg)	Enthalpy of sulphur (kJ/kg)	Earned quantity for hydrogen (kJ/h)	Earned quantity for sulphur (kJ/h)
100	1	1208.02	87.187	1208.02	87.187
80	0.8	966.41	69.75	773.13	55.82
50	0.5	604.01	43.53	302.005	21.76

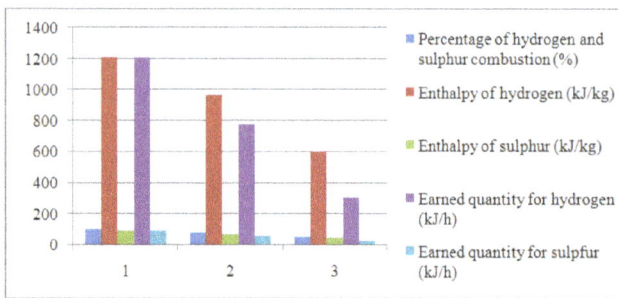

Figure 3.1. *Thermal quantity of hydrogen and sulphur combustion*

4. Balance of Thermal Quantity

Analytical and graphical analysis of hydrogen and sulfur combustion in the oxidation zone, treating the management of the lignite combustion process in boilers in this zone [7,9]. Furthermore, is analyzed in details the thermal quantity gained with combustion of hydrogen and sulpur and is proven to have different values. This changing of thermal quantity is the potential for losses, or thermal quantity without the well managed of process of lignite combustion in boilers. The data related to this issue are presented in table 4.1 and figure 4.1.

Table 4.1. *Changing of thermal quantity eraned with H and S combustion*

Percentage of hydrogen and sulphur combustion (%)	Thermal quantity earned with hydrogen combustion (kJ/h)	Thermal quantity earned with sulphur combustion (kJ/h)	Changing of thermal quantity (kJ/h)
100	1208.02	87.187	1120.833
80	773.17	55.82	717.35
50	302.005	21.76	280.25

Table 4.2. *Remaining quantity of the thermal value of H and S during combustion*

Percentage of H and S combustion (%)	H and S weight (kg)	Remaining thermal quantity of H (kJ/h)	Remaining thermal quantity of S (kJ/h)
100	1	0	0
80	0.8	434.99	31.357
50	0.5	906.05	65.4477

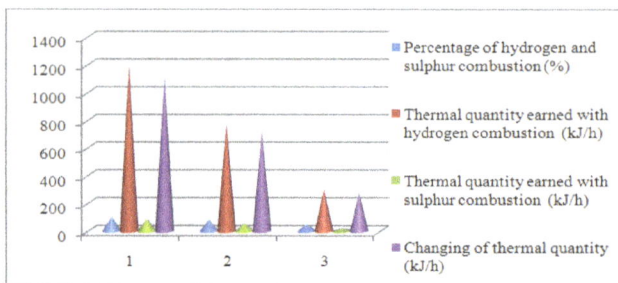

Figure 4.1. *Changing of thermal quantity earned with H and S combustion*

Based on the analysis of sulphur and hydrogen combustion and thermal value gained with their combustion, and the thermal value of the lignite basin of Kosovo is treated the thermal amount gained with complete and incomplete combustion of hydrogen and sulphur. Furthermore is treated the remaining quantity of the thermal value ,which is unused during the technological process of hydrogen and sulphur combustion in oxidizing zone that is potential for losses and for environmental pollution. The data related to this issue are presented in table 4.2 and figure 4.2.

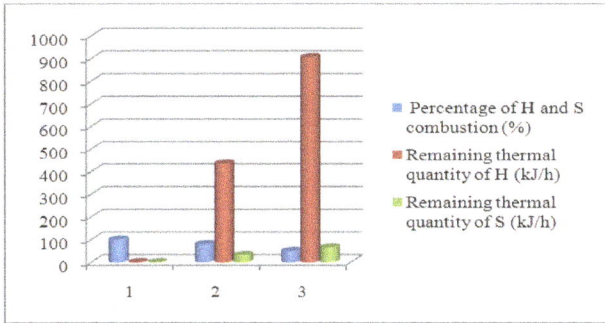

Figure 4.2. Remaining quantity of the thermal value of H and S during combustion 5. Impacts of Sulphur and Hydrogen in the Environment

5. Impacts of Sulphur and Hydrogen in the Environment

Sulphur and hydrogen in Kosova's lignite, during the combustion process release the thermal amount that is managed to be transformed into electricity.

The combustion process of sulphur and hydrogen in boilers is inclined to the formation of gases, especially the oxide gases. [5,8]

Oxide gases of sulphur, occupying a special place in this paper related to the treatment, management and their impact on the environment.

The main characteristics of sulphur and hydrogen in normal conditions

1. Sulphur
- The melting point 115.21 °C,
- The boiling point 444.72 °C,
- The starting point of melting - 31.51 °C,
- The critical point - 1041 °C,
- The powder density - 1960 kg/m^3,
- The molar volume - 1553 cm^3,
- The thermal conductivity - 0.205 W/m K.
2. Hydrogen
- The melting point - 259.2 °C,
- The boiling point - 252.8 °C,
- Density of powder hydrogen 8.99 kg/m^3,
- Molar volume - is associated with air,
- The thermal conductivity – easy combustion

The main parameters of the elements that result in the creation and emission of gases, that polluting the environment are the melting point and thermal conductivity table 5.1 and figure 5.1.

Table 5.1. Melting point and thermal conductivity of S and H

Element	Melting point (°C)	Thermal conductivity (W/m K)
Sulphur	115.21	0.205
Hydrogen	-259.2	Easily combustion

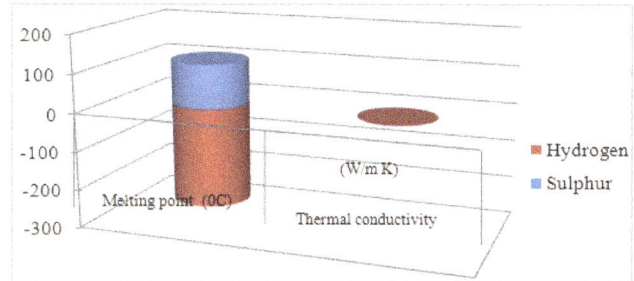

Figure 5.1. Melting point and thermal conductivity of S and H

5.1. Enthalpy of Elementary Sulphur and Hydrogen

The presence of sulphur and hydrogen in the environment as products of coal combustion in boilers, have their multi impacts on the environmental ecology. Based on the physical and chemical properties, and technological process of hydrogen combustion, is minimized the impact of hydrogen on the environment.

Enthalpy for sulphur. Enthalpy of fusion (mixture) is 1.73 kJ mol^{-1}. Enthalpy of evaporation is 9.8 kJ mol^{-1}. Enthalpy of powdering is 279 kJ mol^{-1} (table 5.2 and figure 5.2).

The hydrogen enthalpy. Enthalpy of mixture is minimized. Evaporation enthalpy is minimized. [3, 4]

Enthalpy of powdering is 2416.04 kJ mol-1. (table 5.2).

Table 5.2. Enthalpy of S and H kJ kg^{-1}

Element	Mixture Enthalpy kJ kg^{-1}	Evaporation Enthalpy kJ kg^{-1}	Powder Enthalpy kJ kg^{-1}
Sulphur	0.054	0.30	8.71
Hydrogen	0	0	1208.02

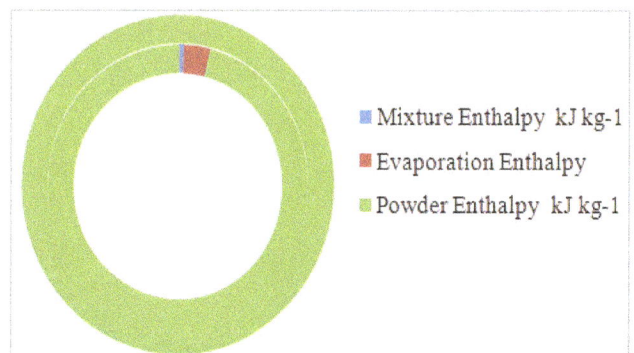

Figure 5.2. Sulphur and hydrogen enthalpy kJ kg^{-1}

6. Conclusion

The paper analyzes the effective thermal value that depends on the percentage of combustion matters and necessary amount of air entered in technological process for the lignite combustion process. In particular in this paper is analyzed the thermal amount that is gained with the complete and incomplete combustion of hydrogen and sulphur, as well and gases released in the oxidation zone during technological process. As a result of analytical and graphical analyzes, led us to conclude that with the complete hydrogen combustion the thermal amount earned is 1208.02 kJ / h, while with the complete sulphur combustion the thermal amount earned is 87. 187 kJ / h. Grounded on the thermal value and its use during the combustion of hydrogen and sulphur, and obtained results will be an overview as follows. With complete combustion of hydrogen use of the thermal value of coal is comprehensive and has an average value of 4.12%. While with the complete combustion of sulphur, use of the thermal value of the coal is weaker and has an average value of 0.297%.

It can be concluded, that with the complete hydrogen combustion, we will have a good use of the thermal quantities during the combustion process of the coal. Complete combustion of sulphur which results with lower thermal value, does not have any major impact on the thermal amount of lignite combustion process in boilers. Sulphur and its oxide gases have huge impact on environmental pollution. The paper as such reflects positively on the management and utilization of the thermal amount in the oxidation zone of boilers and closely is associated to positive effects on the economic and environmental sustainability.

References

[1]   Ekrem B, (1987) Tehnologija sa poznavanje robe. Pejë, 52-109.

[2]   Laboratory of Energetic Corporation of Kosovo

[3]   Feretič D,Tomišič Z, Śkanata D, Čavlina N, Subašič D (2000) Elektrane i okoliš. Zagreb, 213-218.

[4]   Halili A, (1983) Furrat metalurgjike II. Tirane, 207-235.

[5]   Haxhiaj A, Elezi D, Shkololli SH. (2006) "Mjedisi dhe menaxhimi i gazrave termike në zonën e parangrohjes të furrave vatërxhakete në Trepçë" simpoziumi i gjashtë ndërkombëtar Tirane.

[6]   Haxhiaj A, and Haxhiaj E, (2010) "The Thermal Gas Processing in Pre-Heating Zone of "Water-Jacket" Furnaces in "Trepça", TMS Annual Meeting & Exhibition, Seattle Washington, 205-215.

[7]   Haxhiaj A, and Haxhiaj E, (2010) The Optimization of the Coke and Agglomerate Quantity in Lead Production in "Water-Jacket" Furnace, TMS Annual Meeting & Exhibition, Seattle Washington, 249-257.

[8]   Haxhiaj A, and Haxhiaj E, (2010) The Air Pollution from the Port-Piri Furnaces Gases, Journal of International Environmental Application & Science, 357-363.

[9]   Panariti A, Merollari J, (1987) Metalurgjia e gizës-2. Tirane, 7-12.

[10]   Ranko R, (2010) Komercijalno Poznavanje robe. Beograd, 23-43.

17

Forced Convection Heat Transfer Analysis of Square Shaped Dimples on Flat Plates

Jamil Ahmed[1], Hasibur Rahman Sardar[2], Abdul Razak Kaladgi[3, *]

[1]Department of Computer science and Engineering, P.A College of Engineering, VTU, Mangalore, India
[2]Department of Electronics & Communication Engineering, P.A College of Engineering, VTU, Mangalore, India
[3]Department of Mechanical Engineering, P.A College of Engineering, VTU, Mangalore, India

Email address:
jamil.pace@gmail.com (J. Ahmed), hasibpace@gmail.com (H. R. Sardar), abdulkaladgi@gmail.com (A. R. Kaladgi)

Abstract: Dimples play a very important role in heat transfer enhancement of electronic cooling systems. This work mainly deals with experimental investigation of forced convection heat transfer over square shaped dimples on a flat aluminum plate under external laminar flow conditions. Experimental measurements on heat transfer rate and friction characteristics of air (with various inlet flow rates) on a flat plate were conducted. Both staggered and inline arrangements of the dimples were considered for the analysis. From the obtained results, it has been observed that the heat transfer coefficient were high for the plate having dimples.

Keywords: Forced Convection, Dimples, Friction Coefficient, Passive Techniques

1. Introduction

The development of integrated electronic devices with increase level of miniaturization, higher performance and output has increased the cooling requirement of chips considerably. And as the chip temperature increases, the stability and efficiency issues increases so the problem of heat dissipation has become a bottleneck for the development of chips in the electronic industry [1].Passive heat transfer enhancement techniques are used in electronic cooling devices. In these techniques passive augmented devices such as rib-turbulators, concavities (dimples) Extended surfaces or fins, dimples, and protrusions are used. Among these, the dimples (Special concavities) can be considered important because they not only enhances (augument) the heat transfer rate but also produce minimum pressure drop penalties which is important for pumping power requirements[1]. The dimple usually produces vortex pairs, causes flow separation, creates reattachment zones and hence increases the heat transfer rate. And as they do not protrude into the flow so they contribute less to the foam drag, to produce minimum pressure drop penalties [2]. Another advantage is that in dimple manufacture the removal of material takes place and reduces the cost and weight of the equipments.

Kuethe [3] can be considered as the first person to use

dimples on flat surfaces. He observed that the dimples promote rapid or turbulent mixing in the flow, acts as vortex generator & hence enhance the heat transfer. Afanasyev et al [4] carried out an experimental investigation on friction and heat transfer on surfaces having spherical dimples. They used totally ten plates for the investigation and the experiment was carried out for turbulent flow condition. They observed an increase of 30-40% in the heat transfer rate with no significant effect on the hydrodynamics of flow. In an another study, Chyu et al [5] used the transient liquid crystal imaging system to analyze and compare tear drop type and hemispherical dimples to study the heat transfer distribution in the channel. He observed a considerable increment in the heat transfer rate for the surfaces having dimples (about 2.5 times their smooth counterparts $10,000 \leq Re \leq 50,000$). Mahmood et al [6], experimentally investigated the effect of dimples on heat transfer augmentation. They used the flow visualization techniques and concluded that the periodic nature of shedding off of vortices is the main cause of enhancement of heat transfer and is much more pronounced at the downstream rims of the dimples. Mahmood et al [7] studied the effect of Reynolds number, aspect ratio, and temperature ratio & flow structure in a channel having dimples at one wall. They observed through the flow visualization techniques that the vortices that are shed off from the dimples become stronger as the non-dimensional channel height to dimple diameter (H/D) ratio decreases and increases the local Nusselt number in these regions. Xie et al [8] carried out a numerical investigation to study the effect of different types of heat transfer enhancing devices such as circular fins, protrusions, & hemispherical dimples mounted on tip wall. They concluded that though the dimples have a simple geometry but they are best suited for cooling of blade tip especially at low Reynolds numbers.

From the literature, it is very much clear that dimples (vortex generators) have high potential to enhance the heat transfer along with the production of lower pressure drop penalties. The other advantages include low weight and cost and low fouling rates [9], However, most of the researchers conducted numerical or experimental work on spherical dimples of uniform diameter [5, 10]. Also most of the research is confined to flow in the channel or Internal flow, with a very few studies on external flow [10].So the main aim of this project is to experimentally study the effect of square dimples under external laminar forced flow conditions. Both staggered and inline arrangements of the dimples are considered for the analysis.

2. Experimental Setup

The prime objective of the present work is to study experimentally the heat transfer enhancement through square dimples on aluminum flat plate's using forced convection technique. For this to be possible we required a forced convection setup which was fabricated as required. The fabricated setup is shown below.

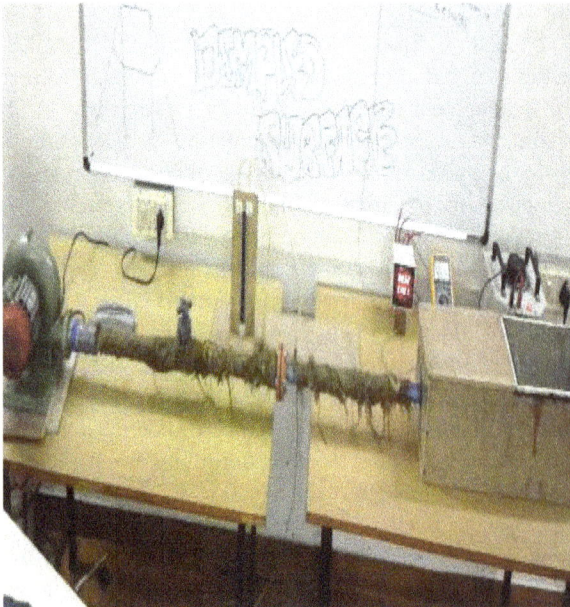

Figure 1. Experimental setup.

The main components of the test apparatus are a test plates of dimensions 100x100x2 mm, a calibrated orifice flow meter, Strip plate heater with capacity of 100 watts, Dimmer stat, Digital temperature, voltmeter, and ammeter with J type thermocouple, a gate valve, and a centrifugal blower. Strip

plate heater was used to provide heat input to the test surface. The provision was made to fix the heater at the base of the test plate & was connected to the blower through an orifice plate of 12mm diameter at inlet and to the atmosphere at outlet. U-TUBE manometer was connected across the orifice plate to indicate the pressure difference in terms of centimeters of water column difference. A PVC pipe was used to connect the blower outlet to the rectangular duct. Next to the blower outlet, flow regulating valve was connected to the pipe to regulate the air flow. Orifice plate was introduced next to the regulating valve to measure the regulated air flow rate in the pipe. Air flows parallel to the dimpled test surface. The strip plate heater fixed at the bottom of the test plate, was connected to power socket through dimmer stat. Dimmer stat provides the required heat input to the test plate. A Calibrated J-thermocouple was used to measure the temperatures. Provisions were made to fix the thermocouples on the test surface. Temperatures of air at inlet and outlet of the heat exchange module are also measured. Digital temperature indicators were used to show the temperature readings (in °C) recorded by thermocouple wires. Only top dimpled surface of the test plate was exposed to the air stream from which the convective heat transfer to the air stream takes place. The regulator was used to vary manometer head to give different trails for different shapes of dimples like Square, Triangular and Circular. The experiment was carried out by varying the inlet flow rates, thus varying the Reynolds number with air as a working fluid.

Figure 2. Schematic representation.

Table 1. Components and Specifications.

Components	Specification
Test plate	10x10x2 cm Aluminium plates
Blower	110W, 0.4BHP, 280rpm
Heater	100W, 4"x4"
Dimmer stat	6A,230V
Digital Temperature Indicator	6 channel,12000C, 230V
Orifice plate	12mm dia.
Manometer	"U-tube" glass manometer
Casing	A wooden casing of size of 8"x8" and 2feet long.
Thermocouple	K-Type, 3000C, 1m long.
Digital Multi-meter	Voltmeter, Ammeter

3. Data Reduction

The study was carried out under Steady laminar external forced convection regime. Steady state values for a given heat flux was used to determine the heat transfer performance parameters.

The various formula used are as follows

- Surface temperature of the plate: $T_s = (T_2+T_3+T_4)/3$
- Film temperature $T_f = (T_\infty + T_s)/2$

Properties of air at "T_f" is used to find parameters like

- Prandtl number(Pr)
- Thermal conductivity(k)
- Density(ρ)
- Kinematic viscosity(ν)
- Reynolds number = u *L/ν

Where "u" is air flow velocity, "L" is the length of aluminum test plate.

- Air flow velocity "u" =Volumetric flow rate/cross sectional area of pipe ,
- Volumetric flow rate $= c_d (\sqrt{2gHw}\ \rho_w/\rho_a -1)/ \sqrt{(1-m^2)}$

From heat transfer data book, Nusselt number can be related by,

- $Nu=0.453\ Re^{0.5}\ Pr^{0.333}$
- Heat transfer coefficient, $h = Nu \times k_{air}/L$

Where "L" is the length of test plate.

- Heat transfer rate, $Q = h\ A_s(T_s - T_f)$

Where "As" is surface area of aluminum test plate.

4. Results and Discussion

Experiments were conducted on Aluminum test plates with square shaped dimples. Inline and staggered arrangements were used for the study. The data obtained were used to find heat transfer parameters like Nusselt number, heat transfer coefficient, heat transfer rate and friction coefficient. And the experimental findings have been plotted in the form of graphs, mainly

- Nusselt number(Nu) vs Reynolds number(Re)
- Heat transfer coefficient(h) vs Reynolds number(Re)
- Heat transfer rate Q vs Reynolds number(Re)

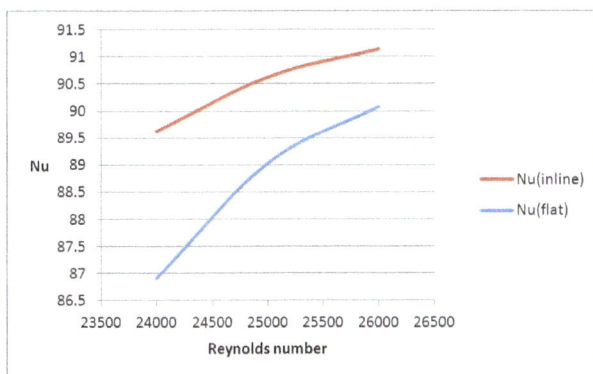

Figure 3. Variation of Nusselt number with Reynolds number(Inline).

Figure 3 & 4 shows variation of Nusselt number 'Nu' with Reynolds number for the square shaped dimples considered. It is obvious that the 'Nu' increases as Reynolds number

increases due to direct flow impingement on the downstream boundary and strengthened flow mixing by vortices at the downstream [1,11]. The formation of vortex pairs which are periodically shed off from the dimples, a large up wash regions with some fluids coming out from the central regions of the dimples are the main causes of enhancement of Nusselt number & is more pronounced near the downstream rims of the dimples [6].It can also be seen that the variation in the Nusselt number is gradual with Reynolds number as expected [12, 13, 16].

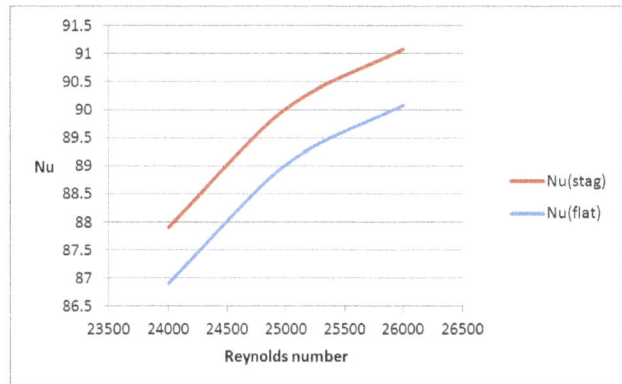

Figure 4. Variation of Nusselt number with Reynolds number(Staggered).

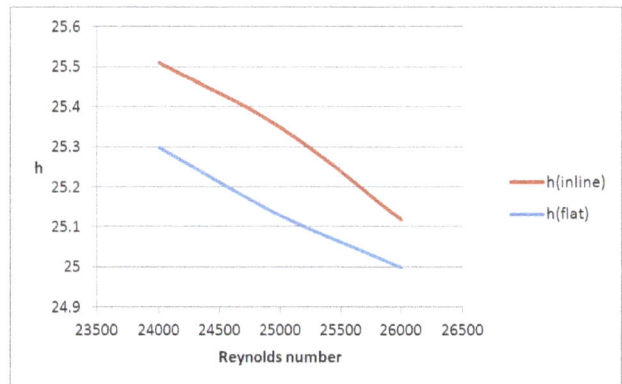

Figure 5. Variation of Heat transfer coefficient with Reynolds number(Inline).

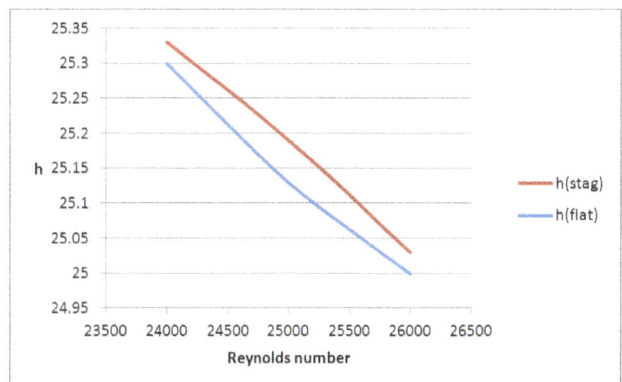

Figure 6. Variation of Heat transfer coefficient with Reynolds number(Staggered).

Figure 5 & 6 shows the variation of heat transfer coefficient 'h' with Reynolds number 'Re' for the square shaped dimple considered. It is obvious that 'h' increases with 'Re' as expected because the development of the thermal boundary layer is delayed or disrupted & hence enhances the local heat transfer in the reattachment region and wake region and increases the heat transfer coefficient [1].

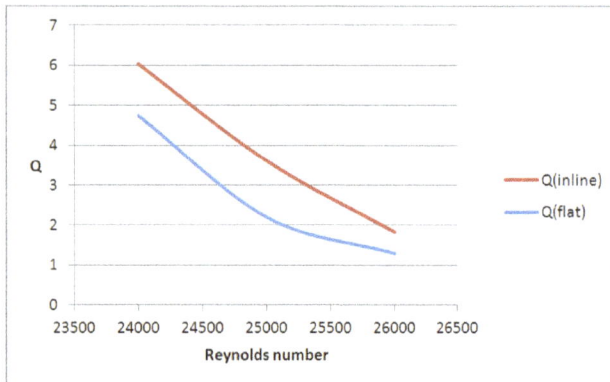

Figure 7. *Variation of Heat transfer rate with Reynolds number(Inline).*

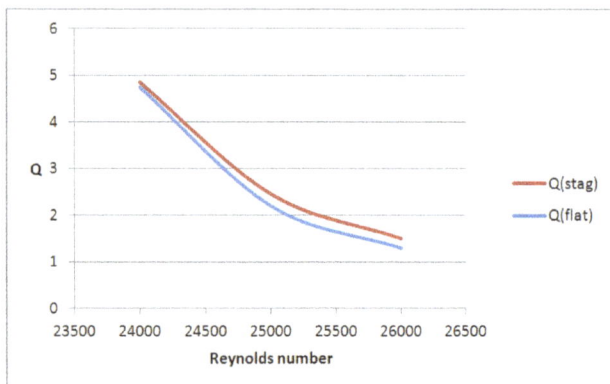

Figure 8. *Variation of Heat transfer rate with Reynolds number(Staggered).*

Figure 7 & 8 shows variation of Heat transfer rate 'Q' with Reynolds number 'Re' for the square shaped dimples considered. It can be seen that again 'Q' increases as 'Re' increases for both the cases i.e. flat plate with dimples and without dimples. It can also be seen that 'Q' is very much higher for plates having dimples because of increased flow area as compared to flat plates without dimples. Hence it can be concluded that dimples helps in better enhancing the heat transfer compared to flat plates without dimples [8].

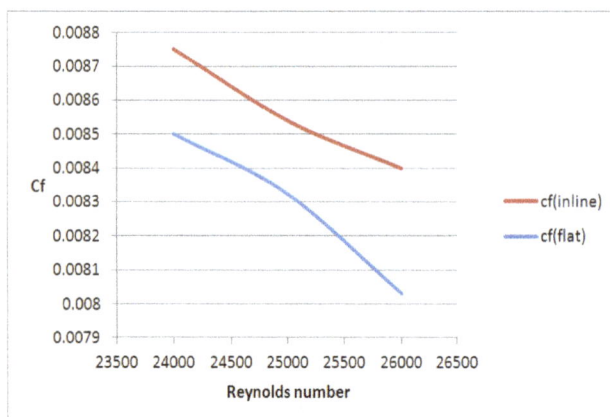

Figure 9. *Variation of friction coefficient with Reynolds number(Inline).*

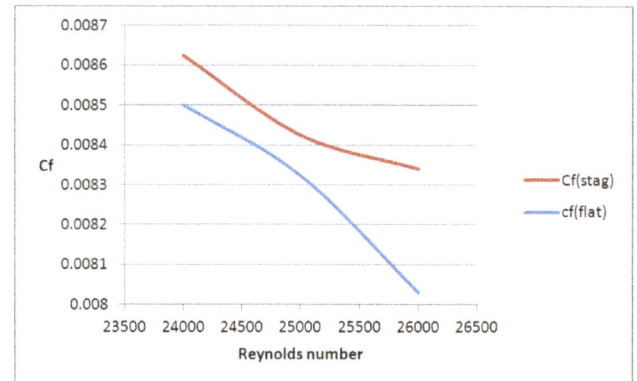

Figure 10. *Variation of friction coefficient with Reynolds number(Staggered).*

Figure 9 & 10 shows variation of average friction coefficient 'Cf' with Reynolds number 'Re' for the square shaped dimples considered. It can be seen that 'Cf' decreases as 'Re' increases for both the cases i.e. flat plate with dimples and without dimples because of the increased flow rates. It can also be seen that 'Cf' is little higher for plates having dimples indicating that the enhancement of heat transfer is at the cost of pressure loss. Hence it can be concluded that dimples helps in better enhancing the heat transfer but the pumping requirement may a little high as compared to flat plates.

5. Conclusion

In this experimental work an investigation of the effect of air flow over a flat plate with square shaped dimples is carried out. The main conclusions of the work are:

- Nusselt number increases with Reynolds number for the dimple arrangement considered due to direct flow impingement on the downstream boundary and strengthened flow mixing by the vortices at the downstream.
- Heat transfer coefficient increases with Reynolds number for the dimple arrangement considered due to the disruption of the thermal boundary layer development & hence enhance the local heat transfer in the reattachment and wake regions.
- It can be seen that 'Q' increases as 'Re' increases for both the cases i.e. flat plate with dimples and without dimples. It can also be seen that 'Q' is very much higher for plates having dimples because of increased flow area as compared to flat plates without dimples. Hence it can be concluded that dimples helps in better enhancing the heat transfer compared to flat plates without dimples However, the Augmentation depends on the configuration[9].
- It can also be seen that 'Cf' is little higher for plates having dimples indicating that the enhancement of heat transfer is at the cost of pressure loss.

References

[1] Zhang, D., Zheng, L., Xie, G., and Xie, Y., An Experimental Study on Heat Transfer enhancement of Non-Newtonian Fluid in a Rectangular Channel with Dimples/Protrusions, Transactions of the ASME, Vol. 136, pp.021005-10,2014.

[2] Beves, C.C., Barber, T.J., and Leonardi, E., An Investigation of Flow over Two-Dimensional Circular Cavity. In 15th Australasian Fluid Mechanics Conference, the University of Sydney, Australia, pp.13-17, 2004.

[3] Kuethe A. M., Boundary Layer Control of Flow Separation and Heat Exchange. US Patent No. 1191, 1970.

[4] Afanasyev, V.N., Chudnovsky, Y.P., Leontiev, A.I., and Roganov, P.S., Turbulent flow friction and heat transfer characteristics for spherical cavities on a flat plate. Experimental Thermal Fluid Science, Vol. 7, Issue 1, pp. 1–8, 1993.

[5] Chyu, M.K., Yu, Y., Ding, H., Downs, J.P., and Soechting, F.O., Concavity enhanced heat transfer in an internal cooling passage. In Orlando international Gs Turbine & Aero engine Congress & Exhibition, Proceedings of the 1997(ASME paper 97-GT-437), 1997.

[6] Mahmood, G.I., Hill, M.L., Nelson, D.L., Ligrani, P.M., Moon, H.K., and Glezer, B., Local heat transfer and flow structure on and above a dimpled surface in a channel. J Turbo mach, Vol.123, Issue 1, pp: 115–23, 2001.

[7] Mahmood, G. I., and Ligrani, P. M., Heat Transfer in a Dimpled Channel: Combined Influences of Aspect Ratio, Temperature Ratio, Reynolds Number, and Flow Structure. Int. J. Heat Mass Transfer, Vol. 45, pp.2011–2020, 2002.

[8] Xie, G. N., Sunden, B., and Zhang, W. H., Comparisons of Pins/Dimples Protrusions Cooling Concepts for an Internal Blade Tip-Wall at High Reynolds Numbers. ASME J. Heat Transfer, Vol. 133, Issue 6, pp. 0619021-0619029, 2011.

[9] Gadhave, G., and Kumar. P. Enhancement of forced Convection Heat Transfer over Dimple Surface-Review. International Multidisciplinary e - Journal .Vol-1, Issue-2, pp. 51-57, 2012

[10] Katkhaw, N., Vorayos, N., Kiatsiriroat, T., Khunatorn, Y., Bunturat, D., and Nuntaphan. A. Heat transfer behavior of flat plate having 45^0 ellipsoidal dimpled surfaces. Case Studies in Thermal Engineering, vol.2, pp. 67–74,2014

[11] Patel, I.H., and Borse, S.H.Experimental investigation of heat transfer enhancement over the dimpled surface. International Journal of Engineering Science and Technology, Vol.4, Issue6, pp.3666–3672, 2012.

18

An Overview on LNG Business and Future Prospect in Bangladesh

Saiful Islam, A. T. M. Shahidul Huqe Muzemder

Department of Petroleum & Mining Engineering, Shahjalal University of Science and Technology, Sylhet, Bangladesh

Email address:

saifulpmre@gmail.com (S. Islam), shahidulpme@gmail.com (A. T. M. S. H. Muzemder)

Abstract: Bangladesh is characterized by both relatively high growth rates in population and expanding economies and a deficiency in domestic fossil fuel energy. Growing population and expanding economies are main causes of increasing energy demand. This study provides an overview of global liquefied natural gas (referred as LNG) technologies, business as it currently exist and examines the future potential growth in this market. In addition, this study examines the prospects for Bangladesh in this sector and the factor behind this potential entry. Over the last fifteen years, world trade in LNG is more than tripled and it is anticipated that this market will continue its rapid expansion with the technological development to meet the demand for energy. The current consumers of LNG are mainly to be found among the energy hungry economies of South East Asia as well as the western European countries. It also expected to exhibit the highest future growth rate, given the underlying economic growth of the countries found in this region and their burgeoning demand for energy. In Bangladesh, natural gas is the main source of energy that accounts for 75% of the commercial energy of the country. Most of the industries and power generation plants are driven by natural gas. Bangladesh currently produces about 2250 million cubic feet per day (mmcfd) natural gas with a shortage of about 450 mmcfd. To overcome this shortage initiative should be taken to find an alternative new source of energy such as- importing LNG, production of coal and renewable energy. For the immediate solution of this problem LNG would be good option but for long term solution all other options should take under consideration to secure the future demand of energy.

Keywords: LNG, Demand, Supply, Bangladesh

1. Introduction

The Liquefied Natural Gas (LNG) is natural gas in liquid state at atmospheric pressure and temperatures around –161°C. The volume is reduced a factor 600times compared to the standard conditions, which allows larger volumes of LNG to transported by sea in refrigerated ships. In the last decades the world-wide LNG has growth significantly to meet the demand of natural gas allowing business opportunities.LNG plants have been designed for high capacities, base load plant, exceeding 150 million cubic feet per day(MCFD) of natural gas. The designs of large capacity LNG plants are focused mainly on the exploration of vast natural gas fields and towards the construction of major facilities, in order to take advantages of economies of scale. Different technologies for liquefying natural gas have been developed; being the most used the technologies of two and three cycles of cooling, with cascade or propane pre-cooling plus mixed refrigerant schemes. At present there are two major technology licensors which have dominated the LNG market for years.

The current consumers of LNG are mainly to be found among the energy hungry economies of South East Asia as well as the western European countries. It also expected to exhibit the highest future growth rate, given the underlying economic growth of the countries found in this region and their burgeoning demand for energy.

In Bangladesh, natural gas is the most important indigenous source of energy that accounts for 75% of the commercial energy of the country. So far 25 gas fields have been discovered of which two of the gas fields are located in offshore area. Currently gas is being produced from 20 gas fields. Oil was tested in two of the gas fields (Sylhet and Kailashtila). To reduce the dependency on natural gas, alternative energy resource including LNG must be explored.

1.1. Properties of Liquefied Natural Gas (LNG)

LNG is the cleanest form of natural gas and contains more than 90% methane therefore, LNG becomes synonyms to methane. It is colorless, odorless, nontoxic and noncorrosive. Its weight is less than one-half that of water. Hazardous include flammability, freezing and asphyxia. The density of LNG is approximately 0.41-0.5kg/L, depending on temperature, pressure and composition, compared to water at 1.0 Kg/L. The heating value depends on the source of gas that is used and the process that is used to liquefy the gas. The higher heating value of LNG is estimated to be 24 MJ/L at 164 °C. This value corresponds to a lower heating value of 21 MJ/L. LNG is produced by cooling natural gas to 161 °C at which it becomes a liquid. This process reduce its volume by a factor of more than 600 similar to reducing the volume of a beach ball to the volume of a ping-pong ball. The ability to convert natural gas to LNG, which can be shipped on specially built to ocean-going ships, provides consumers with access to vast natural gas resources worldwide.

1.2. Global LNG Market

Efforts to liquefy natural gas for storage began in the early 1900s, but it was not until 1959 that the world's first LNG ship carried cargoes from Louisiana to the United Kingdom, providing the feasibility of trains oceanic LNG transport. Five years later the United Kingdom began importing Algerian LNG, making the Algerian state-owned oil and gas company, Sonatrach, the world's first major LNG exporter. The United Kingdom continued to import LNG until 1990, when British North Sea gas became a less expensive alternative. [16].

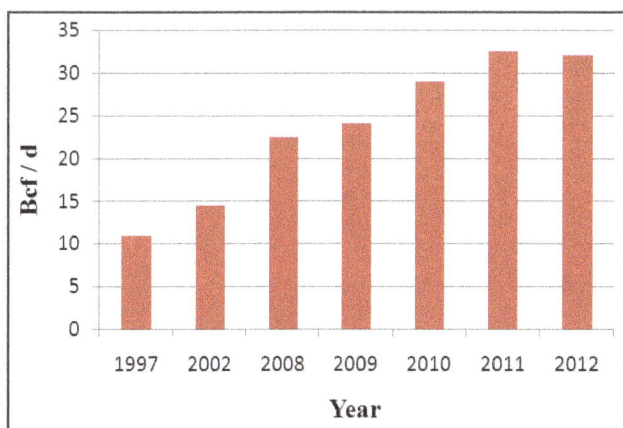

Figure 1. *World LNG Trade. [source: EIA.OGJ, BP Statistical Review of World Energy].*

Japan first imported LNG from Alaska in 1960 and moved to the forefront to the international LNG trade in the 1970s and 1980s with a heavy expansion of LNG imports. These imports into Japan helped to fuel natural gas fired power generation to reduce pollution and relieved pressure from the oil embargo of 1973. Japan currently imports more than 95 percent of its natural gas and as shown in Figure serve as the destination for about half the LNG exported worldwide.

World trade in LNG has grown dramatically over the last fifteen years. As shown in (figure9), trade in LNG more than tripled, growing from just over 10 Bcf/d in 1997 to 32Bcf/d in 2012. It is anticipated that this market will continue its rapid expansion as better production technology means more gas reserves worldwide are available for development while demand for energy, particularly for those with cleaner-burning properties, is expected to grow. [16].

1.3. World LNG Supply

The current suppliers of LNG to worlds markets and those that are expected to emerge as significant suppliers in the future come, not surprisingly, from those countries that are endowed with the largest natural gas reserves. Currently the countries located in Pacific Basin that supply the largest amounts of LNG include Malaysia, Indonesia and Australia (Table 1). Australia has a number of LNG liquefaction projects both under construction and in planning states, which should see it emerges as a much more significant supplier in future years. Also the first exports of LNG from Russia have occurred with the completion of its Shkhalin II project, which is Likely to be the first of several future LNG liquefaction developments in this country. [16]

Table 1. *Global LNG Supply (BCF/D) [Source: Oil and Gas Journal and BP World Energy Statistical Review].*

Exporter	2009	2010	2011	2012
Malaysia	2.8	3	3.22	3.10
Indonesia	2.5	3	2.82	2.40
Australia	2.3	2.5	2.51	2.70
Brunei	0.9	0.8	0.91	0.90
Russia	0.7	1.3	1.39	1.40
Alaska	0.1	0.2	0.2	0.10
Peru	0	0	0.5	0.50
Total Pacific Basin	*9.3*	*10.8*	*11.55*	*11.1*
Qatar	4.8	7.3	9.92	10.20
Oman	1.2	1.1	1.06	1.10
Abu Dhabi	.7	.8	0.77	0.70
Yemen	0	0.5	0.86	.70
Total Middle East	6.7	9.7	12.61	12.7
Trinidad	1.9	2	1.83	1.80
Algeria	2.1	1.9	1.66	1.50
Nigeria	1.5	2.3	2.5	2.60
Egypt	1.2	0.9	.83	.60
Norway	0.30	0.5	0.38	0.50
Equatorial Guinea	0.4	0.5	0.51	0.50
Libya	0.1	0.3	0.01	0.00
Total Atlantic Basin	7.5	8.4	7.72	.5
Total World	*23.5*	*28.8*	*31.88*	*31.30*

Of those countries located in and supplying markets in the Atlantic Basin, Trinidad, Algeria and Nigeria are currently the dominant suppliers. Nigeria, however have a number of LNG liquefaction projects being planned and developed that should increase its relative significance as an LNG supplier in years to come.

Exporters of LNG from the Middle East are dominated by those from Qatar, which is currently the largest exporter of LNG in the World. This status, however, could come under threat from Australia over the next decades as this country has a considerable number of LNG liquefaction projects being planned and developed LNG exports from the Middle East serve markets located in both the pacific and Atlantic basins. [16].

1.4. World Demand for LNG

The Energy Information Administration is expecting world-wide natural gas consumption to increase from 310 Bcf/d in 2010 to 507 Bcf/d in 2040 Figure 11. Much of this increase is due to the anticipated growth in the use of natural gas for power generation as countries take advantage of cleaner burning properties of this fuel. Natural gas consumption is expected to grow considerably faster in developing countries, as the world wide use of natural gas increases, the size of the LNG market will grow as well. While currently about 10% of natural gas produced globally is liquefied, the LNG market will likely account for a growing share of world natural gas trade as worldwide liquefaction capacity increases. [16].

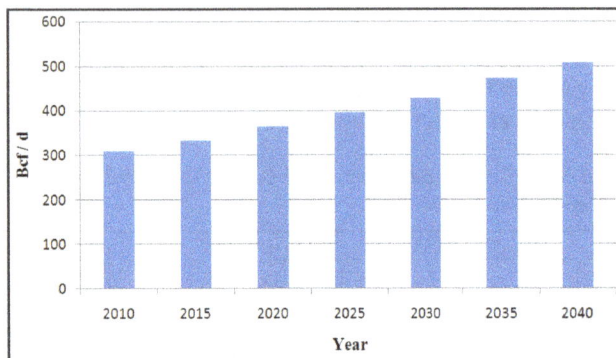

Figure 2. *World Natural Gas Consumption [Source: EIA 2013 International Energy Outlook].*

The current consumers of LNG are mainly to be found among the energy hungry economies of South Asia as well as a number of the more developed Western European countries. Some LNG is also imported into North and South America. The pacific Basin is the Largest Consuming region and is also expected to exhibit the highest future growth rate given the underlying economic growth of the countries found in this region and their burgeoning demand for energy. With respect to South East Asian importers of LNG, Table 2 shows that Japan is by far the largest importer in this region and in fact the whole world with more than 30 LNG import terminals currently in service and several more being planned. Japan currently accounts for over 35% of the entire world wide consumption of LNG and with substantial damage to Japanese nuclear power generating capacity as a result of the Tsunami that occurred in March 2011, LN G imports may grow higher in future to compensate the loss of nuclear power generating capacity. China has recently emerged as a net importer of natural gas. With its almost insatiable demand for Energy, it is expected to become a major importer of LNG in the future. India is also expected to increase its future demand for LNG.

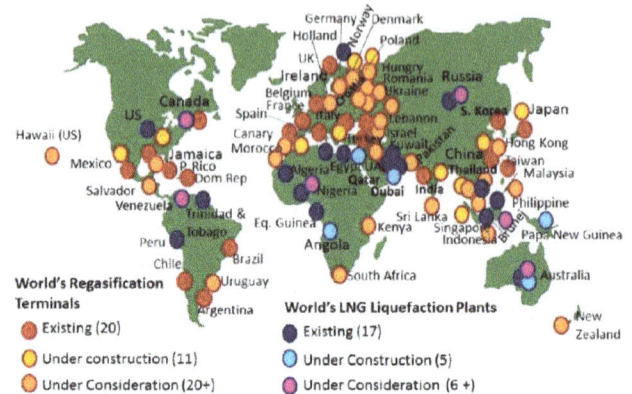

Figure 3. *World's majorLNGexportingandimportingcountries. [Source: BGgroup].*

The developed economies of Western Europe imported 6.6 Bcf/d in2012, down from 8.7Bcf/d in 2011 as economics of the region continue to struggle. These nations are not expected to increase their demand for the LNG anywhere near the same rate as the growing Asian economies. LNG is also consumed in America, but the emergence of shale and gas in North America means that this region is more likely to become a net exporter of LNG rather than an importer. [16].

Table 2. *Imports of LNG (Bcf/d).*

Country	2009	2010	2011	2012
France	1.26	1.34	1.41	1
Spain	2.56	2.66	2.34	2.1
Portugal	0.29	0.3	0.3	0.4
Turkey	0.54	0.77	0.6	0
Belgium	0.62	0.63	0.64	0.4
Italy	0.27	0.88	0.85	0.7
Greece	0.08	0.11	0.13	0
UK	1.02	1.81	2.45	1.3
US	1.2	1.18	0.97	0.5
Puetro Rico	0.07	0.08	0.07	0
Dom. Republic	0.05	0.08	0.09	0
Mexico	0.34	0.55	0.39	0.5
Brazil	0.07	0.27	0.1	0.3
Argentina	0.09	0.16	0.42	0.5
Chile	0.06	0.3	0.37	0.4
Canada	0.08	0.19	0.32	0.2
Kuwait & United Arab Emirates	0.09	0.27	0.43	0.4
Total Atlantic Basin & America	8.771	11.59	11.88	9.4
Japan	8.39	9.04	10.34	11.5
South Korea	3.36	4.3	4.77	4.8
Taiwan	1.07	1.45	1.58	1.6
India	1.18	1.18	1.65	2
China	0.72	1.23	1.61	1.9
Total Pacific Basin	14.71	17.21	19.95	21.8
World Total	23.42	28.79	31.83	31.2

[Source: Oil and Gas Journal and BP World Energy Statistical Review]

1.5. Future Prospect of LNG in Bangladesh

In Bangladesh, natural gas is the most important indigenous source of energy that accounts for 75% of the commercial energy of the country. So far 25 gas fields have been discovered of which two of the gas fields are located in offshore area. Currently gas is being produced from 20 gas fields. Oil was tested in two of the gas fields (Sylhet and

Kailashtila). To reduce the dependency on natural gas, alternative energy resource must be explored.

1.6. Current Scenario of Gas Sector in Bangladesh

The main source of energy of Bangladesh is indigenous natural gas and imported petroleum. At present, daily average gas production is about 2300-2350 mmcf. National Gas companies' produces around 1050-1100 mmcf (42%) and the IOCs produces around 1200-1300 mmcf (58%),off the total production of gas, presently used in Power sector, Fertilizer, Captive Electricity, Industry, CNG, Commercial, Tea Gardens and in Domestic sector (Figure12).

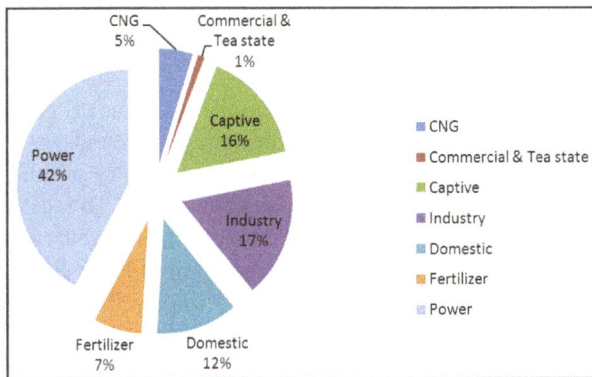

Figure 4. Gas consumption sectors in 2012. [source: Annual Report-2012, Petrobangla].

As per official statistics, there is a suppressed demand of around 500-700 mmcfd of gas. If gas was made available, may be at this time it would have exceeded daily usage of 4000 mmcf. Moreover, during the last few years, gas transmission pipeline has been extended up to Khulna and Rajshahi. When those areas would be brought under full-fledged gas supply without making available sufficient sources of supply, it would be a real difficult situation in maintaining supply to the present consumers what to talk about supplying new customers of the new areas. Therefore, an alternative has to be in place before actual supply starts in the new areas. The only alternative to make available natural gas is import of gas or discovery of a gas reserve like Titas or Bibiyana fields. The means of import of gas is pipe gas or it could be LNG. [15].

1.7. LNG for Bangladesh

Now a days, to meet immediate requirement of natural gas, importing LNG is the main source of supply. The government also took the initiative in 2010 and a delegation headed by the Secretary, Energy visited Qatar in June 2010 and agreed a Memorandum of Understanding which was signed in January 2011. In short term solution various countries adopting the methodology of Floating Storage & Re-gasification Unit (FSRU) instead of land based permanent type. However, considering the cost effectiveness, with the short term solutions one should not forget about the land-based LNG facilities. In Bangladesh, as the port facilities is very limited and specially for LNG vessel because of the

higher requirement of draught, it has been planned to use FSRU near Moheshkhali where required draught is available at about 5-6 km offshore of Moheshkhali coast. The mother vessel carrying LNG would be transferred to the FSRU which would be moored at about 5-6 km off-coast of Moheshkhali. An offshore pipeline would be installed from the FSRU and a delivery point will be stationed on-shore at Moheshkhali Island. From the delivery point, a gas transmission pipeline of about 85-90 km would be installed to bring the gas at the port city of Chittagong. In Chittagong, it would be hooked up with Karnaphully gas system and the gas would be supplied to the customers. As per plan, initially about 500 mmcfd of gas would be supplied for which in total about 4 million ton of LNG would be required annually. [18]

2. Basic Issues of LNG

2.1. Import of LNG & LNG Handling Facilities

The first and foremost requirement of handling LNG vessels is the availability of minimum 15 meter draught. Generally maximum draught available in Chittagong port is 10-12 meter which depends on the tide again. A vessel carrying 30,000 M.Ton of oil could not come to the unloading jetties and need to do lighterage of 10,000 M.Ton and then bring the vessel to the unloading jetties. Draught of 15 meter plus is available in and around 5-6 km off-coast of Moheshkhali. Therefore, to handle LNG vessel an adverse and tough situation had to be faced. As land based LNG terminal installation is not possible, there is no flexibility of choices but to go for FSRU type facilities. The FSRU is nothing but a LNG carrier having re-gasification process facilities on it. The mother vessels would come to the FSRU and transfer the LNG and the FSRU process and converts the LNG into gas and send it to the pipeline. In our case, from the FSRU, offshore pipeline of about 5-6 km would be needed and then from the onshore receiving station gas had to be brought to Chittagong area by connecting a pipeline of about 85- 90 km of appropriate size. Again the underwater current situation of that area needs to be evaluated before installation of offshore pipeline for which bathometry study had to be conducted. As Bangladesh has not handled LNG vessel before, therefore, it is very crucial to know the methodology of handling the mother vessel and guide the mother vessel to the FSRU. Ship to Ship transfer of LNG is another crucial aspect of handling LNG. For all these Bangladesh need to create appropriate technical manpower and those persons need to be trained also. In area like Melaka Straight of Malaysia, it looks like big lake where there is hardly any high wave. In Kuwait, the sea is calm and quiet, and even the Break water can be seen from a distance. But think about the Bay of Bengal in Kutubdia / Moheshkhali Island area. The roughness of the sea especially during monsoon season may cause very difficult situation for handling LNG vessels. So one can easily understand the magnitude of activities involved in handling LNG in our situation. In

Chittagong areas almost every year during monsoon season, we have cyclone and tornedos. Therefore, maintaining the continuous supply of LNG would be a challenging affairs and alternatives arrangement must be in place during such a situation to maintain the chain of continuous supply of LNG. [18].

2.2. Price of LNG

Another basic issue of LNG operation would be the price factor. Gas is already underpriced in our country as it is considered coming from indigenous sources. Presently, in the domestic producer level, three price regimes are existing. First, gas price of Bangladesh Gas Fields Company Limited (BGFCL) and Sylhet Gas Fields Limited (SGFL); second, gas price of Bangladesh Petroleum Exploration & Production Company Limited (BAPEX); and third the gas price of IOCs (International Oil Companies). When import of LNG will start, another price factor would be added. Therefore, a suitable price mechanism needs to be in place so that LNG can be imported on uninterrupted basis. One might argue that present LNG price is tagged with oil price and it would be very costly for Bangladesh to effort. In this respect, according to a businessman of the country who said that instead of keeping the factories or mills in shutdown condition, if the factory can run with higher priced gas, there will be employment opportunity as well as the loan money from the bank which is presently yielding no output resulting non-payment of bank loan, at least it can be returned or process of returning the loan would start. If consider with broader sense and not confining within the issue of price, ultimately country would be benefitted. However, it must be kept in mind that 4 million ton of LNG means about minimum 6 billion dollar plus annually considering the present trend of price of LNG. [18].

3. Long Term Vision for Bangladesh

To start with importing of LNG by Bangladeshi, FSRU seems to be alright. Installation of FSRU and brining the gas into the system would take minimum 24-30 months. Now FSRU is not a proprietary to a particular supplier. While selecting FSRU, an in-depth understanding is required and all out efforts has to be there to minimize the cost as much as possible. Therefore, selection of size of FSRU is very important. Again it must be kept in mind the size of mother vessel of LNG. FSRU of 170,000-175,000 CBM can easily handle full standard size of LNG cargo and it would provide increased security of supply of LNG.A smaller FSRU size can also handle LNG but the flexibility/security would be limited. To meet the ever increasing demand of natural gas, we must plan for a permanent solution and that is installing a land based LNG facilities. Because of the draught restriction, a normal cargo vessel or oil tanker of 30,000 M.Ton cannot come to the port facilities of Chittagong. Government has already initiated "Deep Sea Port Project". It would be appropriate to include LNG facilities with the Deep Sea Port project so that permanent LNG facilities could be installed in due course of time. The main intention would be to reduce costing of LNG and its facilities. Very recently Singapore has taken an initiative to become a LNG hub like Oil hub. As Bangladesh would be signing contract with Qatar for import of LNG on long term basis, it may keep in mind the development of LNG facilities around Bangladesh so that it can take advantage as and when possible. Now in USA, LNG is sold at much lower price than Asia. So importing of LNG from USA may also be kept in mind. Another source of importing LNG may be Australia too. All these options had to be kept open while engaging import of LNG. [18].

4. Discussion

There is a widening gap between the amount of gas supplies available and the growing demand for natural gas. There are a lot of industrial units exists in the country which just cannot go into operation in the absence of gas. To meet this demand it is time think about alternative energy sources like LNG, coal and renewable energy, but to meet immediate requirement of natural gas, importing LNG is the main source of supply as Bangladesh has port facilities. One might argue that present LNG price is tagged with oil price and it would be very costly for Bangladesh to effort. Therefore, a suitable price mechanism needs to be in place so that LNG can be imported on an uninterrupted basis. Finally, it can be said that for the long term solution of the energy crisis in Bangladesh proper step should be taken to reduce the dependency of natural gas in power generation sector where most of the gases are used instead of that coal based plant should be developed.

5. Conclusion

Currently, there are different technological options with the potential to be applied in future LNG developments. These alternative technologies are mainly made up of two and three cycles of cooling, with schemes either cascading process or more pre-cooling with propane plus mixed refrigerants. However, the selection of liquefaction natural gas technology most appropriate depends on the priorities and conditions for each project.

The current consumers of LNG are mainly to be found among the energy hungry economies of South East Asia as well as the western European countries. It also expected to exhibit the highest future growth rate, given the underlying economic growth of the countries found in this region and their burgeoning demand for energy. Bangladesh is also thinking about LNG option to meet the demand of natural gas.

In the last decades the world-wide LNG has growth significantly to meet the demand of natural gas allowing business opportunities. It also expected that this market will continue its rapid expansion as better technologies are available.

References

[1] Vink, K.J. and Nagelvoort, R.K. 1998. Comparison of Baseload Liquefaction Processes, Paper presented at the 1998 Intl. Conference on Liquefied Natural Gas, Perth, Australia, 4–7 May.

[2] Process Evaluation—Research Planning, Liquefied Natural Gas. PERP report by Chem systems Inc, 96/97S2, November.

[3] M.J. Roberrts, J. C. Bronfenbrenner, Yu-Nan Liju - Large Capacity Single Train AP-X Hybride LNG Process- Gastech 2012 Qatar.

[4] R. Nibbelke, S. Kauffman B. Pek – Liquefaction Process Comparison of C3MR and DMR for Tropical Conditions- GPA 81^{1st} annual convention, 2002.

[5] Chabrelie Marie F. LNG the way ahead. En: Fundamentals of the global LNG Industry, pp. 10-14, Londries: Petroleum Economist, 2007.

[6] Coll, Roberto; Carbon, Eudardo; Technology Evaluation Methodology for standard Gas Monetization Options, 19th WPC, 29th June-3rd July, Madrid, 2008.

[7] H. Bauer – A Novel Concept for Large LNG Basehold Plants- AICHE Spring National Meeting, 2001.

[8] Guerrero, Ramiro A. y otors. Processes of liquefied natural gas –state of the art. Caracas Universidad Simon Boliver, 2006.

[9] Nexant LNG: the Expanding Horizons of Liquefaction Technology and Project Execution Strategies. Houston: Nexant, 2007.

[10] PEK B, y otors. Large capacity LNG Plant development. LNG 14, 2014.

[11] Sang gyu Lee, Kun hyung Choe, Young Yang, The state of art LNG Liquefaction Plant Technologies, The 3rd Korean Congress of refrigeration, vol. 3, pp.65-68. 2009.

[12] Seung Taek Oh, Ho Saeng Lee, Jumg In Yoon, Sang Gyu Lee, Development of LNG.

[13] Liquefaction Process, Journal of the SAREK, Vol. 38, No. 3, PP 13-17, 2009.

[14] Annual Report of Petrobangla-2012. <http://energy.gov/sites/prod/files/2013/04/f0/LNG_primerupd.pdf>.

[15] Report on "An Overview of the World LNG Market and CANADA's Potential for exports of LNG", January 2014 <http://www.capp.ca/getdoc.aspx?DocId=237161&DT=NTV > (Accesed19 August 2014).

[16] Wikipedia, Liquefied Natural Gas, (22 August 2014 revision) <http://en.wikipedia.org/wiki/Liquefied_natural_gas> (Accessed 19 August 2014).

[17] Petrowiki, Liquefied Natural Gas, (23 March 2014 revision) <http://petrowiki.org/Liquified_natural_gas_(LNG)> (Accessed 18 August 2014).

[18] <http://ep-bd.com/online/details.php?cid=32&id=17538> (Accessed 20 August 2014).

[19] BG Group, Aworldleader in natural gas. <http://www.bg-group.com/OurBusiness/BusinessSegments/Pages/pgLiquefiedNaturalGas.aspx.>.

[20] A. B. Raheem, A. Hassan, S. A. Samsudin, Z. Z. Noor, A. Adebobajo, Comparative Economic Investigation Options for Liquefied Petroleum Gas Production from Natural Gas Liquids, American Journal of Chemical Engineering. Special Issue: Developments in Petroleum Refining and Petrochemical Sector of the Oil and Gas Industry. Vol. 3, No. 2-1, 2015, pp. 55-69.

[21] Md. Niaz Murshed Chowdhury, Samim Uddin, Sumaiya Saleh. Present Scenario of Renewable and Non-Renewable Resources in Bangladesh: A Compact Analysis. International Journal of Sustainable and Green Energy. Vol. 3, No. 6, 2014, pp. 164-178.

19

Degradation Prevalence Study of Field-Aged Photovoltaic Modules Operating Under Kenyan Climatic Conditions

Macben Makenzi[1], Nelson Timonah[2], Mutua Benedict[3], Ismael Abisai[4]

[1]Mechatronic Department, Jomo Kenyatta University of Agriculture and Technology, Nairobi, Kenya
[2]Physics Department, Jomo Kenyatta University of Agriculture and technology, Nairobi, Kenya
[3]Faculty of Engineering and Technology, Egerton University, Nakuru, Kenya
[4]Ubbink East Africa Ltd, Naivasha, Kenya

Email address:

macbenmm@gmail.com (M. Makenzi)

Abstract: Photovoltaic (PV) modules deployed outdoors can degrade due to exposure to the various elements. This includes exposure to UV light, a range of fluctuating temperatures and humidity and exposure to a range of operating currents and voltages. Different weather conditions have an important influence on degradation rate. Evidence indicates that both degradation and failure mechanisms are location dependent. This paper presents a research investigating the prevalence of various forms of physical degradation experienced by photovoltaic panels which have been in operation in Kenya under various climatic conditions. To study degradation of PV systems, identification and analysis of modules that had been deployed in various locations in Kenya, and which had been in operation for at least the last 2 years was carried out. Imaging instruments were used to study visible signs of weathering and other physical defects. The results indicated that despite the fact that panels are designed to operate in outdoor environment, numerous cases do exist whereby the panels degrade physically, in various ways, and consequently exhibit total failure, diminished performance or just physical manifestation of wear. Apart from manufacturer defects, user ignorance on installation and usage was also proved to contribute to the diminished life span of some panels.

Keywords: Degradation, Photovoltaic, Field-Aged Modules

1. Introduction

Long term performance of photovoltaic (PV) systems is vital to their continuing success in the market place. The gradual energy output loss over long periods of time is a major concern to all renewable energy stakeholders. A wide variety of degradation rates has been reported in literature with respect to technologies, age, manufacturers, and geographic locations [1].

PV modules are degraded by ambient temperature and humidity; moreover, these factors can accelerate the degradation. Moisture can diffuse into photovoltaic (PV) modules through their breathable back sheets or their ethylene vinyl acetate (EVA) sheets [2]. When in service in hot and humid climates, PV modules experience changes in the moisture content, the overall history of which is correlated with the degradation of the module performance [2]. If moisture begins to penetrate the polymer and reaches the solar

cell, it can weaken the interfacial adhesive bonds, resulting in delamination [3] and increased numbers of ingress paths, loss of passivation [4], and corrosion of solder joints [5,6,7]. Significant losses in PV module performance are caused by the corrosion of the cell, that is, the Si-Nx antireflection coating, or the corrosion of metallic materials, that is, solder bonds and Ag fingers [8, 9].

A thorough understanding of the forms of physical degradation experienced by solar modules and their relationship to various climatic conditions, is essential to all stakeholders i.e. private consumers, utility companies, integrators, investors, and researchers alike. Financially, degradation of a PV module or system is equally important, because a higher degradation rate translates directly into less power produced and, therefore, reduces future cash flows [10]. Technically, degradation mechanisms are important to understand because they may eventually lead to failure [11].

It is not well known how long modules last after their installation. Manufacturer warranties often guarantee 80% maximum power for 25 years; however relatively few PV modules have been in use for that long. As a result, there is a limited amount of data on PV module lifetimes. Different methodologies have been used to study degradation effects including both indoor and field measurements. Photovoltaic module manufacturers usually make efforts to eliminate the impact of short-term and long-term environment-induced degradation, but the difficulty in correlating indoor with outdoor testing at local conditions, poses a great challenge.

Degradation studies are more often than not based on various scientific studies that are typically grounded on general conditions. Manufacturers usually expose their products to accelerated tests which rarely depict the actual environment that the panel will eventually end up operating in after outdoor deployment.

As such, there exists a huge knowledge gap on the degradation forms and mechanisms of solar panels over medium to long term usage. A larger gap tending towards almost total exclusion of this knowledge, for the Kenyan environment, does exist. The present article presents an empirical study of PV panels, in their natural operating environments and having been in operation over an extended period, to investigate the degradation forms, their prevalence and their correlation between various climate conditions experienced in Kenya.

2. Procedures

Solar panels were sourced from different regions in the country, implying that each of the panels was exposed to a specific set of climatic conditions. The panels also varied in age. The oldest panel analysed had been in operation for 27years whilst the youngest had an operation lifetime age of 9 years as at the time of analysis.

Precipitation and temperature data information for the various areas from which the photovoltaic modules were sourced from was obtained and used for zoning sample source regions.

A Nikon D800, 36.3 megapixels FX-format HD-SLR was used for visual imaging of defects and degradation manifestations.

3. Results

A brief description of the seventeen PV-module samples used in the investigation is outlined in Table 1 below. Information on location, age and climate of the area is listed.

Table 1. Information on Solar Modules Samples.

Module I.D	Installation Location	Age (Years)	Temperature Zone	Rain Zone
S 1	Bunyore	27	Zone 2	zone 1
S 2	Tala	23	Zone 2	Zone 3
S3	Sultan Hamud	21	Zone 3	Zone 3
S 4	Voi	17	Zone 3	Zone 3
S 5	Homa Bay	17	Zone 2	Zone 1
S 6	Kisii	20	Zone 2	Zone 1
S 7	Kisii	22	Zone 2	Zone 1
S 8	Migori	15	zone 3	zone 1
S 9	Uasin Gishu	10	zone 1	Zone 1
S 10	Nanyuki	11	Zone 1	Zone 3
S 11	Nakuru	13	Zone 1	Zone 1
S 12	Nakuru	9	Zone 1	Zone 1
S 13	Nairobi	12	zone 2	Zone 3
S 14	Kapsabet	12	Zone 1	Zone 1
S 15	Kericho	21	Zone 2	Zone 1
S 16	Kericho	11	Zone 2	Zone 1
S 17	Mwingi	20	Zone 3	Zone 3

Table 2 provides the zoning criteria developed to facilitate grouping of the panels into common temperature and precipitation regions. Three zones were defined for temperature locations and two zones for precipitation.

Table 2. Climate Zoning of Samples Origin.

Temperature zone	Annual Temperature Range(°C)	Precipitation Zone	Annual Precipitation Range (mm)
Zone 1	≤17	Zone 1	>1000
Zone 2	<21, >17		
Zone 3	≥ 21	Zone 3	< 1000

Visual inspection was carried out on all of the PV module samples and each was found to exhibit several degradation effects as outlined in Table 3.

Table 3. Forms of degradation exhibited by solar module samples.

Module I.D	Degradation Forms
S1	Electrochemical Corrosion; Browning, Hot spots, Delamination, junction box damage, cell discoloration, cell interconnect degradation
S2	Browning, Cell discoloration, Cell fracture, Hot spots, Delamination, cell interconnect degradation
S3	Browning, Cell fracture, Hot spots, Delamination, cell interconnect degradation
S4	Browning, Cell discoloration, Hot spots, Cell fracture, cell interconnect degradation junction box damage, Hot spots, Delamination
S5	Electrochemical Corrosion; Cell discoloration Browning, hot spots , Delamination
S 6	Electrochemical Corrosion; Browning, Cell fracture, hot spots, Delamination
S7	Electrochemical Corrosion; Browning, Cell discoloration, Hot spot ,cell interconnect degradation, Delamination
S8	Cell fracture, cell interconnects degradation, junction box damage Hot spots, Delamination
S9	Hot spots, Browning, Delamination
S10	Hot spots, Browning, Delamination
S11	Front glass crack, cell interconnects degradation, junction box damage Hot spots, Delamination
S12	Browning and cell discoloration
S13	Browning
S14	Front glass crack, Browning, Delamination
S15	Cell fracture, cell interconnect degradation, Hot spots, Browning, Cell discoloration Delamination
S16	Browning, Cell discoloration
S17	Hot spots, Browning, Delamination, cell interconnect degradation

4. Discussion

Table 4 shows the magnitude of recurrence of each observed degradation form in the samples studied. Delamination of the encapsulate was seen to be the most predominant degradation effect, experienced by 82 % of the samples. Hot spots and encapsulant browning were also noted to be common effects, being observed in 76 % and 71 % of the modules respectively.

Physical damages e.g front glass fracture and deformed junction box were observed to occur very minimally.

Table 4. *Percentage prevalence of degradation forms in studied solar modules.*

Degradation Effect	Percentage prominence in samples
Browning	71
Cell discoloration	41
Electrochemical corrosion	24
Delamination	82
Hotspots	76
Cell fracture	35
Interconnects degradation	24
front glass fracture	6
deformed junction box	12

The primary degradation forms observed in the samples were those that affect the encapsulant i.e. delamination and encapsulant browning. Module delamination occurs as a result of the disintegration of bond between the encapsulant and other material layers that make up the PV module [12].

Figure 1. *Destruction of laminate by water ingress.*

Most of the delamination observed in the field has occurred at the interface between the encapsulant and the front surface of the solar cells in the module. A common observation has been that delamination is more frequent and more severe in hot and humid climates (Temperature zone 3 and Precipitation Zone 1).

The browning of EVA encapsulant used in PV modules with outdoor exposure was observed in several samples. Formulations of EVA that undergo yellowing/browning has been shown to produce acetic acid, with UV exposure which

corrodes solder bonds and electrical contacts [13]. This also corresponds to increased leakage current through the encapsulant [12].

Hotspots were also observed as a common degradation form in many of the sample modules investigated. No climatic relation can be conclusively associated with the hotspots observed in the samples investigated in this research. However, the severe hotspots, for example those shown in Figure 2 below, were observed on panels which had not been ideally located and allowed some form of partial shading to the panel during the day.

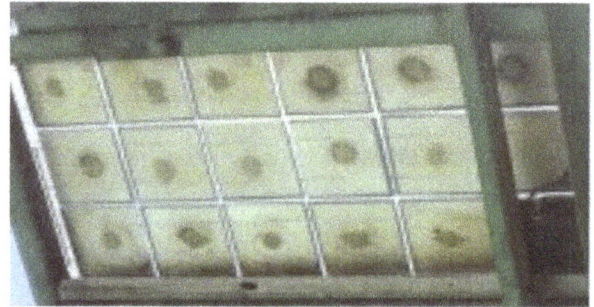

Figure 2. *Hotspots on PV Module.*

A number of samples were observed to have open-circuited inter-cell connections, broken solder joints and broken interconnects which may result from mechanical stresses. Mechanical stress on a module can sever the delicate ribbons and solder bonds. Reducing the number of available solder joints increases the series resistance and degrades performance which increases the possibility of hot spots and burn marks to form at solder joints, on the back sheet, and in the encapsulant [14].

Figure 3 shows junction box from one of the sample which had undergone warping of the cover, probably due to the high temperature conditions of the region. It was also installed directly on the corrugated iron sheets without allowing flow of air. As such, the junction box was exposed to elevated temperatures which consequently weakened the plastic cover

Figure 3. *Junction box defects.*

Glass breakage is an important degradation factor of PV modules. They occur in most of the cases during installation, maintenance, and especially during the transportation of modules to their installation sites [14].

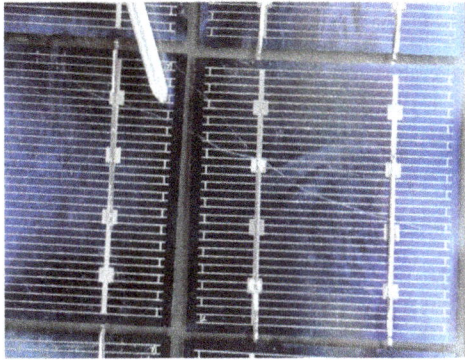

Figure 4. *Cracked front glass of a solar module.*

The broken modules or with cracks may keep functioning correctly. However, the risk of an electrical shock and of a moisture infiltration increases. Breakages and cracks are usually followed by other degradation types such as corrosion, discoloration and delamination [12].

Analysis was done to show the prevalence of each degradation form in each age cluster of the samples studied. Figure 5 (a-e), illustrates the results observed. Encapsulant browning was seen to be the most prevalent form of degradation in the 9-12 Years age clusters as illustrated in Figure 5(a). Figure 5(c) and 5(d) indicates that hot spots were observed to take place more commonly in the 17-20 and 21-24 Years age cluster. As only one panel existed in the 25-28 age clusters, the severeness of deformations in this group could not be determined. This panel exhibited all the forms of deformations observed in all the other panels.

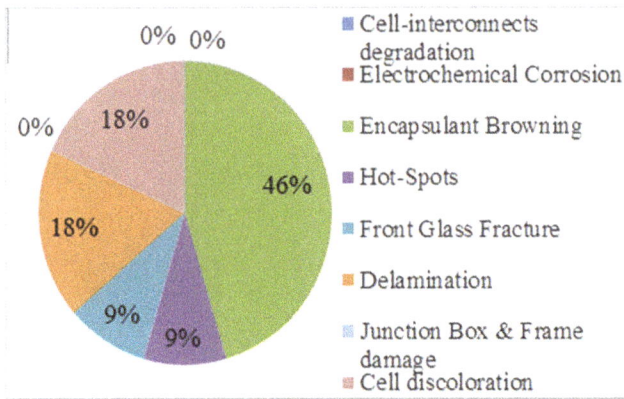

(c) 17 - 20 Years

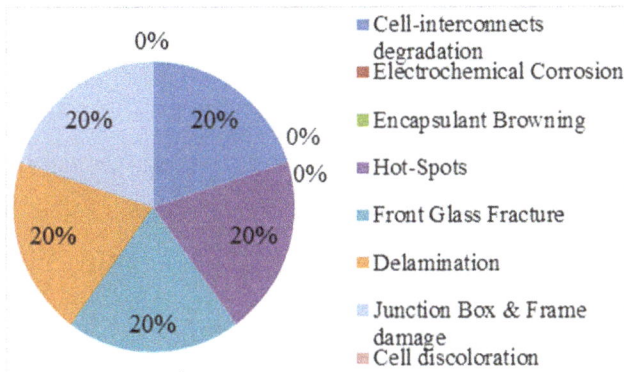

(d) 21 to 24 Years

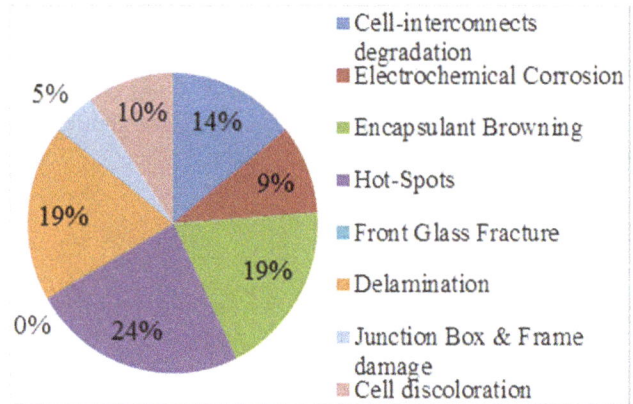

(e) 25 to 28 Years

Figure 5. *Correlation between Degradation prevalence and age of solar panels.*

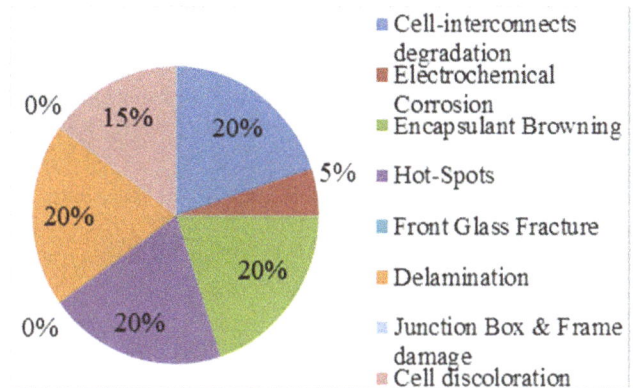

(a) 9 - 12 Years

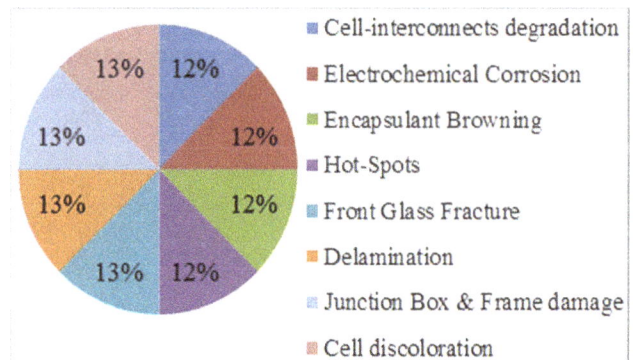

(b) 13 - 16 Years

5. Conclusion

The following conclusions were drawn after analysis of the results:

i). Main forms of degradation observed were defects on the encapsulant i.e. delamination and browning.

ii). Photovoltaic modules operating in regions of elevated temperatures experience more degradation forms.

iii). Photovoltaic modules operating in regions of higher annual mean precipitation exhibit more degradation forms as compared to those operating in areas of less humidity and moisture.

iv). A majority of solar systems which had been been

installed more than nine years ago were poorly sized for the load they are feeding. Wrong balance of system components were used in the installations and there is almost complete absence of any maintenance routines carried out by the users of these systems.

References

[1] D.C Jordan and S.R Kurtz, "Photovoltaic degradation rates an analytical review," Progress in Photovoltaics: Research and Application, 2011.

[2] M. D. Kempe, "Modeling of rates of moisture ingress into photovoltaic modules", Solar Energy Materials and Solar Cells, vol. 90, no. 16, p. 27202738, 2006.

[3] K. Morita, T. Inoue, H. Kato, I. Tsuda, and Y. Hishikawa, "Degradation factor analysis of crystalline-Si PV modules through long-term field exposure test", in Proceedings of the 3rd World Conference on Photovoltaic Energy Conversion, p. 19481951, 2003.

[4] E. E. van Dyk, J. B. Chamel, and A. R. Gxasheka, "Investigation of delamination in an edge-defined film-fed growth photovoltaic module," Solar Energy Materials and Solar Cells, vol. 88, no. 4, p. 403411, 2005.

[5] N. G. Dhere and N. R. Raravikar, "Adhesional shear strength and surface analysis of a PV module deployed in harsh coastal climate," Solar Energy Materials and Solar Cells, vol. 67, no. 1-4, pp. 363{367, 2001.

[6] X. Han, Y.Wang, L. Zhu, H. Xiang, and H. Zhang, "Mechanism study of the electrical per formance change of silicon concentrator solar cells immersed in de-ionized water," Energy Conversion and Management, vol. 53, no. 1, pp. 1-10, 2012.

[7] D. Polverini, M. Field, E. Dunlop, and W. Zaaiman, "polycrystalline silicon PV modules performance and degradation over 20 years," Progress in Photovoltaics: Research and Applications, vol. 21, no. 5, pp. 1004-1015, 2013.

[8] M. Kontges, V. Jung, and U. Eitner, "Requirements on metallization schemes on solar cells with focus on photovoltaic modules," in Proceedings of the 2nd Workshop on Metallization of Crystalline Silicon Solar Cells, 2010.

[9] C. Dechthummarong, B. Wiengmoon, D. C. and C. Jivacate, and K. Kirtikara, "Physical deterioration of encapsulation and electrical insulation properties of PV modules after longterm operation in thailand," Solar Energy Materials and Solar Cells, vol. 94, no. 9, pp. 1437{1440, 2010.

[10] W. Short, D. Packey, and T.Holt, "A manual for the economic evaluation of energy efficiency and renewable energy technologies," Report NREL/TP-462-5173, 1995.

[11] L. Meeker WQ, Statistical Methods for Reliability Data. Jo. John Wiley & Sons: New York, 1998.

[12] M.A Quintana, D.L King and T.J McMahon and C.R Osterwald, "Commonly observed degradation in field-aged photovoltaic modules", IEEE Photovoltaic Specialist Conference, 2002.

[13] M. D. Kempe, G. J. Jorgensen, K. M. Terwilliger, T. J. McMahon, C. E. Kennedy, and M. D. Kempe, "Acetic acid production and glass transition concerns with ethylene-vinyl acetate used in photovoltaic devices", Solar Energy Materials and Solar Cells, 2007.

[14] D.W Cunningham, P. Monus and J. Miller, "Long term reliability of PV modules",In: Proc. 20th European Photovoltaic Solar Energy Conference, 2005.

20

Design and Manufacture Three Solar Distillation Units and Measuring Their Productivity

Rusul Dawood Salim[1], Jassim Mahdi AL-Asadi[1], Aqeel Yousif Hashim[2]

[1]Physics Department, College of Education, University of Basrah, Basrah, Iraq
[2]Technical institute of Basrah, Southern Technical University, Basrah, Iraq

Email address:
Rosalh7975@Gmail.com (R. D. Salim)

Abstract: In this research, a comparison is made among three units of solar water distillation (M1, M2 and M3), where M1 is a passive solar still, M2 is a passive solar still coupled with flat plate collector and M3 is similar to M2 but it was provided with copper tube inside the still to realize the latent heat. Only the daily productivity factor was used to make a comparison among these units. The results shows that the productivity of M2 is more than it for M1 by 42%, While the productivity of M3 is more than it of M1 by 52%.

Keywords: Solar Collector, Solar Energy, Solar Distillation, Solar Water Heating, Solar Still

1. Introduction

Searching the energy sources become the main issue in the world, environmental pollution encoding world to search for clean energy sources, the sun is the most important source of make clean energy so that solar radiation can be transformed through systems and equipments which innovated by scholars and researchers it can be invested in different fields in the factories, domiciles and academic institutions, etc.

The need of clean, pure drinkable water in many countries. Often water sources are brackish and or contain harmful bacteria and therefore cannot be used for drinking. In addition, there are many coastal locations where seawater is abundant but potable water is not available. Pure water is also needful in some industries, hospitals and schools. Distillation is one of many processes that can be used for water purification. Solar radiation can be the source of heat energy. In this process, water is evaporated and thus separating water vapour from dissolved substances, and is then condensed as pure water [1].

The process for solar distillation in common with all other process of distillation as well as evaporation and condensation process, but it differ from other energy production processes in terms of not having to recurrence costs, and the major cost of this process can be included in the cost of solar collectors.

Solar still is characterized by having simple design, installation and maintenance as well as the ease of operation. Research has indicated that solar stills has an economic use to provide small amounts of water in the same time it can be expensive to produce water for an arid region.

The glass has the property of transmitting incident short-wave solar radiation which passes through the glass, the glass being a medium of transfer of heat, into the still to heat the brine. However, the re-radiated wavelengths from the heated water surface are infra-red and very little of it is transmitted back through the glass as it is shown in figure (1).

Water is an abundant natural resource that covers three quarters of the earth's surface. However, around 97% of the water in the world is in the ocean, only about 3% of all water sources are potable. Less than 1% fresh water is available within human reach and even this small fraction (ground water, lakes and rivers) is believed to be adequate to support life and vegetation on the earth and the rest is permanent snow cover, ice and permafrost in polar region. About 25% of the world does not have access to good quality and quantity of fresh water and more than 80 countries face severe water problem [3].

The rapid increasing need for energy and environmental concerns has focused much attention on renewable energy resources. Nowadays pollution in rivers and lakes by industrial effluents and sewage disposal has resulted in scarcity of fresh water in many big cities around the world [4].

The use of solar energy is more economical than the use of

fossil fuel in remote areas having low population densities, low rain fall and abundant available solar energy. Solar stills are highly reliable and they can easily provide us with the necessary daily amount of drinking water for the water scare

and drought areas. Also it is simple and has no moving parts and maintenance free. The problem of solar stills is the low productivity. Different techniques were used to enhance the output of the stills.[5].

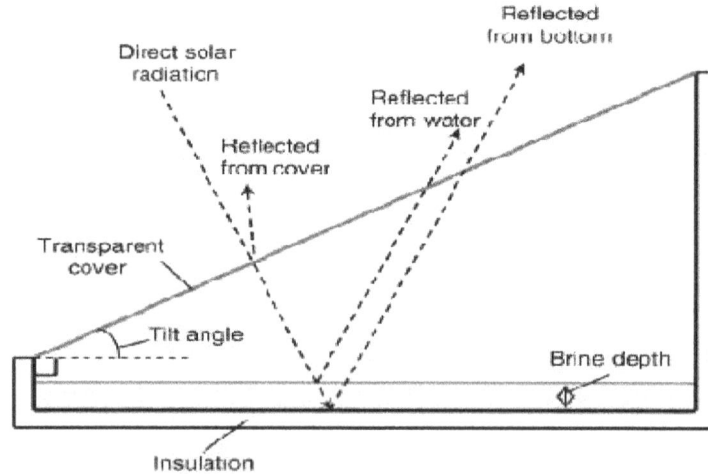

Fig (1). Schematic diagram of a basin-type solar still [2].

Solar energy desalination is generally the collecting of solar thermal energy that is used for desalination directly in solar still or indirectly (solar collector, photovoltaic (PV), ... etc.) [6].

Solar distillation is one of many processes that can be used to produce fresh water by using the heat of the sun directly in a simple equipment to purify water. The equipment, commonly called a solar still [7]. Solar stills are more attractive for smaller required output than the other techniques of desalination [8].

An experimental investigation on a solar still with an integrated flat plate collector was studied. Rajaseenivasan et al Experiments are carried out by varying the water depth in the basin and using the wick and energy storing materials in basins of both stills. The flat plat collector coupled with basins still gives about 60% higher distillate than the conventional still for the same basin condition. Economic analysis shows that the cost of distilled water for the FPCB still is lower than that for the conventional still [9, 10].

The general objective of this work is to select the optimum solar distillation unit in Basrah City which is located in south of Iraq (longitude 47°45' 06.45"E, latitude 30°33' 56.26"N), and several attempts have been made to use cheaper available materials, where these materials are assembled in the laboratory of Solar Energy in Physics Department – University of Basrah.

2. Experimental Setup

In this work, three asymmetric single basin solar stills with a same sizes were manufactured. The shallow basin made up of aluminum plate was placed inside the outer box. The outer box made by glass. These solar stills are similar in measurements and sizes but they are differ in the way of supplying their basins with brackish water. Also two flat plate

collectors were constructed. Three different units of solar distillation (M1,M2 and M3) were obtained by assembling the stills and solar collectors as shown below :

The first solar unit M1 (asymmetrical solar still); the first solar unit for distillation water in this work is an asymmetric slope basin type solar still. This still was constructed from glass where the base of the solar still has a thickness of (0.6) cm , length of (66) cm, and width of (60) cm. A hole of (0.8) cm diameter was drilled in the right side of base at a distance (4) cm from the edge for outlet the produced distill water through a copper tube. The walls and the covers of the stills are made of glass thickness of (0.4) cm and with dimensions. A small copper tubes diameter of (0.95) cm are fixed at the base and the back glass cover in order to outlet and inlet water to the solar stills. The front cover has a hole to insert thermometer to measure the temperature of the water in the basin. A basin was manufactured from aluminum plate thickness of (0.05) cm, length of (50) cm, width of (50) cm and high of (2) cm. This basin was installed inside the still at height of (0.4) cm from the still base to allow the condensing water to flow under it to the hole and then to the distilled water vessel. A matt black paint is used to paint the basin to increase the temperature of the water in the basin by absorbing all the incident solar radiation. This still is provided with brackish water directly from external tank. Idle water cooler was used as external tank with a capacity of (20) liter in order to control the level of water at the basin, where the depth of water at the basin is (1) cm, as shown in figure (2).

The second unit (M2) (M1 coupled with flat plate collector); this unit is a unit M1 coupled with flat plate collector (Z1). In this unit, the brackish water is coming from the external tank goes through flat plate collector (Z1) and then goes to the basin. The flat plate collector has been constructed from a zigzag copper tube (diameter of 0.9525 cm and length of 757cm) installed on copper plate (with length of 100 cm , width of 50 cm and thickness of 0.05 cm) inside insulated plywood box (thickness of 1cm, length of 112 cm , width of 62 cm and high of 15 cm) as shown in figure (3). Inclination angle of the solar collector was changed manually between 15 degrees in summer and 45 degrees in winter.

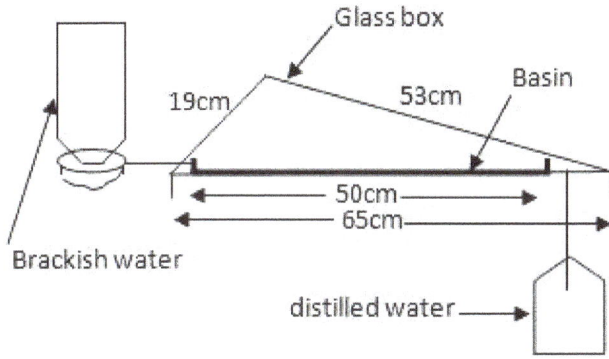

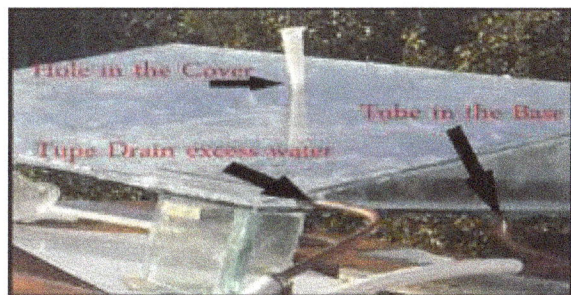

(a)

(b)

(c)

Fig. (2). *(a) Schematic diagram of an asymmetric basin type solar still and experimentally setup of M1. (b) A photograph for distillation unit M1. (c) Shows hole in the cover, drain excess water and tube in the base.*

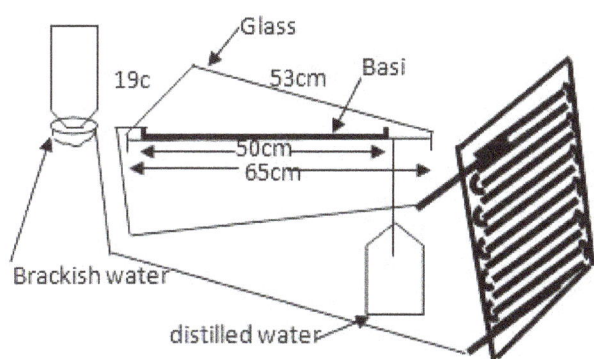

(a)

(b)

Fig. (3). *(a) Schematic diagram of an asymmetric basin type solar still and experimentally setup of M2. (b) A photograph for experimentally setup of solar distillation M2.*

The third unit (M3); This solar distillation unit is different from that for (M2) by placing an internal condenser from copper tube diameter of (0.79) cm and length of (285) cm on the inner glass walls of the still to enhance the condensing rate by releasing the latent heat of evaporation as shown in figure (4). In this type the water flow from external tank to the condensing tube and go through the flat plate collector and finally to the basin of the still, as shown in figure (4).

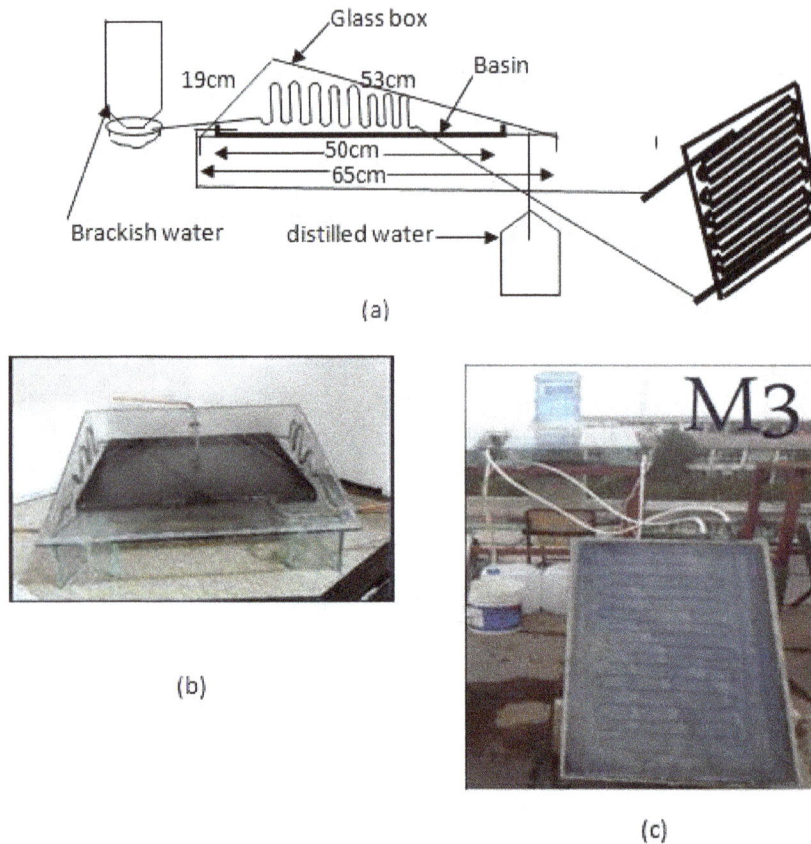

(a)

(b)

(c)

Fig. (4). *(a) Schematic diagram of an asymmetric basin type solar still and experimentally setup of M3. (b) A photograph shows the copper tube inside the still. (c) A photograph for the experimentally setup of M3.*

3. Results and Discussion

Figures (5) & (6) shows the monthly mean productivity for

solar distillation units in the period of measurements which begin from (December 2013) to (May 2014). It is clearly shown that the productivity of unit (M3) is the best one, the unit (M2) was in the second rank and M1 was at the last rank .

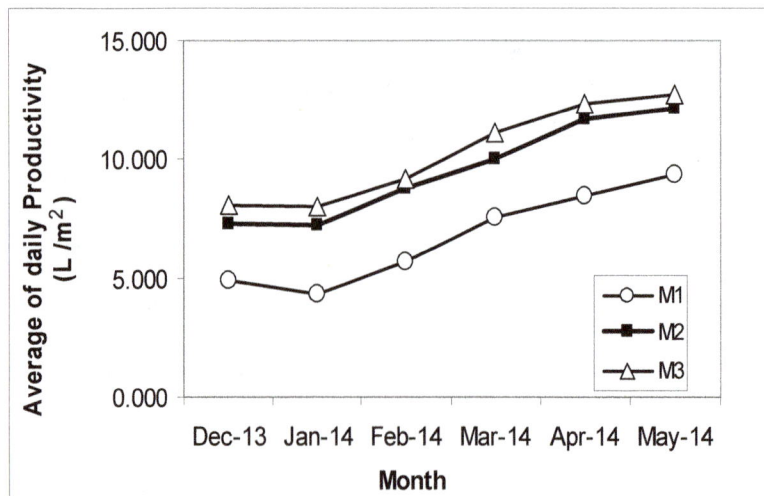

Fig. (5). *The average of daily productivity for the Month (December- 2013 to May-2014).*

From figures 5 and 6 , it is clear that the second unit M2 to be more productive than the first unit M1 due to the effect of

adding the flat plate solar collector to the solar still, which lead to rise the basin water temperature, so there is increasing in the evaporation, condensation and production rate with respect to the basin water temperature means the higher basin water temperature, higher rate of evaporation and higher the distillate output from the solar still. This high productivity is gained by the solar still coupled with flat plate collector. This is explained by fact that when temperature of water inside the tube will heated by the incoming solar radiation, hence the water temperature will be increased and such kind of preheated water will be used in solar still, so less heat is required to evaporate the water into steam and higher

distillate output is enhanced by solar still.

In the M3 unit in addition to what has been mentioned in M2, the copper tube which have been installed inside the solar still helps to realize the latent heat of vaporization and acquired by the water passing through it, which helps to increase the condensation process within the still and then the hot water goes to the solar collector and increases its temperature more then enters the basin, which became ready to evaporation process. The average daily productivity of unit M3 is more than M2 by 8% and by 52% more than M1.

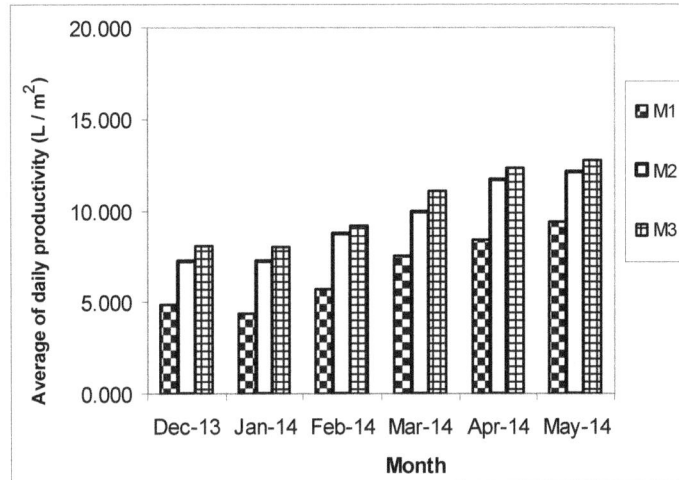

Fig. (6). Average daily Production for different months of the year for the three solar distillation (M1, M2 and M3)

References

[1]   Hitesh N. Panchal et al, A comparative analysis of single slope solar still coupled with flat plate collector and passive solar still, IJRRAS 7 (2) May 2011.

[2]   Khalifa, A.J.N., A.M. Hamood Performance Correlations for Basin-type Solar Stills. Desalination 249 (2009) 24-28

[3]   Omar O. Badran, Mazen M. Abu-khader, "Evaluating thermal performance of a single slope solar still" , Heat And Mass Transfer, No.43, pp. 985-995, 2007.

[4]   Al Hayek I and Bardan O O, "The effect of using different designs of solar stills on water distillation" , Desalination, No.150, pp.230-250, 2004.

[5]   M. Koilraj Gnanadason et al, "Comparison of Performance Analysis between Single Basin Solar Still made up of Copper and GI", International Journal of Innovative Research in Science, Engineering and Technology Vol. 2, Issue 7, July 2013.

[6]   Al-Kharabsheh S.& Goswami D. Y. Theoretical Analysis of a Water Desalination System Using Low Grade Solar Heat. Journal of Solar Energy Engineering MAY, Vol. 126, 2004.

[7]   Tiwari A.K. & Tiwari G.N. Thermal modeling based on solar fraction and experimental study of the annual and seasonal performance of a single slope passive solar still: The effect of water depths. Desalination 207,pp. 184–204, 2007.

[8]   Kalogirou S. A. Seawater desalination using renewable energy sources. Progress in Energy and Combustion Science 31,pp. 242–281,2005.

[9]   T. Rajaseenivasan, P. Nelson Raja, K. Srithar. An experimental investigation on a solar still with an integrated flat plate collector, Desalination 347 (2014) 131–137.

[10] Mohamed A. Eltawil , Z.M. Omara. Enhancing the solar still performance using solar photovoltaic, flat plate collector and hot air, Desalination 349 (2014) 1–9.

21

Production of Smokeless Briquette Fuel from Sub-bituminous Coal for Domestic and Industrial Uses

Izuchukwu Francis Okafor[*], Cosmas Ngozichukwu Anyanwu

National Centre for Energy Research and Development, University of Nigeria Nsukka, Enugu State, Nigeria

Email address:

izuchukwu.okafor@unn.edu.ng

Abstract: This study highlighted on coal production and reserves in Nigeria, chemical characteristics of air dried coal and utilization potential of the Nigerian coal in the domestic and industrial sectors. It discussed coal briquetting technology and processes involved in producing smokeless briquette fuel - coal drying, screening, crushing and carbonization to remove obnoxious volatile matters. The coal briquetting press developed at Energy Research Center University of Nigeria, Nsukka was used to produce three different briquette samples (B), (C) and (D) using a sub-bituminous coal obtained from Onyeama Mine in Enugu. The coal sample was pulverized, sieved and carbonized at the temperature of about 550°C, and mixed with 5%, 10% and 15% starch-binder concentrations respectively and compacted under pressure of 9 N/mm^2 and allowed to stay for five minutes before ejection. The briquette D possessed better handling quality than briquette B and C, but has lower calorific value, higher ignition time and lower burning rate. The moisture contents and volatiles of the briquettes tend to increase with increase in starch-binder concentrations, while the ash content decreased with an increase in starch-binder concentration. The calorific values and fixed carbon content of the produced briquettes decreased with an increase in the starch-binder concentrations and this could be due to decrease in the coal content of the briquettes, which has a higher calorific value and fixed carbon.

Keywords: Coal, Briquetting Processes, Coal Briquettes, Starch-Binder

1. Introduction

Like other parts of the world, coal was the oldest commercial fuel to be exploited by Nigeria [1]. Coal mining in Nigeria began in 1916. Average production in the first decade was >150,000 tons annually. This reached around 300,000 tons/year by the time World War II broke out. From 1940s to mid-1960s, production averaged > 600,000 tons a year, until the Nigerian civil war (1967–70) disrupted the mining activities [2]. Coal production in Nigeria reached the peak in 1950s [3]. In 1976/77, production began to decline rapidly, reaching as low as 53,500 tons annually in 1983. Figure 1 shows the coal production in Nigeria for the period of 1980 to 2012 [4]. The decline was attributed to the advent of the "oil boom" and the shift of attention from coal. The "dieselization" undertaken by the Nigerian Railway Corporation (NRC) (major customer of the coal industry), switching to natural gas and oil by several Power Stations; another major consumer of coal in Nigeria and the disruptive consequences of the civil war also led to the decline in coal consumption in Nigeria [2]. Table 1 shows the coal reserves and coal fields in Nigeria, indicating great potential and Table 2 shows the chemical characteristics of the air dried Nigerian sub-bituminous and lignite coals.

Despite the abundance of coal resources in Nigeria, it still remains the smallest contributor to the overall fuel energy mix in the country. In 2001, coal accounted for only about 0.2% of Nigeria energy consumption [5]. However, efforts are currently being made by Nigerian Government to increase the country's level of coal utilization to help reduce over dependence on crude oil and to stem the loss of forests to domestic fuelwood harvesting. With about 100 million m^3 of wood consumed annually, Nigeria's forests are under severe pressure from harvesting fuelwood for cooking [6]. The environmental consequences of deforestation– the erosion of watersheds, flooding, destruction of farmlands and desertification, are devastating. This study, therefore focused on the processes involved in producing smokeless coal briquettes suitable for use in brick factories, foundries, and laundry and bakery industries and as substitute for fuel wood in domestic cooking to reduce household dependence on firewood, kerosene and liquid petroleum fuels, which are not only very expensive, but also sometimes very scarce.

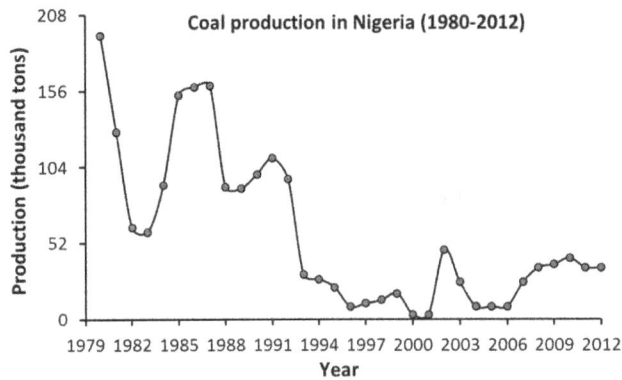

Figure 1. Coal production in Nigeria (International Energy Statistics, 2015).

Table 1. Coal reserves and production potential in Nigeria.

Coal Reserves in Nigeria		
1	Proven coal reserves	About 639 million metric tones
2	Inferred reserves	2.75 billion metric tonnes
Coal Production Potential in Nigeria		
1	Underground potential	200,000 to 600,000 tonnes per year
2	Surface production potential	400,000 to 800,000 tonnes per year (World Bank, 1983)
(c) Coal Fields in Nigeria		Location
1	22 Coal fields	Spread to over 13 States
2	Consist of approximately: 49% sub-bituminous coal 39% bituminous coal And 12%Lignite coal	

Source: Nigerian Coal Corporation (nd) [7].

Table 2. Chemical characteristics of air dried Nigerian coal.

Chemical Characteristic of Air-dried Nigerian Coal		
	Characteristic	Sub-bituminous Coal	Lignite
1	Caloric Value, MJ/kg	22.3 – 28.8	22.9 – 23.4
2	Moisture content, %	2.9 – 11.1	8.1 – 12.5
3	Ash content, %	3.9 – 26.0	7.8 – 10.8
4	Volatile matters, %	32.5 - 41.8	38.9 – 42.7
5	Fixed carbon, %	38.6 – 46.3	37.8 – 41.4

Source: Enibe (1998) [8].

2. Research Methodology and Materials Used for the Study

This study highlighted on the reserves, production and utilization potential of the Nigerian coal, processes involved in producing smokeless coal briquette fuel - coal drying, screening, crushing and carbonization to remove obnoxious volatile matters - the production sequence for coal briquetting. The sub-bituminous coal sample used for this study was obtained from Onyeama Mine in Enugu. The briquetting press developed at the Energy Research Center, University of Nigeria Nsukka, was employed for producing the sample briquettes used for evaluating the characteristics (bulk density, porosity index and handling durability etc.) of the briquettes produced in this study. The briquettes were dried using solar dryer developed at the National Centre for Energy Research and Development. The calorific values of the briquettes were determined using an oxygen bomb calorimeter model

XRY-1A. The MB35 Halogen Ohaus moisture analyser was used to determine the moisture contents of the briquettes. The performance of the briquettes (ignition time and burning rate) was carried out by conducting a simple water boiling test using a briquette stove also developed at the National Centre for Energy Research and Development.

3. Utilization Potential of the Nigerian Coal

Nigerian coal is environment-friendly due to its low sulphur and ash contents [8], has high calorific values of about 22.3 MJ/kg to 28.4 MJ/kg [9] and therefore has wider areas of applications. The potential areas where coal is used as source of fuel include cement production, brick factories, foundries, and laundry and bakery industries. Other areas of potential uses are as the important raw material base in tyre and battery manufacture, substitute for fuel wood in domestic cooking. In metallurgical industry, Nigerian coal is suitable in producing coke of acceptable strength and character for the steel plants. Up to 200,000 metric tons of Nigerian coal will be required annually when Ajaokuta Steel Plant goes into full production [7]. Okpara and Onyeama Coal in Enugu State have the potentials of providing sources of future coal supply to the steel sector. In power generation, Nigerian coals are suitable as energy fuel for power generation for the abandoned Oij Power Station in Enugu State and other proposed power stations at Kogi, Benue, Anambra and Delta States. With the current deregulation of power generation in the country which was facilitated by the lack of sufficient power generating capacity by PHCN is a fertile ground for the Independent Power Producers. Also coal and its derivatives, smokeless coal briquettes, have been demonstrated as the cheapest, safest and therefore most suitable substitute to fuel wood. About 2 million metric tonnes of coal/briquettes equivalent are required annually to substitute fuel wood utilization. More than 20 carbonization/briquetting plant of 100,000 tons per annual capacity is required to be sited within Okaba, Ogboyoga, Oba/Nnewi, Gombe/Bauchi and Azagba/Ogwashi coal/lignite fields in Nigeria [7]. These plants are yet to be accomplished.

4. Coal Briquetting Processes

Coal is often described as a ''dirty'' fuel emitting smoke that is offensive and containing constituents that are harmful to human health. But with suitable briquetting processes and carbonizing when necessary, coal can be put into a relatively clean, compact, and stable form for uses in the domestic, commercial and industrial sectors. Coal briquetting has a long history. In the late 1800s relatively worthless fine coal or slack was compressed to form a "patent fuel" or briquette [10]. Coal briquetting has been researched worldwide in places like Germany, USSR, Korea, India, and the United States [11]. Reasons for the extensive research are as follows: (i) All coals are not alike, and often research has been aimed at developing an" improved briquetting process for a particular coal. Briquetting can be done with or without an additive (binder) to help in agglomerating and giving cohesive strength to the briquette. (ii) To obtain suitable binders, as well as processes by which briquetting can be performed with or without a binder. (iii) To improve the properties of the briquettes, such as maintaining ignitability while keeping volatile matter low as well as reducing smoke and sulfur emissions upon burning [10]. In coal briquetting, a set of parameters such as temperature, pressure, pressing time, pretreatment, etc can be varied to produce unique briquetting processes. Also, it is very essential that the Run-Off-Mine (ROM) coal is subjected to cleaning and preparation processes [12], which include sorting out and screening of different sizes of coal pieces, removal of impurities such as sulphur, dirt, and mud etc, before dispatch for briquetting. The processes involved in producing "smokeless" fuel briquettes are closely tied to the nature of the feed coal [13] and such processes are discussed as follows:

(i) Coal drying: Methods used in coal drying are direct drying (a flash dryer using hot gas) and indirect drying (a disc dryer using steam heat). This stage is very crucial as moisture content has critical impact on the strength of briquettes. Excess moisture physically militates again development of the maximum contact area that would otherwise be established at a particular pressure. In addition to drying the coal to the appropriate moisture content, it is essential to ensure that moisture is uniformly distributed; and in commercial operations, coal leaving the dryer should be very slowly cooled before dispatch to the briquette press.

(ii) Crushing: This can be done before or after screening. It is generally essential to crush the coal to <4mm, with 60-65% < 1mm to facilitate close compaction of coal particles during compression [14]. Crushing coal to this particle size results in a stronger briquette and better ignition properties. The equipment required for crushing include rotary crusher, double roller crusher and hammer mills etc.

(iii) Binders: Coal briquetting can be performed with or without binder. Briquetting without prior addition of binders are required to ensure that the briquette has adequate strength to withstand normal handling. Briquetting is usually conducted at temperature (35-65°C) at which advantage can be taken of the enhanced plasticity of small particles of and as bulk of binderless lignite briquettes is used for residential heating, the preferred press is Exter extrusion press [14]. In the modern version of this equipment, coal is compacted under approximately 1400 kg/cm^2 by reciprocating ram that pushes it into, and through, a ventrui type channel by repeated forward strokes and simultaneously admits additional coal during each return stroke. Although the high pressures required for satisfactory compaction make production of binderless fuel briquettes from more mature coals uneconomical, smaller quantities are produced for other purposes such as preparation of shaped activated carbon. This operation extends to lignite briquetting by using much more finely comminuted coal – commonly 100% <75μm and pressure on the order of 2100 – 2800kg/cm^2, requires special mold designs and fairly complex mold-charging cycles [14].

Briquetting coal with a binder, the production sequence can be represented in Figure 2 as follows:

Figure 2. *Production sequence for coal briquetting.*

The broken lines indicate alternative sequences. Also close attention must be given to the quality of the binder and the manner of introducing into the coal and the overall process is more flexible than briquetting without binder. In most cases the coal is dried to <4% moisture, which experience has shown to minimize binder requirements, and crushed to <1mm (with 75 – 80% < 0.5mm) [14]. After addition of a binder, the mixture is homogenized by tempering it in a pug where it is heated to 95 – 100°Cby injection of superheated steam and agitated by rotating paddles. And before dispatch to the presses, it is cooled to a few degrees above the softening point of the binder in order to endow it with appropriate plasticity for compaction. The binder is usually a pitch (crushed to 65 -95% < 0.5mm) or a bitumen, and is either introduced into coarse coal, ie between coal drying and grinding or added after grinding stage. Alternatively, it is melted and sprayed into the coal. Bitumen is always applied in liquid form. The binder proportion depends upon the nature of the coal and ranges from 5 to 8 wt. % in the case of pitch or from 8 to 10 wt. % if bitumen is used. Other binder types include coke oven pitch, petroleum asphalt, and ammonium lignosulphorate, starch and molasses and addition rate is 5 to 15% by weight [10].

(iv) Devolatilizing: This is only applicable to low- rank high-volatile coals. The equipment used is a retort or a beehive type coke oven. This involves driving off excess gas or volatile matter through carbonization process prior to briquetting in order to produce a "smokeless" fuel.

(v) Sulphur removal from coal: The most objectionable impurities in coal is sulphur. Organically bound sulphur in coal constitutes about 70% of sulphur in coal and this cannot be removed by physical sorting, rather by chemical cleaning processes [12]. Several steps can be taken to remove sulfur to make the final smokeless coal lump or briquette clean from sulphur oxide emissions. Washing coal with water prior to pyrolysis is a simple, inexpensive, and effective step. High sulphur coals are washed and chemically treated for sulphur removal. Water soluble sulphate and pyritic are removed by washing the coal. The sulphur content of coal can also be reduced by carbonization processes by converting it to hydrogen sulphide gas.

(vi) Coal Carbonization: Burning raw or untreated briquetted coal produces emissions that can have serious health effects. It releases volatile organic matter, sulphur compounds, nitrogen oxides, particulates, and trace elements. Each of these has potential adverse effects on human health. Carbonizing coal to produce a soft smokeless coke can significantly reduce the adverse health effects of burning raw coal or untreated coal briquettes. Carbonization describes the

process for producing fuels – solid, liquid and gaseous – from coal and other carbonaceous materials such as biomass and organic wastes, by heating them in an oxygen-deficient atmosphere [15]. During the heating the volatile content of coal is reduced, and gases and tars containing some other harmful constituents are given off. The result is a product high in fixed carbon, which can burn with little or no smoke and emissions not worse than the present practice of burning coal and firewood. Carbonization can be performed before or after briquetting – or even without briquetting. One of the important aspects of coal carbonization is that many of the potentially harmful constituents of coal are removed. A certain level of volatiles is left in the soft coke (appx. 1% to 20%) to promote its easy ignition, but this level is deemed safe and does not allow the fuel to smoke significantly [14]. Particulates are also considerably reduced by the process of carbonization. Carbonization is only needed for coals below the rank of anthracite, that is, bituminous, sub-bituminous, and lignite coals. Anthracite is sufficiently low in volatile and ash contents and high in fixed carbon that carbonization is unnecessary for a, smokeless fuel.

Carbonization processes can be classified into three based on temperature range: (i) Low Temperature Carbonization (LTC) (450 – 700°C) is used on small scale to produce smokeless solid fuel, and by-product tars as essential feedstocks to chemical industry or refined to synthetic motor gasoline, heating oil, and lubricants [16]. Preferred coals – As a rule, are lignite or sub-bituminous and high-volatile bituminous coals, when pyrolyzed at temperatures between 600 and 800°C furnish porous chars or so-called semicokes [14]. LTC Equipment: Vertical and Horizontal retorts – with process heats supplied directly by circulating hot combustion gas through the charge or indirectly by passing it through flues between contiguous retorts – were used for batch as well as continuous processing. (ii) Medium Temperature Carbonization (MTC): If LT carbonization is extended into the 700 – 900°C range, it is termed medium temperature processes. This reflects the pronounced physical changes that coal undergoes between 600 and 800°C. (iii) High Temperature Carbonization (HTC): HT carbonization is normally carried out at or above 900°C. If coal is carbonized at temperatures above 650- 700°C, LT chars will become less reactive through devolatilization and loss of porosity [14]. At this point, the chars tend to lose the properties that make them useful domestic and industrial boiler fuels (difficult to ignite). In practice, HTC is therefore limited to the production of metallurgical cokes, which acquire much of their industrial utility through processing at high temperatures.

5. Coal Briquetting Press

Different types of coal briquetting presses are already in existence ranging from simple to complex ones and they include roll types and hydraulic piston press, pellet press, screw-press and extruder types etc. These machines are either very expensive or not readily available for local production of briquettes in Nigeria. The mechanical construction of the

machines determines such important characteristics as reliability, ease of maintenance and cost of operation. Depending on the configuration of the mould installed in the press, briquettes are produced in different shapes such as: (i) pillow shape (most common), (ii) circular, polygonal, or irregular in cross-section; (iii) they may be solid or concentrically perforated to form a hollow cylinder or polygon.

Figures 3 (a) and (b) show a briquetting press designed and fabricated at the Energy Research Center, University of Nigeria Nsukka, indicating different parts of the press. The press was designed to be manually operated to circumvent power supply issues, for it to be used in a remote location and to minimize cost for the local users. However, it can still be automated to increase production rate. It uses two vertically mounted 200kN capacity hydraulic presses to generate its maximum compaction pressure of 160N/m^2 and has twelve cavities of 70 mm x 100 mm cross-sectional area, where the coal-binder mixture is to be loaded for compaction. The high pressure ram of the press is used in compressing the briquettes and the ejector ram is for ejecting the compacted briquettes.

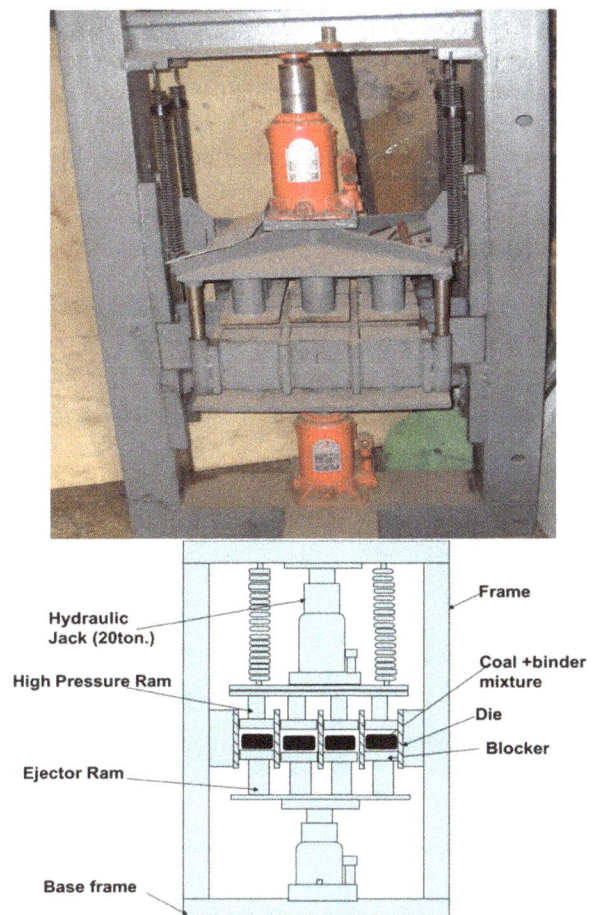

Figures 3. (a) Pictorial view of the briquetting press; (b) Sectional view of the briquetting press.

The press is designed to handle maximum load up to 7 kg of coal-binder or coal-biomass-binder mixtures in a single batch operation and can produce more than 2.8 tonnes of briquettes per day. The frame structure of the press was designed with metal steel to provide the required rigidity to withstand high pressure, ensure durability and efficient performance.

6. Performance Evaluation of the Samples Briquettes Produced by the Press

A sub-bituminous coal sample obtained from the Okpara mine in Enugu was used in producing the briquettes. The coal sample struck into small chunks was washed in a container filled with water to remove impurities and then sun-dried for three days to reduce moisture content. The washed sample was pulverized with a grinding machine and sieved to particle sizes of not more than 1mm and then carbonized at the temperature of about $550^{\circ}C$. The pulverized coal-binder (cassava starch) mixtures were compacted at constant pressure of 9 N/mm^2 and allowed to stay for 5 minutes (dwell time) before ejecting the compacted briquettes. The starch-binder is normally used in briquetting industries. The briquettes produced with starch-binder ignite easily and they burn with less ash deposits compared to that produced with asphalt as binder. The three briquettes samples B, C and D shown in Figure 4 were produced with the briquetting press in Figure 3. The briquettes were dried using solar dryer developed at the National Centre for Energy Research and Development. The calorific values of the briquettes were determined using an oxygen bomb calorimeter model XRY-1A. The MB35 Halogen Ohaus moisture analyser was used to determine the moisture contents of the briquettes.

Figure 4. *Sample B briquettes Sample C briquettes Sample D briquettes*

(a)   (b)

Figure 5. *(a) Burning briquettes in the stove and (b) briquette stove used for the water boiling test.*

The bulk density of the briquettes was determined from ratio of the mass and volume of each sample of the produced briquettes. The porosities of the briquettes were determined from the ratio of the mass of water absorbed by the sample briquettes when immersed in water for 25 minutes to the mass of each sample of the briquettes immersed in the water [17]. The handling durability (shattering index) of the briquettes was determined from ratio of the weight of briquettes still remaining in the sieve after dropping each sample from a height of 2 m on a concrete floor to the weight of the briquettes before the dropping [18]. Also, the ignition time for the sample briquettes was determined as the time it took each sample briquettes to catch fire when it was lit at its edge and this was repeated three times for each sample and the average time was taken as the ignition time. Figure 5 (a) shows the burning briquettes.

The performance of the briquettes was also carried out by conducting a simple water boiling test using a briquette stove shown in Figure 5 (b) also developed at the National Centre for Energy Research and Development. In this test, the stove was loaded with four pieces of sample B briquette used for heating 2.5 liters of water to boiling and 3 mL of kerosene were used to ignite the briquettes. The weight of the burnt briquette was obtained by calculating the difference in the initial weight of the stove loaded with the briquettes and the final weight when the water started boiling and the pot with the boiling water removed from the stove. The burning rate was determined. The processes were repeated for sample C and D briquettes.

7. Results and Discussion

The produced briquette samples B, C and D weighed average 198.5 g, 159.67 g and 149.45 g respectively after five days drying using a passive solar dryer. The briquettes exhibited very insignificant linear expansions while ejecting them from the mould. Table 3 shows the proximate analysis for the pulverized raw coal (sample A) and the three sample briquettes produced with the press. It indicates that the calorific values, ash contents and fixed carbons of the briquettes tend to decrease with an increase in starch-binder concentrations, while the moisture contents and volatiles increased with an increase in the starch-binder concentrations. The higher calorific value for sample B briquette indicates that they could release more heat during combustion than the briquettes C and D. The lower ash content of sample D indicates that it contains less noncombustible materials than briquettes B and C. The higher ash content of sample B could retard the burning rate of the briquette.

Table 3. *Characteristics of the pulverized raw coal (sample A) and produced briquettes.*

Briquette Sample	Calorific Value(MJ/kg)	%Moisture Content	% Volatiles	%Ash Content	% Fixed Carbon
A	27.46	7.20	36.14	18.53	38.13
B	26.36	8.11	37.83	16.95	37.11
C	25.23	9.52	39.12	15.03	36.33
D	24.82	11.23	40.84	13.37	34.56
A-Pulverized Raw Coal (100 %), B – Coal (95 %) + Binder (5%), C- Coal (90 %) +Binder (10%) and D-Coal (85 %) +Binder (15%)					

Table 4. Performance analysis of the produced sample briquettes.

	B – Coal (95%) + Binder (5%)	C- Coal (90%) + Binder (10%)	D-Coal (85%) + Binder (15%)
Bulk density (g/cm³)	1.03	1.01	0.91
Porosity index	0.24	0.19	0.14
Handling durability	0.061	0.052	0.043
Ignition time (min)	1.31	1.56	2.18
Burning rate (g/min)	2.65	2.30	1.91
Smoke	Burnt with little white smoke on starting the fire and vanished when fully ignited.	Burnt with little white smoke on starting the fire and vanished when fully ignited.	Burnt with little white smoke on starting the fire and vanished when fully ignited.
Odour	Burnt with no odour, except on starting the fire.	Burnt with no odour, except on starting the fire.	Burnt with no odour, except on starting the fire.

Table 4 indicates that the bulk density and porosity index of the briquettes decreased with an increase in the binder concentration. The ignition time increased with an increase in the binder concentration and this could be due to decrease in the porosity of the briquette. The lower porosity index exhibited by sample D could reduce the heat and mass transfer rate during combustion due to a decrease in voids in the briquette matrix as a result of the increase in the binder concentration. The briquette D also exhibited best handling quality due to its lower shattering index when dropped from a height of 2 m on a cemented concrete floor due to its higher binder concentration. The briquette B has a higher burning rate due to its higher volatile content and porosity index and lower moisture content.

8. Conclusion

Coal was the oldest commercial fuel to be exploited by Nigeria, but later neglected due oil boom and other factors. Through coal briquetting, efforts are being made to increase the country's level of coal utilization to help reduce over dependence on crude oil and to stem the loss of forests to domestic fuelwood harvesting. Coal briquetting involves processing and compacting pulverized coal mixed with or without binder in briquetting press to produce solid fuels. Carbonizing coal to produce smokeless briquettes fuel reduces the volatile matters that adversely affect health from burning raw coal or untreated coal briquettes. The calorific values and fixed carbon content of the produced briquettes decreased with an increase in the starch-binder concentrations. The moisture and volatile contents of the briquettes tend to increase with an increase in the starch-binder concentration, but the ash content decreased with an increase in starch-binder concentration. The handling quality of the briquettes increased with an increase in the starch-binder concentration, but this lowers the burning rate and increased the ignition time due to decrease in the calorific values and porosity index of the briquettes.

References

[1]   Nigeria-Coal and Lignite-OnlineNigeria.com, (2006). www.onlinenigeria.com/aboutus.asp.

[2]   Oyeyinka, Management of Technological Change in Africa: The Coal Industry in Nigeria International Development Research Centre, pp. 3-29, (2004).

[3]   Dayo, F. B., Clean Energy Investment in Nigeria: The domestic context. Published by the International Institute for Sustainable Development, 161 Portage Avenue East, 6th Floor Winnipeg, Manitoba Canada R3B 0Y4, pp. 24 – 26, 2008.

[4]   Pallavi, S., Nigeria Coal Production 1980-2012. International Energy Statistics, 2015.

[5]   EIA, Country Analysis Brief – Nigeria, U. S. Energy Information Administration, 2005. www.eia.gov/emeu/ecbs/nigeria.html.

[6]   ICEED, Improved woodstoves Workshop and Exhibition. International Centre for Energy, Environment and Development, 2007. http://www.iceednigeria.org/project.

[7]   Nigerian Coal Corporation (nd). Coal Reserves and Production Potential in Nigeria. Nigerian Coal Corporation, Okpara Avenue, Enugu, Nigeria.

[8]   Local Sourcing of Raw Materials: Solid Mineral Deposits in Nigeria Profile (1999): Nigeria Investment Commission, Maitama District Abuja Nigeria. www.nipc-nigeria.org/solidmin.htm

[9]   Enibe, S. O. Power Plant Engineering Lecture Notes, 1998. Department of Mechanical Engineering, University of Nigeria, Nsukka. Unpublished.

[10]   Walters, A. D. (nd). Coal Preparation. www.ilo.org/safework_bookshelf/english? content&nd.

[11]   Komarek, K.R. The Briquetting Process and Compacting Process. Inc. 548 Clayton Ct. Wood Dale, IL, 60191, USA. 2009. http://www.komarek.com/briquette-process.html

[12]   Rao, S. and Parulekar, B., Energy Technology Nonconventional, Renewable and Conventional. Khanna Publishers, 2-B Nath Market, Nai Sarak, Delhi – 110006, pp.742- 744. 2004.

[13]   Krug and Nauendorf "Briquetting of brown coal Volume 1, section on Drying, VEB Deutscher Verlag fur Grundstoffindustrie, Leipzig, 1984, 1st Edition.

[14]   Barkowitz, N., An Introduction to Coal Technology. Second edition. Academic Press INC. 525 B Street, Suite 1900, San Diego, California 92101 – 4495, pp. 205-209, 1994.

[15]   Skodras, G. and Amarantos, P., Overview of Low Temperature Carbonisation. Center for Research and Technology Hellas Institute for Solid Fuels Technology and Applications, 2004.

[16]   King, J., GLow-Temperature Carbonisation. 1st World Petroleum Congress, London, UK, 2003.

[17] Ikelle I. I. and Mbam. J. The Study of Briquettes Produced With Bitumen, CaSO4 and Starch as Binders. American Journal of Engineering Research (AJER) Volume-03, Issue-06, pp-221-226, 2014.

[18] Davies R. M. and Abolude D. S., Mechanical Handling Characteristics of Briquettes Produced from Water Hyacinth and Plantain Peel as Binder Journal of Scientific Research & Reports 2(1): 93-102, 2013.

22

Behaviour of industrial machinery foundation on pre stressed geogrid-reinforced embankment over soft soil under static load

Masih Allahbakhshi[1, *], Habib Sadeghi[2]

[1]Department of Civil Engineering, Mazandaran University of technology, Babol, Iran
[2]Department of Chemical Engineering, Isfahan University, Isfahan, Iran

Email address:

masih2768@gmail.com (M. Allahbakhshi)

Abstract: Results of parametric study to investigate the applicability of finite element method for analyzing industrial machinery foundation on pre stressed-reinforced embankment over soft soil are investigated in this paper. Model tests were carried out using model footing of 1 m in diameter and geogrids. Particular emphasis is paid on the reinforcement configurations including number of layers, spacing, layer length and depth to ground surface on the behavior of industrial machinery foundation on reinforced silty sand embankment on peat and soft clay under static load is determined. A series of finite element analyses were performed on a slope using two-dimensional plane strain model using the computer code Plaxis. Soil was represented by non-linear hardening soil model, which is an elasto-plastic hyperbolic stress-strain model while reinforcement was represented by elastic elements. Test results indicate that the inclusion of geogrid layers in sand not only significantly improves the footing performance but also leads to great reduction in the depth of reinforced sand layer required to achieve the allowable settlement. However, the efficiency of the sand–geogrid system increases with increasing number of geogrid layers and layer length. Based on the theoretical results. In this paper we can see the effect of pre stressed geotextile is more than that unreinforced and reinforced (without pre stress) embankment.

Keywords: Bearing Capacity, Industrial Machinery Foundation, Pre stressed - Reinforced Embankment, Soft Soil, Finite Element Analyses

1. Introduction

Soil can resist pressure and shear forces very well, but it is not able to tolerate tensile forces. Reinforced soil is composite material that contains components that can easily stand tensile forces. Nowadays reinforcing materials is widely used to overcome technical problems. Reinforced soil is used in stabilizing embankment (slope), fill dams, retaining walls, foundation and in-situ slope for increasing the shear resistance of soil layer in different earth structures. The subject of reinforcing soil beneath footings has gained considerable attention in the past few years (e.g. Dash et al., 2003; Boushehrian and Hataf, 2003; Ghosh et al., 2005; Bera et al., 2005; Patra et al., 2005, 2006).This paper is interested in the many situations where footings are constructed on/or adjacent to soft clay sloping surfaces under static load such as industrial machinery footings on sloping embankments. In

this case, two major problems arise; the low bearing capacity of soft clay and the potential failure of the slope itself. Therefore, over some years, the subject of stabilizing earth slope has become one of the most interesting areas for scientific research and several techniques have been suggested to improve the stability of earth slope and hence improve the bearing capacity. Typical examples include modifying the slope surface geometry, chemical grouting, using soil reinforcement, or installing continuous or discrete retaining structures such as walls or piles. Geosynthetics recognized as synthetic materials are used in soil. The specific families of Geosynthetics are the following: Geotextiles, Geogrids, Geomembranes and Geocomposites. When synthetic fibers are made into a flexible, porous fabric by standard weaving machinery or are matted together in woven and nonwoven manner, the product known as "Geotextile".

Geogrids are plastics formed into a very open netlike configuration. Geotextiles and Geogrids are used usually as reinforcing material for soil improvement. These reinforcing materials are not susceptible to corrosion, have relatively low stiffness and flexible enough to tolerate large deformation. These factors make them to be superior to steel reinforcing materials in soils. As use of geotextile in reinforcing embankment is growing. Several case studies described the successful use of geogrids to reinforce a weak subgrade such as variable soft clay (Tsukada et al., 1993; Khing et al. (1993); British Rail Research, 1998; Omar et al. (1993); Dashet al., 2003; Yetimoglu et al., 1994; Sitharam and Sireesh (2004).

Tsukada et al. (1993) investigated the use of geogrids for roadway foundation and reported that settlement response and pressure distributions were directly related to the thickness and configuration of the geogrid-reinforced foundation. Khing et al. (1993) conducted model tests on a strip footing supported by a sand layer reinforced with layers of geogrid. The test results show that the maximum benefit of geogrid reinforcement in increasing the bearing capacity was obtained when the ratio of the depth of the first reinforcing layer to the foundation width was less than unity. British Rail Research (1998) has demonstrated that geogrid inserted in the ballast where tracks lie over soft ground can help extend maintenance intervals. Omar et al. (1993) presented the results of the laboratory model tests for strip and square foundations supported by sand reinforced with geogrid layers. The test results demonstrate that for the development of maximum bearing capacity, the depth of reinforcement is about 2B for strip foundation and 1.4B for square foundation, where B is the width of the footing. The maximum depth of placement of the first layer of geogrid should be less than about B to take advantage of reinforcement. Dash et al. (2003) performed model tests in the laboratory to study the response of reinforcing granular fill overlying soft clay beds and showed that substantial improvements in the load carrying capacity and reduction in surface heaving of the foundation bed were obtained. Yetimoglu et al. (1994) conducted laboratory model tests to investigate the bearing capacity of rectangular footings on geogrid-reinforced sand. For a single layer of geogrid reinforcement, the optimal placement depth was 0.3 times the footing width. Sitharam and Sireesh (2004) conducted laboratory model tests to determine the bearing capacity of an embedded circular footing supported by sand bed reinforced with multiple layers of geogrids. The test results demonstrate that the ultimate bearing pressure increases with embedment depth ratio of the foundation.

2. Prototype Study

2.1. Finite Element Analysis

A series of two-dimensional finite element analyses (FEA) on a prototype footing-slope system was performed in order to understand the deformations trends within the soil mass. The analysis was performed using the finite element program Plaxis software package (professional version 8, Bringkgreve

and Vermeer, 1998). Plaxis is capable of handling a wide range of geotechnical problems such as deep excavations, tunnels, and earth structures such as retaining walls and slopes. The software allows the automatic generation of six or fifteen node triangle plane strain elements for the soil, and three or five node beam elements for the footing while three or five node elastic elements were used for the geotextile elements. Initial step for analyzing the models to create the geometry of the model. The geometry characteristics such as embankment height, slope and crest width. The other geometry which should be defined is under laying soil profile such as thickness of the soft layer. The second step is to provide the material properties of the embankment and the under laying soil. For present investigation the main model with 4m height,8m crest width,1:3(V:H) slope and is placed on a peat layer of 3m thickness and soft clayey layer of 3m thickness and the vibrating source is an industrial machinery founded on a 0.2 m thick concrete footing of 1 m in diameter. In addition to the weight of the footing, the weight of the industrial machinery is assumed 5 kN/m², modelled as a uniformly distributed load and special boundary conditions have to be defined to account for the fact that in reality the soil is a semi-infinite medium.

2.2. Finite Element Modeling

The non-linear behavior of sand was modeled using hardening soil model, which is an elasto-plastic hyperbolic stress–strain model, formulated in the framework of friction hardening plasticity. The foundation was treated as elastic beam elements based on Mindlin's beam theory with significant flexural rigidity (EI) and normal stiffness (EA). A basic feature of the hyperbolic model is the stress dependency of soil stiffness. The interaction between the geogrid and soil is modeled at both sides by means of interface elements, which allow for the specification of a reduced wall friction compared to the friction of the soil. The limiting state of stress are described by means of the secant Young's modulus (E_{50}^{ref}), tangent stiffness modulus for primary compression (E_{oed}^{ref}), Poisson's ratio (ν), effective cohesion (c), angle of internal friction (Φ), angle of dilatancy (ψ), failure ratio (R_f) and interface reduction factor (R_{int}). The modeled boundary conditions were assumed such that the vertical boundaries are free vertically and constrained horizontally while the bottom horizontal boundary is fully fixed. The software allows the automatic generation of six node triangle plane strain elements for the soil, and three node beam elements for the footing and the geogrid. The number of element used in reinforced tests are 250 element while in unreinforced tests the number is 160. The analyzed model slope geometry, generated mesh, and the boundary conditions are shown in Fig. 1. An internal angle of friction and secant Young's modulus (E_{50}^{ref}) representing dense sand conditions derived from a series of drained tri axial compression tests were used for the sand. A value of 10 kN/m² to the undrained cohesion (c) for the peat and 25 kN/m² for the soft clay derived from undrained tri axial compression tests was used. Then hyperbolic parameters for the sand, peat and clay were taken from database provided by the software manual as shown in Table 1.

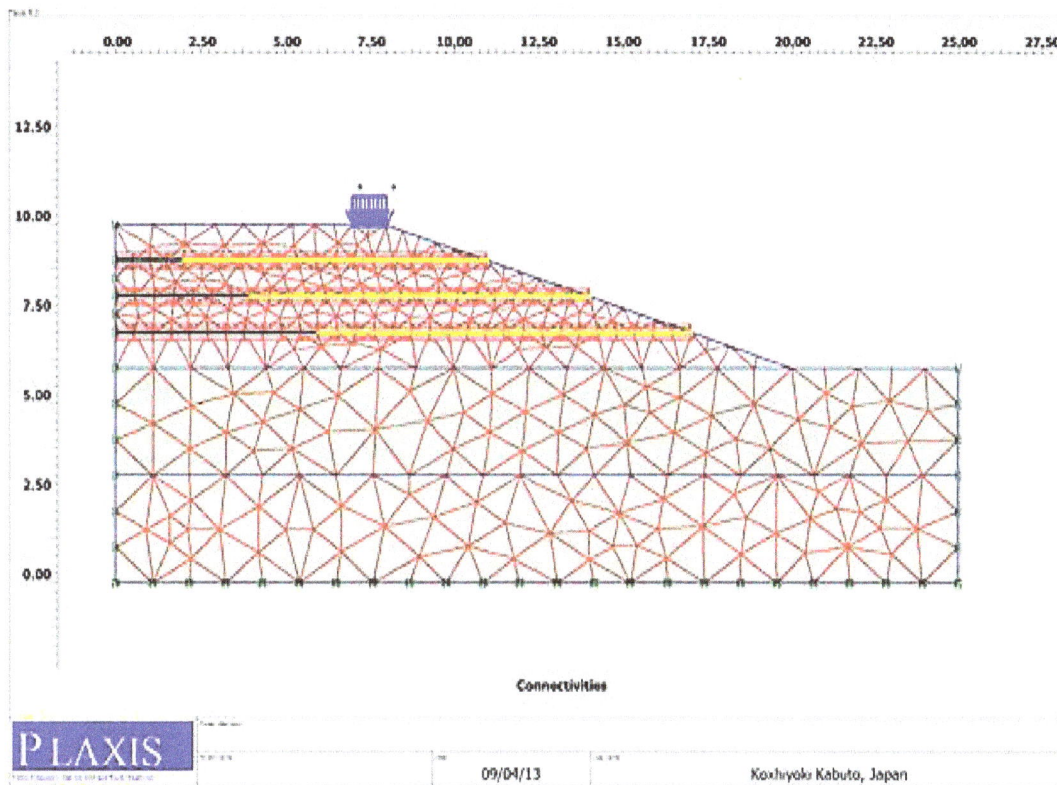

Fig. 1. *Prototype slope geometry, generated mesh, and boundary conditions.*

Table 1. *Hardening soil-footing model parameters used in the finite element analysis.*

Parameter	sand	peat	soft clay	Footing	Geogrid
Primary loading stiffness(E_{50}^{ref}) (kN/m^2)	45000	10000	15000	-	-
Cohesion (c) (kN/m^2)	0.00	10	25	-	-
Friction angle (ϕ)	35	5	2	-	-
Dilatancy angle (ψ)	10	0.00	0.00	-	-
Soil unit weight (γ) (kN/m^3)	20	13.5	18	-	-
Poisson's ratio (ν)	0.30	0.35	0.35	-	-
Failure ratio (R$_f$)	0.90	0.90	0.90	-	-
Interface reduction factor (R$_{int}$)	0.80	0.50	0.30	-	-
EA of the footing (kN/m)	-	-	-	7600000	-
EI of the footing (kNm^2/m)	-	-	-	24000	-
EA of the geogrid (kN/m)	-	-	-	-	2500

3. Results and Discussion

A total of 54 model tests were carried out on model plane strain footing supported on sand pads overlying peat and soft clay ground slope. The effect of geogrid parameters on the ultimate load and displacement were obtained and discussed. An additional numerical study on the effect of reinforcing the sand pad on the behavior of a model footing was carried out using the finite element model.

3.1. Bearing Capacity Behavior

The BCI of the footing on the reinforced sand is represented using a non-dimensional factor, called BCI factor. This factor is defined as the ratio of the footing ultimate pressure with the slope reinforced (q$_{u\ reinforced}$) to the footing

ultimate pressure in tests without slope reinforcement (q$_u$). The footing settlement (S) is also expressed in non-dimensional form in terms of the footing width (B) as the ratio (S/B, %). The ultimate bearing capacities for the model footing when located on non-reinforced and reinforced sand layer obtained from the FEA are 35 and 50kPa respectively is determined from the load–displacement curve. The measured and calculated ultimate loads for footing supported on both reinforced and non-reinforced slopes for the different studied parameters are given in Tables 2–4. These results are discussed in the following sections.

Table 2. *Results of footings located near to reinforced slopes.*

Test results	x/B							
	0	0.5	1	1.5	2	2.5	3	3.5
q (kPa) FEA	35	37	40	43	45	47	50	50

Table 3. Results of footings located near to reinforced slopes.

Test results	L/B							
	0	3	5	7	9	11	13	15
q (kPa) FEA	35	37	39	41	43	44	45	46

Table 4. Results of footings located near to reinforced slopes.

Test results	L/B							
	0	3	5	7	9	11	13	15
q (kPa) FEA	35	37	39	41	43	44	45	46

Table 5. Results of footings located at different locations.

	Non-reinforced						Reinforced					
Test results	b/ B						b/ B					
	0	1	2	3	4	5	0	1	2	3	4	5
q (kPa) FEA	35	45	50	85	95	95	78	90	94	157	174	176

3.1.1. Effect of Number of Geogrid Layers

A series of studies were carried out in order to study the effect of varying the number of geogrid layers on the Footing- slope performance. In this series, geogrid length, location, and spacing, was kept constant but the number of geogrid layers was varied. To assess the effect of presence of number of geogrid layers, initially embankment is modeled without geogrid. In second step, one layer of geogrid is introduced at level of 1m below the foundation. For third step of analyses, two layers of geogrid are considered at level of 1mand 2m below the foundation respectively. For last step of analyses, three layers of geogrid are considered and they are placed one between the embankment base and the soft layer and remaining two others in the body of embankment at level of 1m, 2m and 3m below the foundation respectively. Typical variations of q obtained from numerical analysis against settlement ratios (S/B) for a footing located at the slope crest are shown in Figs. 2. For the same displacement ratio, the figure demonstrates that the inclusion of geogrid layers resulted in an increase in the load capacity of the model footing. Also, for the same footing load, the settlement ratio decrease significantly with increasing the number of geogrid layers. This increase in footing ultimate load can be attributed to reinforcement mechanism which derived from the passive earth resistance, interlocking in front of the transverse members, and adhesion between the longitudinal/transverse geogrid members and the sand. The mobilized passive earth resistance of soil column confined in the geogrid apertures along with the interlocking limit the spreading of slope and lateral deformations of sand particles. The mobilized tension in the reinforcement enables the geogrid to resist the imposed horizontal shear stresses built up in the soil mass beneath the loaded area and transfer them to adjacent stable layers of soils leading to a wider and deeper failure zone. Therefore, sand pad–geogrid interaction not only result in increasing the bearing capacity due to developed longer failure surface but also results in widening the contact area between sand and soft clay. As a result, the developed acting net stress due to footing load decreased leading to decreasing the consolidation settlement of soft clay. Fig. 3 presents comparisons of the variations of the calculated and measured BCI for a footing located at the slope crest for varying values of N.

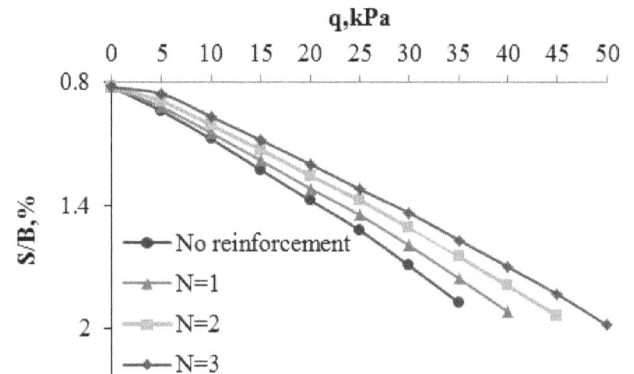

Fig. 2. Variations of q with S/B for prototype slope for different N.

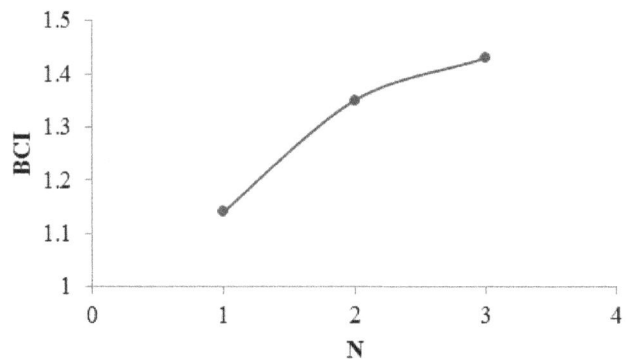

Fig. 3. Variations of BCI with number of geogrid layer, N.

3.1.2. Effect of Geogrid Layer Length

The effect of length of the geogrid layer L/B is studied using only one layer of geogrid placed in dense sand at 2m beneath of the footing. In order to determine how far to extend the geogrid layers into the soil mass to provide an adequate anchorage length for each geogrid layer, seven tests were carried out to study the effect of varying the layer length on the footing behavior. Fig.4. shows the variations of BCI with the geogrid length for model slope-footing system. The BCI increases with increasing geogrid length. This behavior illustrates that sufficient anchorage lengths must be provided to maximize the reinforcing effect through full mobilization of pullout capacity of the reinforcements. With short layers of geogrid, the anchorage length of geogrid in sand is insufficient and the mobilized lateral resistance by passive resistance, interlocking and friction in the stable mass of soil is less than the transferred horizontal shear stresses and the geogrid layers will move down with the soil movement underneath the footing. For longer layers, sufficient anchorage length mobilizes larger lateral resistance than that built up underneath the footing and therefore with footing settlement the geogrid will not move down with supporting soil but mobilize greater resistance up to maximum pullout capacity of geogrid layer after which the system fails.

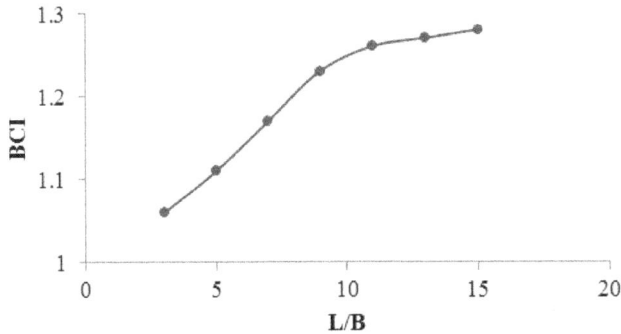

Fig. 4. Variations of BCI with geogrid layer length L/B.

3.1.3. Effect of Footing Location Relative to Slope Crest

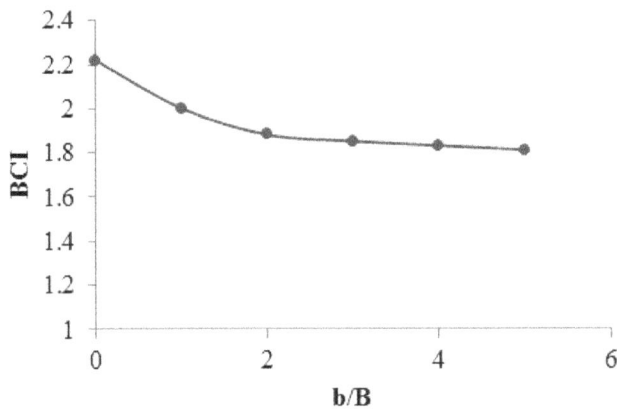

Fig. 5. Variations of BCI with footing location b/B.

In order to study the effect of the proximity of a footing to the slope crest (b/B), a series of tests were carried out on industrial machinery foundation resting on reinforced sand fill overlying soft clay slopes. While the first was carried out on non-reinforced sand fill, the second was carried out on 3-layer of geogrid- reinforced sand. Fig.5.shows the variation of the BCI against the b/B ratios for model results. It can be seen that, while the bearing capacity load significantly decrease as the footing location moves closer to the slope crest, the effect of soil reinforcement on the bearing capacity significantly increase. Also, the figure clearly shows that maximum benefit of slope geogrid reinforcement is obtained when footing is placed at slope crest. This change in bearing capacity of the footing with its location relative to slope crest can be attributed to soil passive resistance from the slope side and reinforcement effect. When, the footing is placed far away of the slope, the passive resistance from the slope side to the failure wedge under the footing increases. Also, using geogrid reinforcement decreases lateral displacements and

results in wider and deeper failure zone as discusses in previous sections, leading to increasing the bearing capacity load.

3.2. Effect of Depth to Top Layer

The effect of depth of the geogrid layer to the ground surface x/B is studied using only one layer of geogrid placed in dense sand at different depths of ground surface. Seven tests were carried out on model footing using FEA. Fig.6 shows the variation of the BCI of the footing against the normalized depth x/B for model footing. Graph clearly show that the BCI initially increases with increasing the depth until it attains a maximum value after which the BCI comes down with increasing the depth of geogrid layer. Also, the variation of BCI with x/B reported by Selvadurai and Gnanendran (1989) and Yoo (2001) for reinforced sand slope are similar to that obtained from the present investigation. This can be explained as follows; at shallow depths under the footing, both the vertical and horizontal soil displacements are greater. Maximum benefits could be obtained when soil reinforcement are placed at these depths where mobilized lateral resistances for soil lateral displacements are maximum. When the depth of geogrid layer increases, both lateral and vertical soil displacements in the zone between the footing and the geogrid layer increase and hence the bearing capacity decreases.

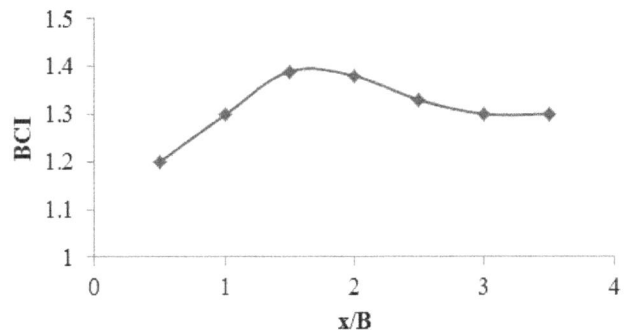

Fig. 6. Variations of BCI with depth of geogrid layer x/B.

3.3. Effect of Pre stressed GeoGrid

To assess the effect of presence pre stressed geogrid layers, initially three types of embankments are modeled: in first step consider an embankment without geogrid, in second step three layers of geogrid are introduced between the embankment and the under laying soft soil and for last step of analyses, three layers of pre stressed geogrid are considered and they are placed between the embankment to investigate the degree of improvement generated by pre stressing the geosynthetic layer for several embedment depths of a footing resting on a reinforced sand bed. The addition of pre stress to the geogrid reinforcement results in significant improvement to the settlement response and the load-bearing capacity of the foundation. Fig.7 the beneficial effects of the pre stressed geogrid configuration were evident, in comparison with unreinforced and reinforced (without pre stress) counterparts.

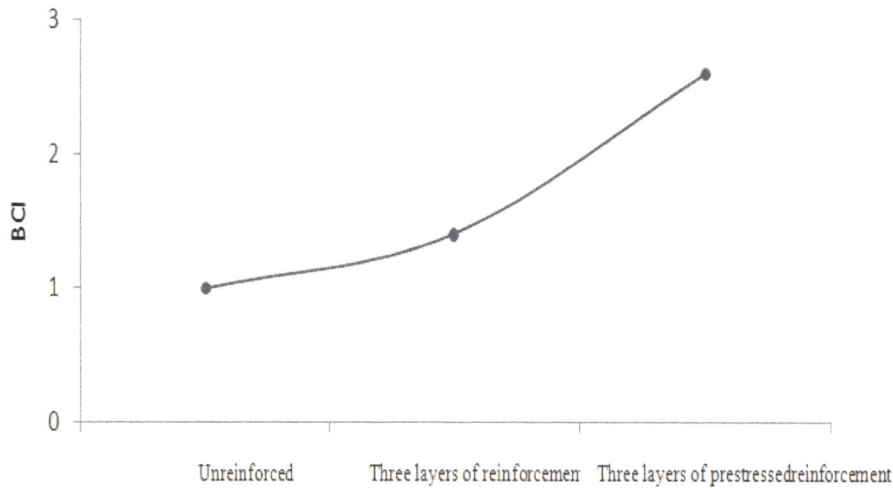

Fig. 7. *Variations of BCI with number of geogrid layer N.*

4. Displacement Vectors

Fig.8.and Fig.9 presents the failure pattern and deformed mesh for a footing placed at the crest of both the non-reinforced and three-pre stressed geogrid layers reinforced slope, respectively. The figure clearly shows the tendency of the footing rotation toward the slope face on reinforced test while in tests on non-reinforced slope, the footing tend to fail by punching shear failure. Typical plots of the displacement vectors obtained from the FEA are also presented. Comparing the plastic flow between these two cases, it can be observed that displacement vectors at failure for non-reinforced slope are concentrated underneath the footing toward the slope face while for the reinforced slope, the displacement vectors are widely distributed underneath the footing for greater width and depth than that in the non-reinforced case. It is clear that the geogrid layers prevent the soil particles from lateral movement toward the slope face and pushes them downward for greater depth and hence spreads the footing load wider and deeper into the soil, which in turn meant a longer failure surface and greater bearing capacity.

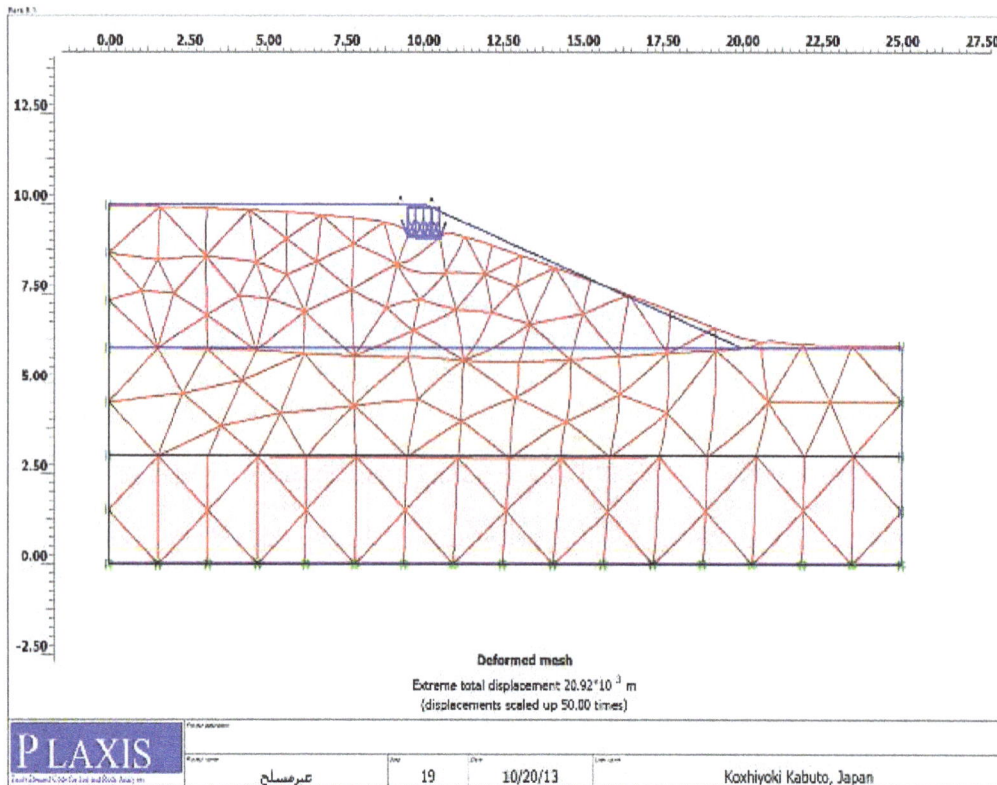

Deformed mesh

Extreme total displacement 20.92*10^{-3} m

(displacements scaled up 50.00 times)

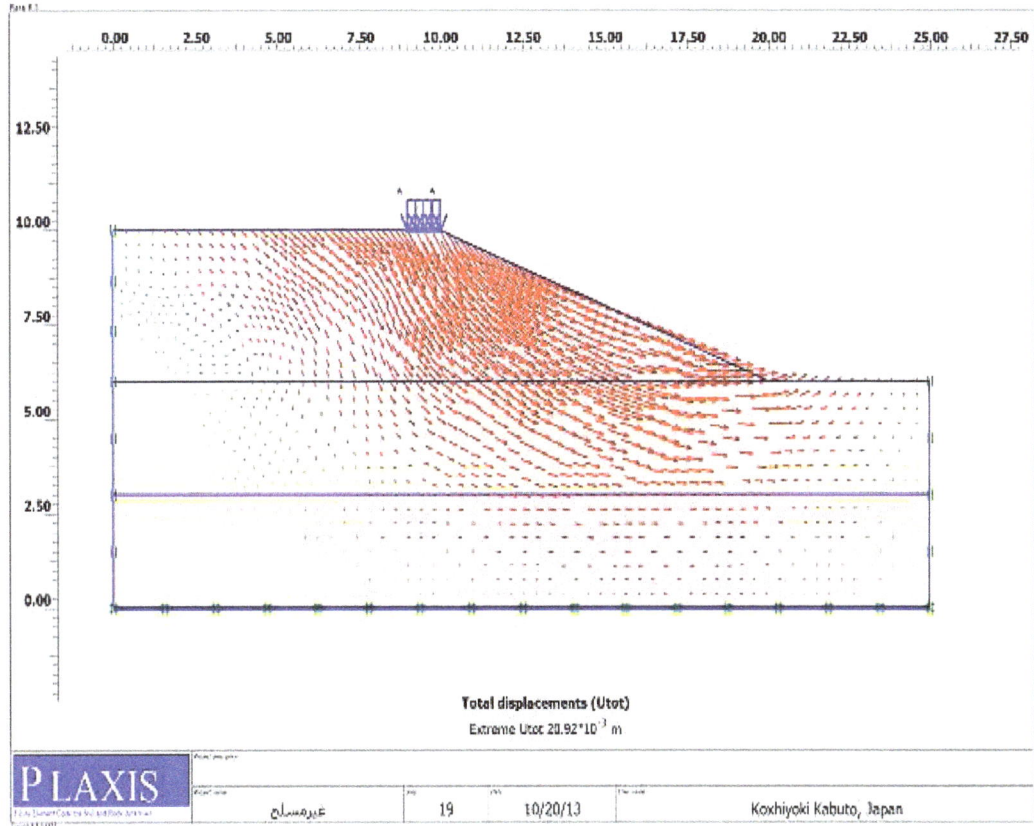

Fig. 8. *Failure pattern and displacement vectors plot for non-reinforced slope.*

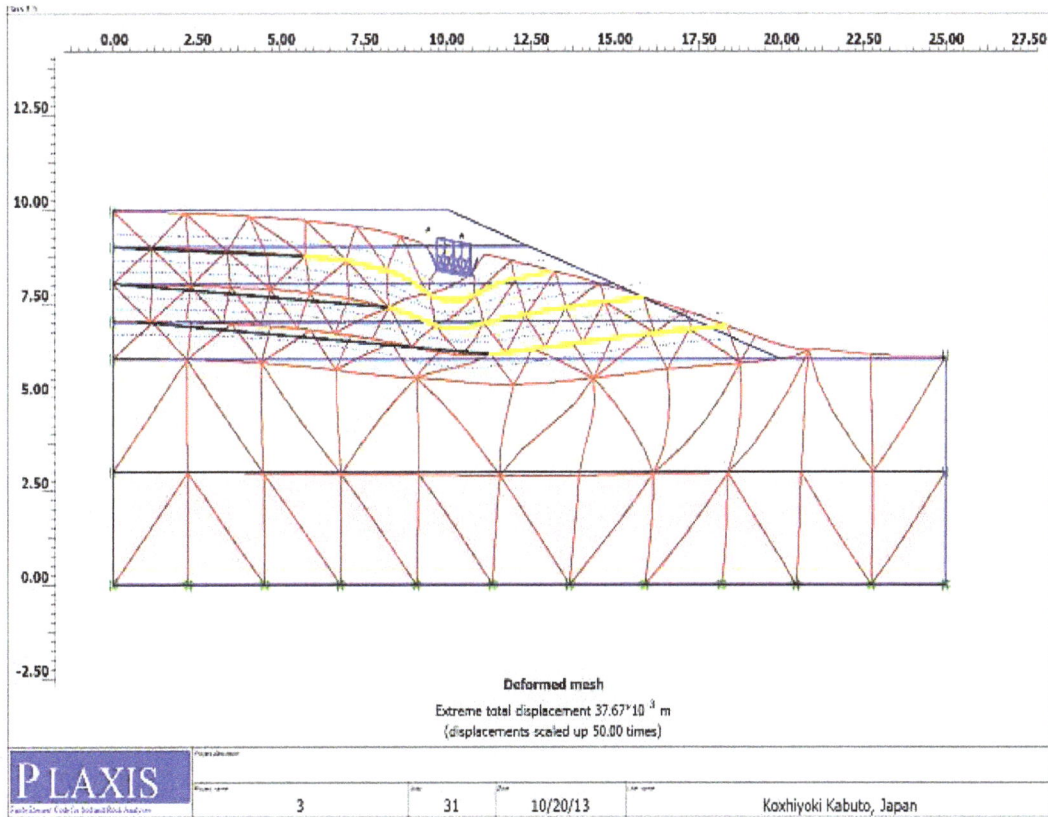

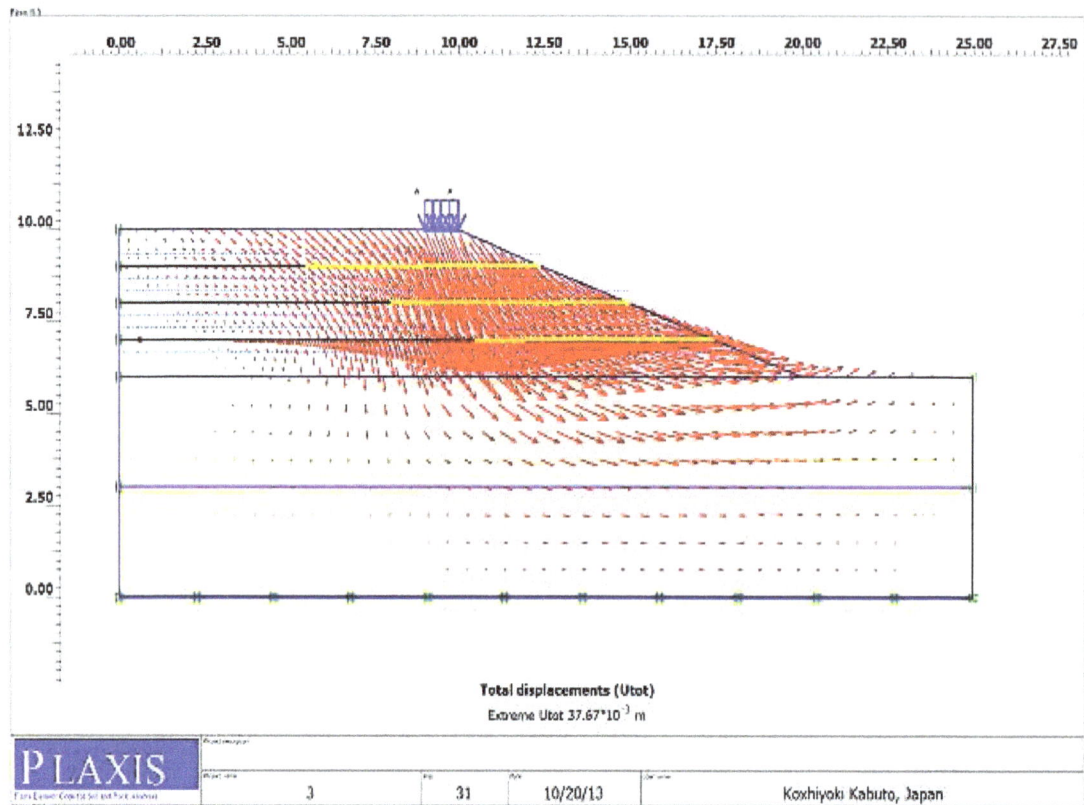

Fig. 9. Failure pattern and displacement vectors plot for reinforced slope.

5. Conclusions

The bearing capacity behavior of industrial machinery foundation resting on reinforcement sand layer constructed on a soft clay slope was investigated. Also, the effect of inclusion of pre stressed geogrid reinforcement on the footing response was studied theoretically. Wide ranges of boundary conditions including footing location and the geogrid parameters were considered. Based on the results from this investigation, the following conclusions can be drawn:

(1)The inclusion of soil reinforcement not only improves the footing behavior but also leads to significant reduction of footing settlement, at the same load levels.

(2)The effect of geogrid reinforcements on the footing performance is dependent on the footing location relative to slope crest. In terms of BCI, geogrid is most effective when the footing is placed on the slope crest rather than any distance away from the slope crest.

(3) For a footing located at slope crest, an adequate anchorage length for each geogrid layer should be provided along with an optimum number of geogrid layers should be used.

(4) For the studied slope geometry and conditions, the maximum benefit of geogrid reinforcements is dependent on reinforcement configuration. The BCI initially increases with increasing the depth until it attains a maximum value after which the BCI comes down with increasing the depth of geogrid layer.

(5)Using three pre stressed layers of geotextile between the embankment base and the under laying soft layer decrease vertical and horizontal displacement and increase BCI and the beneficial effects of the pre stressed geogrid configuration were evident, in comparison with unreinforced and reinforced (without pre stress) counterparts

References

[1] Bringkgreve, R., Vermeer, P., 1998. PLAXIS-Finite Element Code for Soil and Rock Analysis. Version 7 Plaxis B.V., the Netherlands.

[2] British Rail Research, 1998. Supporting roll—geogrids provide a solution to railway track ballast problems on soft and variable subgrades. Ground Engineering 31 (3), 24–27.

[3] Das, B., Khing, K., Shin, E., Puri, V., Yen, S., 1994. Comparison of bearing capacity of strip foundation on geogrid-reinforced sand and clay. In: Proceedings of the Eighth International Conference on Computer Methods and Advances in Geomechanics, Morgantown, WA, USA, pp. 1331–1336.

[4] Dash, S., Sireesh, S., Sitharam, T., 2003. Model studies on circular footing supported on geocell reinforced sand underlain by soft clay. Geotextiles and Geomembranes 21 (4), 197–219.

[5] El Sawwaf, M., 2005. Strip footing behavior on pile and sheet pilestabilized sand slope. Journal of Geotechnology and Geoenvironmental Engineering 131 (6), 705–715.

[6] Ghosh, A., Ghosh, A., Bera, A.K., 2005. Bearing capacity of square footing on pond ash reinforced with jute-geotextile. Geotextiles and Geomembranes 23 (2), 144–173.

[7] Boushehrian, J.H., Hataf, N., 2003. 241 Experimental and numerical investigation of the bearing capacity of model circular and ring footings on reinforced sand. Geotextiles and Geomembranes 21 (4), 241–256.

[8] Otani, J., Ochiai, H., Yamamoto, K., 1998. Bearing capacity analysis of reinforced foundations on cohesive soil. Geotextile and Geomembranes 16, 195–206.

[9] Patra, C.R., Das, B.M., Bhoi, M., Shin, E.C., 2006. Eccentrically loaded strip foundation on geogrid-reinforced sand. Geotextiles and Geomembranes 24 (4), 254–259.

[10] Tsukada, Y., Isoda, T., Yamanouchi, T., 1993. Geogrid subgrade reinforcement and deep foundation. In: Raymond, Giroud (Eds.), Proceedings of the Geosynthetics Case Histories. ISSMFE, Committee TC9, pp. 158–159.

[11] Vesic, A., 1973. Analysis of ultimate loads of shallow foundations. Journal of Soil Mechanics and Foundations Division—ASCE 94 (SM3), 661–688.

[12] Yetimoglu, T., Inanir, M., Inanir, O., 2005. A study on bearing capacity of randomly distributed fiber-reinforced sand fills overlying soft clay. Geotextiles and Geomembranes 23 (2), 174–183.

[13] Huang, C.C., Menq, F.Y., 1997. Deep footing and wide-slap effects on reinforced sandy ground. Journal of Geotechnical and Geoenvironmental Engineering, ASCE 123 (1), 30–36.

[14] Ismail, I., Raymond, G.P., 1995. Geosynthetic reinforcement of granular layered soil. In: Proceedings of Geosynthetics '95 Conference, Nashville, TN, vol. 1. Industrial Fabrics Association International, Roseville, MN, USA, pp. 317–330.

[15] Khing, K.H., Das, B.M., Puri, V.K., Cook, E.E., Yen, S.C., 1993. The bearing capacity of a strip foundation on geogrid-reinforced sand. Geotextiles and Geomembranes 12 (4), 351–361.

[16] Koerner, R.M., 2005. Designing with Geosynthetics, fifth ed. Prentice Hall, New Jersey, USA. Kurian, N., Beena, K.S., Kumar, R.K., 1997. Settlement of reinforced sand in foundations. Journal of Geotechnical Engineering 123 (9), 818–827.

[17] Madhav, M.R., Poorooshasb, H.B., 1988. A new model for geosynthetic-reinforced soil. Computers and Geotechnics 6, 277–290.

[18] Meyerhof, G.G., 1963. Some recent research on bearing capacity of foundations. Canadian Geotechnical Journal 1 (1), 16–26.

[19] Shukla, S.K., 2002. Geosynthetics and their Applications. Thomas Telford, London. Shukla, S.K., Yin, J.H., 2006. Fundamentals of Geosynthetic Engineering. Taylor and Francis, London.

[20] Shukla, S.K., 1995. Foundation model for reinforced granular fill – soft soil system and its settlement response. Ph.D. thesis, Department of Civil Engineering, Indian Institute of Technology, Kanpur, India.

[21] Alawaji, H., 2001. Settlement and bearing capacity of geogrid-reinforced sand over collapsible soil. Geotextiles and Geomembranes 19, 75–88.

[22] Bera, A.K., Ghosh, A., Ghosh, A., 2005. Regression model for bearing capacity of a square footing on reinforced pond ash. Geotextiles and Geomembranes 23 (2), 261–286.

[23] Ovesen, N.K., 1979. The use of physical models in design: the scaling law relationship. In: Proceedings of the Seventh European Conference on Soil Mechanics and Foundation Engineering, vol. 4, pp. 318–323.

23

Optimization and Utilization of Wastepaper for Bio-energy Production

Solomon Hailu[1], Solomon Kahsay G. Mariam[1], Tesfay Berhe[1, 2]

[1]Department of Biological and Chemical Engineering, Mekelle Institute of Technology, Mekelle University, Tigray, Ethiopia
[2]Department of Chemical Engineering, Kombolcha Institute of Technology, Wollo University, Amhara, Ethiopia

Email address:

tesadi@gmail.com (T. Berhe)

Abstract: Bioenergy future depends on an increased share of renewable energy, especially in developing countries. Bioconversion of lignocellulosic based biomass to ethanol is significantly hindered by the structural and chemical complexity of biomass, which makes these materials a challenge to be used as feed stocks for cellulosic ethanol production. Bioethanol is one of the most important alternative renewable energy sources that substitute the fossil fuels. Wastepaper has a content of cellulose and hemicelluloses, which make it suitable as fermentation substrate when hydrolyzed. The objective of this work is ethanol production from wastepaper by fermentation process. Eight laboratory experiments were conducted to produce bioethanol from wastepaper. By using Design Expert 7 software, it was formulated the dilute acid hydrolysis step to investigate the effects of hydrolysis parameters on yield of ethanol and optimum condition. All the three hydrolysis parameters were significant variables for the yield of ethanol. The optimum combinations of the three factors chosen for optimum ethanol yield 10.86 ml/50 g sample were 92.59°C hydrolysis temperature, 30 minutes hydrolysis time and 1%v/v acid concentration.

Keywords: Bioethanol, Distillation, Fermentation, Hydrolysis, Wastepaper

1. Introduction

Energy consumption has increased steadily over the last century as the world population has grown and more countries have become industrialized. Crude oil has been the major resource to meet the increased energy demand (Campbell et al, 1998). As energy demand increases the global supply of fossil fuels cause harm to human health and contributes to the greenhouse gas emission. At the same time, increasing waste generation linked to rising population and living standards is a worldwide challenge to waste management systems. World energy consumption is predicted to increase by 50% to 2030 according to the United States Energy Information Agency (EIA, 2009). The combustion of fossil fuel is responsible for 73% of the CO_2 emission. In this scenario, renewable sources might serve as an alternative. The alternative fuel must be technically feasible, economically competitive, environmentally acceptable, and readily available. Numerous potential alternative fuels have been proposed, including bioethanol, biodiesel, methanol, hydrogen, boron, natural gas, liquefied petroleum gas (LPG), Fischer–Tropsch fuel, p-series, electricity, and solar fuels. Biofuels which include bioethanol, vegetable oils, biodiesel, biogas, biosynthetic gas (bio-syngas), bio-oil, bio-char, Fischer–Tropsch liquids, and biohydrogen offer many advantages over petroleum-based fuels (Demirbas, 2008). These advantages include biofuels are easily available from common biomass sources, they represent a CO_2-cycle in combustion, biofuels have a considerable environmentally friendly potential, there are many benefits to the environment, economy and consumers in using biofuels, and they are biodegradable and contribute to sustainability (Ngunzi, 2014).

Bioethanol have the major applications in automobile, beverage, pharmaceuticals industry etc. due to economically as well as ecofriendly (Galbe and Zacchi, 2007). Bioethanol is appropriate for the mixed fuel in the gasoline engine due to its high octane number, due to its low cetane number and high heat of vaporization impede self-ignition in the diesel engine. Absolute ethanol and 95% ethanol are themselves act as good solvents, somewhat less polar than water and used in perfumes, paints and tinctures. Ethanol is used in medical wipes and in most common antibacterial hand sanitizer gels at a concentration of about 62% (Cardona and

Sanchez, 2007).

Unlike fossil fuels, ethanol is a renewable energy source produced through fermentation of sugars. Currently, the technology for lignocellulosic ethanol production relies mainly on pretreatment, chemical or enzymatic hydrolysis, fermentation and product separation or distillation (Balat, 2010). An appropriate pretreatment strategy is essential for the efficient enzyme hydrolysis of lignocellulosic biomass as lignin hinders the saccharification process (Gaur, 2006). Efficient saccharification of cellulose is of great importance from a viewpoint of the disposal of cellulosic wastes and the utilization of cellulose as a renewable resource in relation to the preventing of a greenhouse effect due to carbon dioxide.

Various raw materials like sugarcane juice and molasses, sugar beet, beet molasses, sweet sorghum (Stokes, 2005) and starchy materials like sweet potato, corn cobs and hulls, cellulosic materials like cocoa, pineapples and sugarcane waste coffee husk and milk/cheese/whey using lactose hydrolyzing fermenting strains has available in literature. The yield of bioethanol from corn and sugar are high and the techniques are mature. However, it increases the risk of causing global food shortage (MoFED, 2010). Therefore, another alternative material has been used which contain lignocellulosic such as weed, grass, saw dust, municipal solid waste, woody biomass, and paper mill waste.

Wastepaper consists of a considerable share of municipal and industrial waste even though recycling efforts have been strengthened in recent years by legal provisions like the packaging directive (Forum for Agricultural Research in Africa, 2008). However, the recycling rate of wastepaper is low and the recycled wastepaper has a low grade paper product because of shorted fiber length. Since the shortening of paper fibers decreases the quality of paper, the maximum ratio of paper-to-paper recycling is said to be 65% (Ikeda et al., 2006). This means that a certain fraction of paper would always be sent to disposal. Still, wastepaper is considered as one of the prospective and renewable biomass materials to produce bioethanol.

The enzymatic hydrolysis of wastepaper is desirable from the standpoint of green and clean processing, although the process presents such challenges as a slow reaction rate and low process efficiency, mainly due to the high crystallinity of cellulose, the presence of some lignin, low specific surface area of the material, and the complexity of celluloses as multicomponent enzyme systems, etc (Hahn-Hagerdal et al., 2006).

Moreover, this alternative outlet for waste papers could help reduce pressure on other waste management options (i.e. recycling, incineration and landfill) from the increasing waste generation due to rising population and ultimately we create aesthetically very attractive city (Halidini et al, 2014). Therefore, the aim of this work is to investigate the possibility of using and transforming different wastepaper to ethanol by fermentation using S. Cereviciae thereby contributing towards alternative energy supply as well as creating an employment opportunity. Also to study the optimum hydrolysis of different waste papers with diluted sulfuric acid.

2. Material and Methods

2.1. Materials

Wastepaper: The wastepaper was collected from the campus area of Mekelle University, Ethiopia. The wastepaper was then dried in oven at 60°C for 48 hours). The dried paper was placed in a mortar crusher and the maximum particle sizes of 3 mm. The sample of larger particle size than 3 mm was cut over and over again until all particle size became 3 mm. The cut material was kept at low temperature until the next stage of experiment.

Chemical: Sulfuric acid (98% H_2SO_4), sodium hydroxide (NaOH) solution, dry instant yeast (*saccharomyces cerevisiae*), yeast extracts (agar), urea, dextrose sugar, ($MgSO_4.7H_2O$), distilled water, and potassium dichromate were used during the experiment.

Equipment: Scissor, oven, balance, digital pH meter, shaking incubator, centrifuge, flasks of different volumes, graduated cylinders of different volumes, autoclave, alcoholmeter, shaker, rotary evaporator, computer, aluminum foil, and cotton were used during the experiment.

2.2. Methods

Steam Pretreatment: The purpose of pretreatment is to remove lignin, reduce cellulose crystallinity, and increases the porosity of the materials (Prasad, 2003). Pretreatment must meet the following requirements: improve the formation of sugar, avoid the degradation or loss of carbohydrate, avoid the formation of by-product inhibitors and must be cost effective.

Procedure in Steam Pretreatment: First distilled water was prepared and then 50 g of the cut sample was soaked in a distilled water of 500 mL in conical flasks for 24 hours. The conical flasks were capped with the help of aluminum foil. Then the lignocellulosic biomass was rapidly heated at 121°C by high-pressure steam without addition of any chemicals in autoclave. The biomass/steam mixture was held for 15 minutes to promote hemicellulose hydrolysis, and terminated by an explosive decompression. After finishing the given pretreatment time and temperature the sample in autoclave was allowed to cool and the soluble portion was separated from the non-soluble portion. The non-soluble portion was hydrolyzed in the next steps and the soluble solution was placed in another conical flask.

Dilute Acid Hydrolysis: The carbohydrate polymers in lignocellulosic materials need to be converted to simple sugars before fermentation, through a process called hydrolysis. Various methods for the hydrolysis of lignocellulosic materials have recently been described. The most commonly applied methods can be classified in two groups: chemical hydrolysis and enzymatic hydrolysis (Silverstein, 2004). Even though there are many types of hydrolysis types, dilute acid hydrolysis is an easy and productive process and the amount of alcohol produced in case of acid hydrolysis is more than that of alkaline hydrolysis. This process is conducted under high temperature and pressure, and has a reaction time in the range of seconds or minutes, which facilitates continuous processing.

The three-parameter and two-level ($2^3 = 8$) factorial design was applied to hydrolysis step of the experimentation. Three experiments were carried out in series for each sample: hydrolysis, fermentation, and distillation. Because of the series nature of the three experiments, it was not simple to assess the contribution of change in parametric values of each experiment on the final result (Sun and Cheng, 2002). For that reason the factorial design was applied only to the hydrolysis experiment and parameters were changed at two levels to see their directional effects on the response parameter (ethanol yield). The parametric values for the rest of the experiments were kept constant for the experiment to be sharply justified. Therefore, the optimum values of main variables in hydrolysis process (time, acid concentration, and temperature) which give high ethanol yield were set. Table 1 shows the hydrolysis parameters and their respective maximum and minimum.

Table 1. Minimum and maximum value of parameters.

S.No	Experiments	Minimum	Maximum
1	Hydrolysis temperature (°C)	80	100
2	Hydrolysis time(minutes)	30	60
3	Acid concentration(% by volume of distilled water)	1	5

The tabulated numeric representation of the factorial design of this research is shown in Table 2. For each one the hydrolysis parameters (acid concentration, hydrolysis time and hydrolysis temperature) there are four low and four high levels in the design shown in the table (Table 2). 50 g of ground wastepaper was used for each experiment and the factors for hydrolysis were time (30 to 60 minutes), hydrolysis temperature (80 to 100°C), and acid concentration (1 to 5%).

Table 2. Numeric values of parameters in hydrolysis according to factorial design.

Sample Number	Acid concentration (% by volume)	Hydrolysis time(min)	Hydrolysis temperature(°C)	Coded representation of parameters
1	1	30	80	- - -
2	5	30	80	+ - -
3	1	60	80	+ -
4	5	60	80	+ + -
5	1	30	100	- +
6	5	30	100	+ - +
7	1	60	100	+ +
8	5	60	100	+ + +

Procedures in Dilute Acid Hydrolysis: Sulfuric acid (by volume to water) was diluted to 1% and 5% concentration. The 50 g ground wastepaper of 10% Weight per Volume to the prepared dilute acid solution (1 to 5%) was then added into the glass vessel. Then the prepared sample was hydrolyzed in autoclave with the vessels unlidded between 80 and 100°C for 30 to 60 minutes. Centrifugation and then filtration was used to separate the solid particles from the liquid in the hydrolyzate (remove the non-fermentable lignin portion). The diluted hydrolyzed samples were conditioned to temperature of 30°C before fermentation step was started. This was the temperature at which all fermentation experiments were carried out.

Fermentation: The supernatant from dilute acid hydrolysis of lignocelluloses can contain both six-carbon (hexoses) and five-carbon (pentoses) sugars (if both cellulose and hemicellulose are hydrolyzed). Depending on the lignocelluloses source, the hydrolysate typically consists of glucose, xylose, arabinose, galactose, mannose, fucose, and rhamnose. Microorganisms can be used to ferment all lignocellulose-derived sugars to bioethanol.

Microorganisms: S. cereviciae was purchased from local market and it was used in all experiments throughout this work. Media which is favorable for the yeast was prepared.

Procedures in Media Preparation: First, 100 ml media containing 10 gm Sugar (Dextrose), 0.2 gm yeast extract, 1.0gm urea and 1.0g $MgSO_4.7H_2O$ was prepared. The 100 ml media was sterilized in autoclave. Next to the 100 ml media, 0.5 gm of yeast, *Saccharomyces cerevisiae* was added in a 250 ml conical flask and then properly covered with aluminum foil. The conical flasks were then placed in a shaking incubator for 24 hours at a temperature of 30°C and 200 rpm.

Adjustment of pH: Before addition of any microorganism to the diluted hydrolyzed sample, pH of these samples had to be adjusted. Otherwise the microorganism will die in hyper acidic or basic state. A pH of around 5.0 - 5.5 was maintained. The hydrolyzed samples were primarily checked for pH using a digital pH meter. The pH then was adjusted to 5.0 - 5.5. When the pH went below 5.0 - 5.5, sodium hydroxide solution was added drop wise to the flask with constant stirring until the pH reaches to a range of 5.0 - 5.5. When the pH went beyond 5.0 - 5.5, concentrated sulfuric acid was added drop wise to maintain the pH in the range.

Procedures of the Experiment: Fermentation was taken place in shaking incubator. The shaking incubator was set at 30°C and the prepared samples were dipped into the water-filled-beaker until a set temperature and temperature of the water in the beaker became equal. The yeast, *Saccharomyces Cerevisiae* culture was added with the proportion of 1:10 to the hydrolyzed sample. The vessel was lidded with a piece of ginned cotton covered with aluminum foil. Fermentation was let take place. After 72 hours of fermentation, the sample was taken out and distilled. The parameters of fermentation i.e. incubation time, yeast concentration (yeast proportion) and fermentation temperature was set to be at 72 hours, 10% and 30°C respectively.

Distillation: A distillation system was used to separate the bioethanol from water in the liquid mixture. All distillation experiments were carried out at a temperature of 85°C and a distillation time of 3 hours by rotary evaporator.

Identification of Bioethanol: About 5 ml fermented sample was taken and pinch of potassium dichromate and a few drops of H_2SO_4 were added. Color change from pink to green indicated the presence of bioethanol.

Determination of Ethanol Concentration: The ethanol concentration of the samples collected every 3 hours interval by rotary evaporator of fermented solution was measured by alcoholmeter. An alcoholmeter is a hydrometer which is used for determining the alcoholic strength of liquids. It only measures the density of the fluid. Alcohol meters have scales marked with volume percent of potential alcohol, based on a pre-calculated specific gravity.

Procedure: Each sample was poured in to a graduated cylinder leaving enough space for the alcoholmeter to disperse the liquid. Then the alcoholmeter was dipped in to the sample and spinned to dislodge any bubbles that may have been in the liquid, and could potentially cause the alcoholmeter to float slightly higher, resulting in a low reading. Finally a reading was taken from the mark where the liquid level crosses the stem of the alcoholmeter. This was the percentage alcohol by volume of the sample.

Data Analysis: Design expert® 7 software was used to determine the effect of three hydrolysis parameters (time, temperature, and acid concentration) on ethanol yield, and to optimize these hydrolysis parameters. The response variable was ethanol yield. Significance of the result was set from analysis of variance (ANOVA).

3. Result and Discussion

Effect of Temperature and Time:

Figure 1(a) & (b) show effect of temperature and time on yield of ethanol when acid concentration was at the center point. Ethanol yield increased with increasing hydrolysis temperature when hydrolysis time was at low level. Similarly, ethanol yield increased with increasing hydrolysis time when hydrolysis temperature was at low level. This is because at low temperature and time cellulose might not be converted to fermentable sugars and at high temperature and time the fermentable sugars might be converted to non-fermentable molecules. Hence both temperature and time have interaction effect, in addition to main effect for the yield of ethanol production.

Figure 1(c) shows the contour plot graph showing predicted response of ethanol yield as a function of hydrolysis time and hydrolysis temperature. Ethanol yield increased as hydrolysis time increases at lower level temperature and it decrease when the hydrolysis time and temperature became higher and higher.

The response surface figure 1 (d), obtained from hydrolysis temperature and time shows that ethanol yield increased with increasing time when hydrolysis temperature was at low level and with increasing hydrolysis temperature when time was at low level.

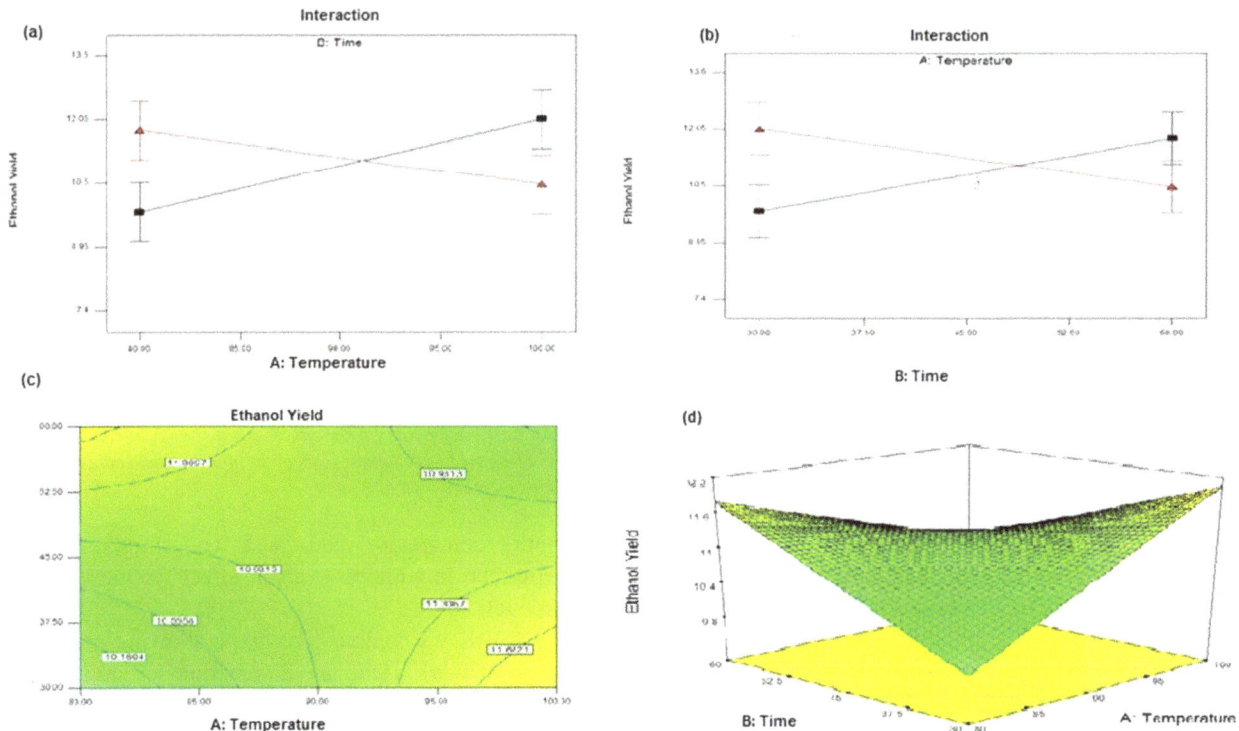

Figure 1. *Effect of temperature and time on yield of ethanol when acid concentration was at the center point (a, b, c & d) (a) The effects of temperature and time (fixed) on the yield of ethanol, when the concentration was at the center point (b) The effects of temperature(fixed) and time on the yield of ethanol, when the concentration was at the center point (c) Contour plots of the effects of time and temperature on the yield of ethanol (d) Response surfaces plot of the effects of temperature and time on the yield ethanol.*

Effect of Acid concentration and time:

Figure 2 (a) & (b) show effect of acid concentration and time on yield of ethanol when hydrolysis temperature was at the center point. Ethanol yield increased with increasing acid concentration when hydrolysis time was at low level. Similarly, ethanol yield increased with increasing hydrolysis time when acid concentration was at low level. This is because at low concentration and time cellulose might not be converted to fermentable sugars and at high concentration and time the fermentable sugars might be converted to non fermentable molecules. Hence both time and acid concentration have interaction effect, in addition to main effect for the yield of ethanol production.

Figure 2 (c) shows the contour plot graph showing predicted response of ethanol yield as a function of hydrolysis time and acid concentration. Ethanol yield increased as hydrolysis time increases at lower level acid concentration and it decrease when the hydrolysis time and acid concentration became higher and higher.

The response surface figure 2 (d), obtained from hydrolysis time and acid concentration shows as hydrolysis time increases at lower level of acid concentration and as increase level of acid concentration and lower level of time gives a positive effect on the yield of ethanol.

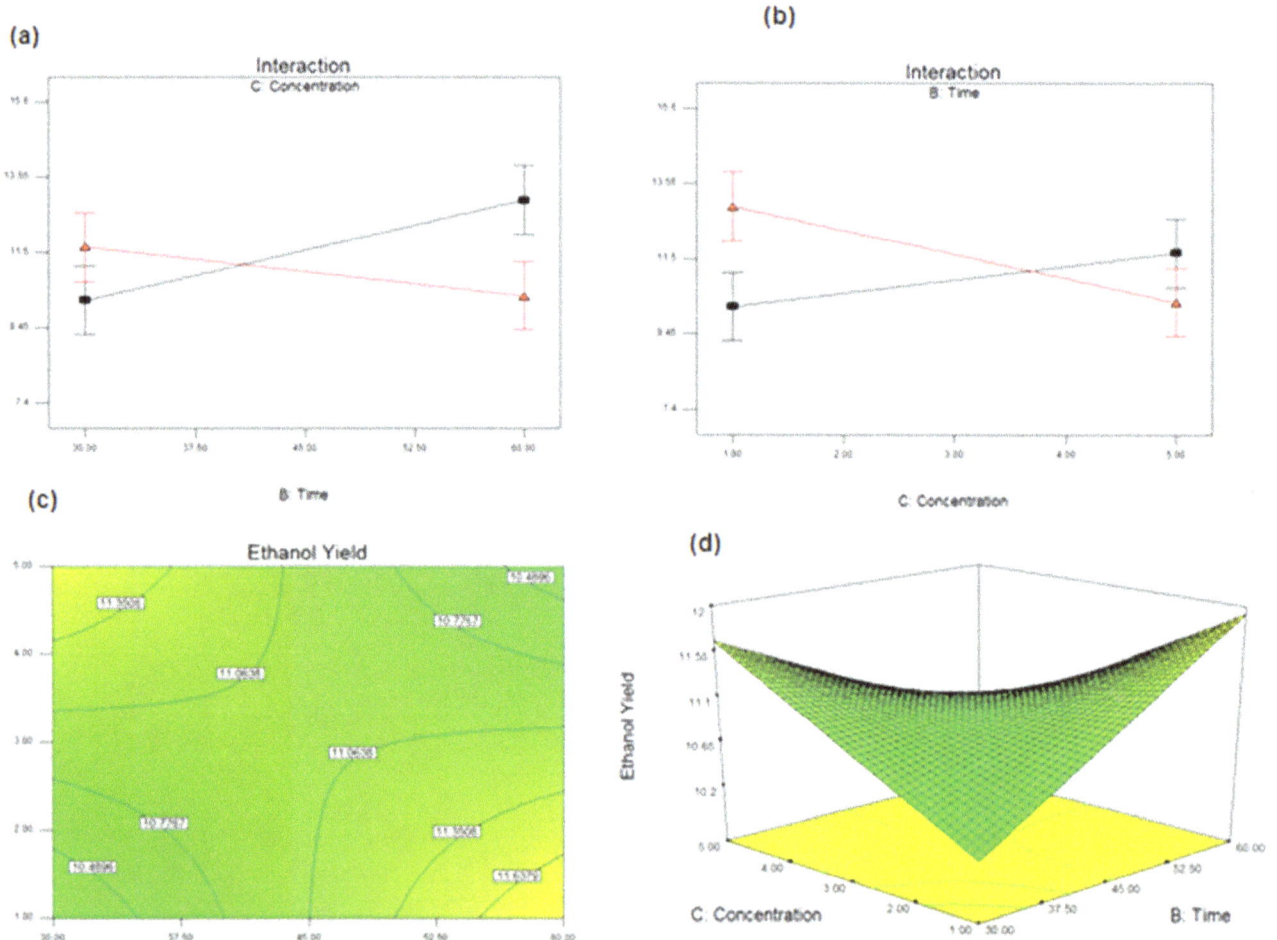

(a)

(b)

(c)

(d)

Figure 2. *Effect of acid concentration and time on the yield of ethanol when temperature was at the center point. (a)The effects of time and acid concentration (fixed) on the yield of ethanol, (b) The effects of time (fixed) and acid concentration on the yield of ethanol, (c) Contour plots of the effects of acid concentration and time on ethanol yield, (d) Response surface plots of the effects of acid concentration and time on ethanol yield.*

Effect of temperature and acid concentration

Figure 3 (a) & (b) show effect of acid concentration and temperature on yield of ethanol when hydrolysis time was at the center point. Ethanol yield increased with increasing acid concentration when hydrolysis temperature was at low level and with increasing hydrolysis temperature when acid concentration was at low level. At lower temperature and acid concentration the cellulose might not be hydrolyzed to fermentable sugars and at higher acid concentration and time the cellulose might be converted to non fermentable molecules.

Hence both temperature and acid concentration have interaction effect, in addition to main effect for the yield of ethanol production.

Figure 3 (c) shows contour plot graph showing predicted response of ethanol yield as a function of hydrolysis temperature and acid concentration. Yield of ethanol increase with increasing acid concentration at low level of hydrolysis temperature and with increasing hydrolysis temperature at low level of acid concentration.

The response surface figure 3 (d), obtained from hydrolysis

temperature and acid concentration shows ethanol yield increased with increasing acid concentration when hydrolysis temperature was at low level and with increasing hydrolysis temperature when acid concentration was at low level. This was consistent with the study on ethanol production from mango and banana peel reported by Taye in 2009.

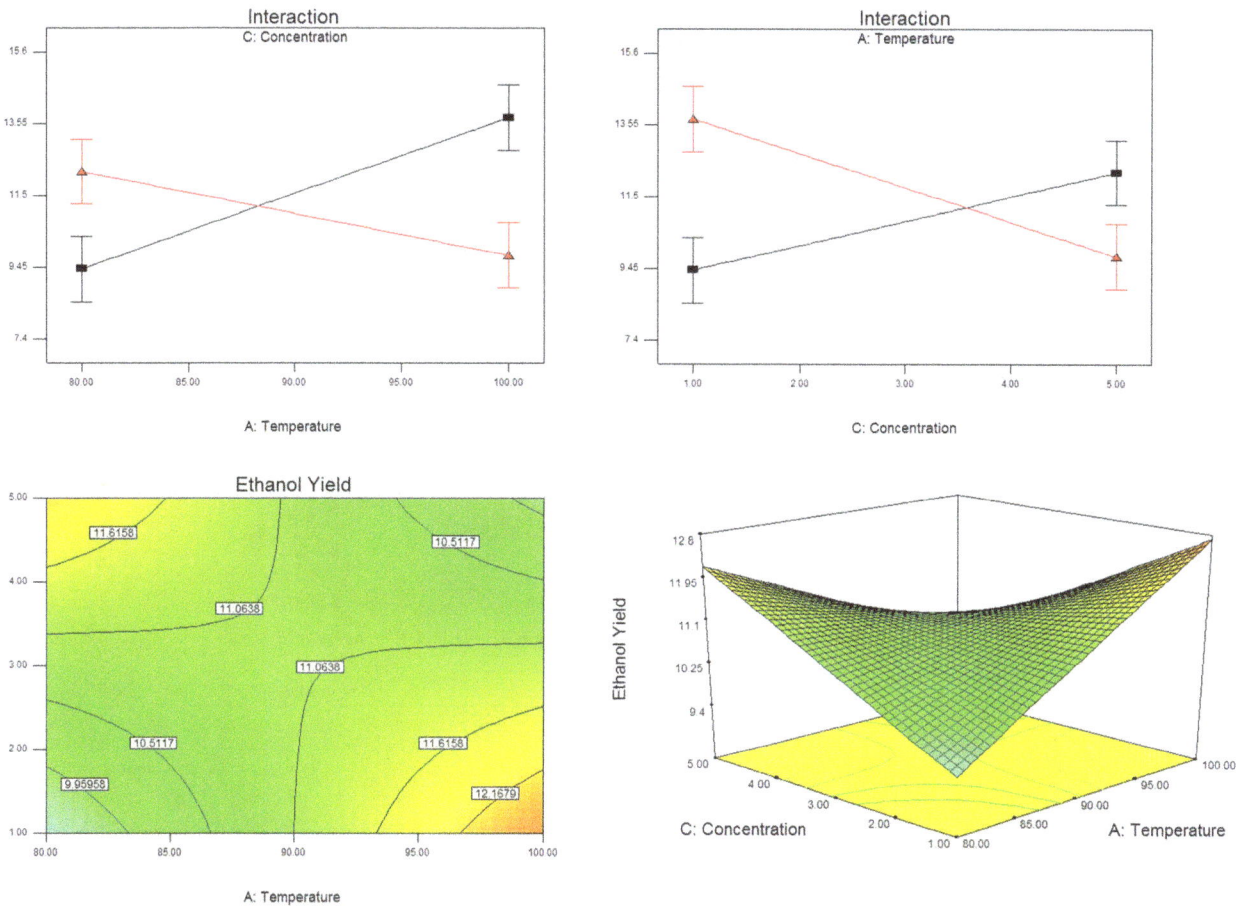

Figure 3. *Effect of temperature and acid concentration on the yield of ethanol when time was at the center point, (a) The effects of temperature and acid concentration (fixed) on the yield of ethanol, (b) The effects of temperature (fixed) and acid concentration on the yield of ethanol, (c) Contour plots of the effects acid concentration and temperature and (d) Response surface plots of the effects acid concentration and temperature.*

Therefore, if it is required to increase ethanol yield with increasing acid concentration, low level of hydrolysis temperature is required. Or else, increased temperature while keeping acid concentration at low level depending on whether economic optimum or technical optimum is the choice or the combination. Technically, optimum operations do not necessarily provide economically optimum performances. Variables such as environment, socio-economic situations and accessibility of relevant technologies must be evaluated to systematically combine the two optimums. From environmental point of view, pollution-free or closed-loop systems are preferred to environmental unfriendly and open end processes. Taking in to consideration the economy of production, using as low quantity of chemicals as possible is beneficial for environment in that treatment and disposal costs are significant portions of production costs. Therefore, using as minimum acid as possible is advisable if possible. Increasing hydrolysis temperature and reducing acid concentration could be economically feasible and environmentally beneficial as heat for steam generation could be derived from non-fermentable lignin-rich fraction of the

waste without the threat of any polluting emissions.

When the above results were compiled to one, high hydrolysis time and high hydrolysis temperature would yield maximum ethanol yield at low acid concentration. This conclusion was consistent with the actual data at 1% acid concentration, 60 minutes hydrolysis time and 100°C hydrolysis temperatures. The maximum ethanol yield found was 13.515 at (1% v, 60minutes, 100°C) of acid concentration, hydrolysis time and hydrolysis temperature respectively.

Table 3. *Optimization criteria for optimum yield of ethanol.*

Parameter	Purpose	Minimum Value	Maximum value
Acid concentration (%)	Minimize	1	5
Temperature (°C)	Minimize	80	100
Time (min)	Minimize	30	60
Ethanol Yield (mL/50g sample)	Maximize	7.42	13.515

Optimization of Hydrolysis Parameters: Design expert® 7 software was used for optimization of hydrolysis parameters. The optimization of hydrolysis criteria for ethanol production from wastepaper using dilute acid treatment are summarized in Table 3.

Design expert® 7 software calculates 15 optimum possible solutions for ethanol production using different hydrolysis parameters as shown in Table 3. The optimum combinations of the three factors chosen for optimum ethanol yield (10.8538) were 92.5°C (hydrolysis temperature), 30 minutes (hydrolysis time) and 1%v acid concentration. This was global optimum combination of the factors. The local optimization usually requires hundreds or perhaps thousands of experiments. The

gross choice between these two factors values do not purely rely on the technical success of the experiment. It also had to consider the economic variables of ethanol production and also whether the production is pilot scale or large scale. The Figure 4, Figure 5 and Figure 6 show contours plot and response surfaces plot generated by Design expert® 7 software for the optimum combinations of the three factors chosen respectively.

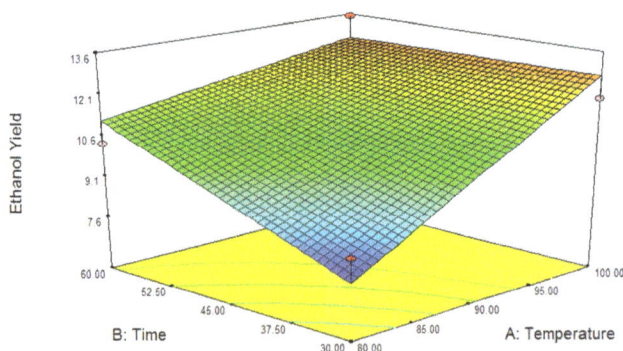

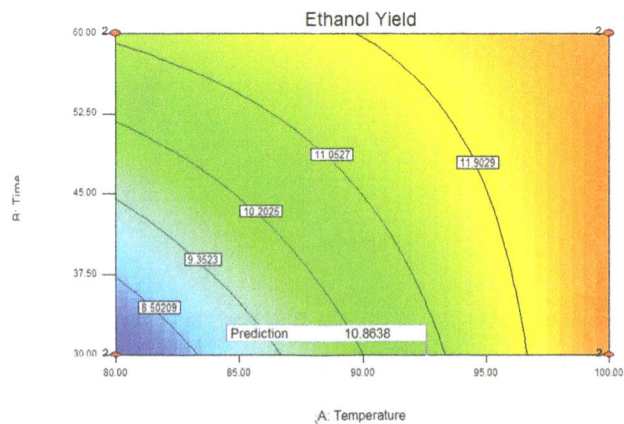

Figure 4. *Optimization of time and temperature (a) response surface method, (b) counter plot.*

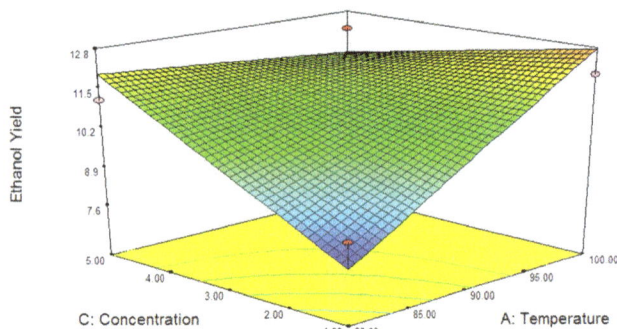

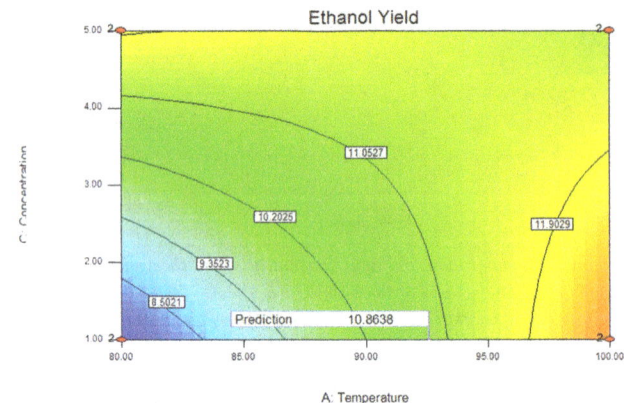

Figure 5. *Optimization of concentration and temperature (a) response surface method, (b) counter plot.*

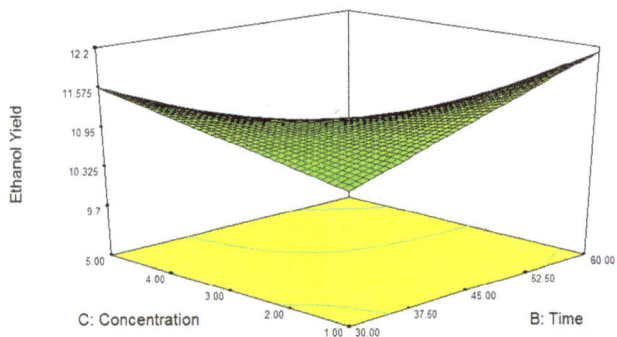

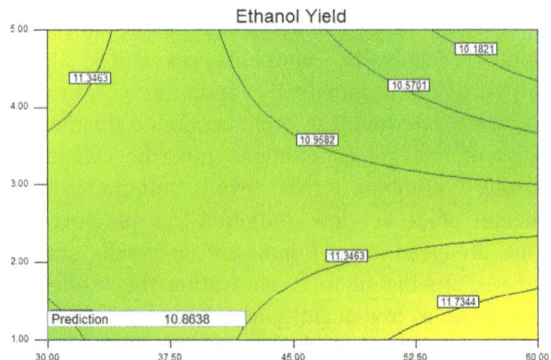

Figure 6. *Optimization of concentration and time (a) response surface method, (b) counter plot.*

To see the consistence between the theoretical ethanol yield at theoretical combination of parametric values and the actual result at that point, an experiment with hydrolysis acidic concentration, temperature and time were conducted at the

optimized conditions. The actual result of ethanol yield at theoretical combination (10.45) was slightly lower than what was expected (10.86).

4. Conclusion

It is concluding that wastepaper is promising lignocellulosic feedstock for bioethanol production. One of the most important factors in the acid treatment of lignocellulose is the determination of optimal conditions required to provide the maximum yield of fermentable sugars and the least amount of inhibitors. All the three hydrolysis parameters were significant variables for the yield of ethanol. Yield of ethanol decreases at very high and low hydrolysis temperature, hydrolysis time and acid concentration. The optimum combinations of the three factors chosen for optimum ethanol yield (10.86 ml/50 g sample) were 92.59°C (hydrolysis temperature), 30 minutes (hydrolysis time) and 1% v acid concentration. Ethanol production from wastepaper is doubtlessly an attractive business from economic and environmental point of view.

References

[1] Balat H. (2010). Prospects of biofuels for a sustainable energy future: A critical assessment. Energy Educ Sci Technol Part A; 24:85–111

[2] Campbell, C.J., Laherrere, J.H., (1998). The end of cheap oil. Sci. Am. 3, 78–83.

[3] Demirbas. (2008). Biofuels sources, biofuel policy, biofuel economy and global biofuel projections Energy Convers Manage, pp. 2106–2116.

[4] EIA. International Energy Outlook. (2009). Energy information administration office of integrated analysis and forecasting US Department of Energy. DOE/EIA-0484;

[5] Forum for Agricultural Research in Africa (FARA). (2008). Bioenergy value chain research and development Stakes and Opportunities. Burkina Faso, FARA Discussion Paper; April 08.<www.fara-africa.org/media/.../Bioenergy_Discussion_Paper. [Accessed November 2009].

[6] Galbe, M. and Zacchi, G. (2007). A review of the production of ethanol from softwood. *Appl. Biochem. Biotechnol.* 59: 618-628.

[7] Gaur, K. (2006). *Process optimization for the production of ethanol via fermentation.* Master thesis, Thapar Institute of Engineering and Technology, Patiala.

[8] Hahn-Hagerdal, B., Galbe, M., Gorwa-Grauslund, M., Liden, G. and Zacchi, G. (2006). Bioethanol – the fuel of tomorrow from the residues of today. *Tre. Biotechnol.* 24: 549-556.

[9] Halidini SarakikyaIkeda, Y., Park, E. Y., & Naoyuki, O. (2006). Bioconversion of waste office paper to gluconic acid in a turbine blade reactor by the filamentous fungus Aspergillus niger. Bioresource Technology, 97, 1030–1035.).

[10] Ministry of Finance and Economic Development (MoFED). (2010). *Ethiopia: Country Report on the Implementation of the Brussels Program of Action (BPOA).* Addis Ababa.

[11] Prasad S, Singh A, Joshi HC. (2007). Ethanol as an alternative fuel from agricultural, industrial and urban residues. Resour, Conserv Recycl 50:1–39.

[12] Stokes H. (2005).Alcohol fuels (ethanol and methanol): safety. In: Presentation at ETHOS conference, Seattle, Washington.

[13] Sun, Y. and Cheng, J. (2002). Hydrolysis of lignocellulosic materials for ethanol production: a review. *Biores. Technol.* 83: 1-11.

[14] *Taye, A. (2009). Conversion of banana and mango peel to ethanol. Addis Ababa.*

[15] Veronica Kavila Ngunzi. Analysis of Energy Cost Savings by Substituting Heavy Fuel Oil with Alternative Fuel for a Pozzolana Dryer. Case.

[16] Study of Bamburi Cement. *American Journal of Energy Engineering.* Vol. 3, No. 6, 2015, pp. 93-102. doi: 10.11648/j.ajee.20150306.13.

24

Extensive study on hydrodynamics and heat transfer of laminar mixed convection

De-Yi Shang[1], Liang-Cai Zhong[2]

[1]136 Ingersoll Cres., Ottawa, ON, Canada K2T 3W9
[2]Department of Ferrous Metallurgy, Northeastern University, Shenyang 110004, China

Email address:

deyishang@yahoo.ca (De-Yi Shang)

Abstract: Through extensive study on hydrodynamics and heat transfer, calculation correlations of heat transfer for laminar free/forced mixed convection on a vertical flat plate is obtained. It contains the following three research investigations: (i) local-similarity analysis and transformation based on our developed new similarity analysis method replacing the traditional Falkner-Skan type transformation; (ii) New governing local-similarity mathematical model, which is first applied in study of laminar free/forced mixed convection. It is more conveniently obtained and applied compared with that based on the Falkner-Skan type transformation for investigation of free/forced mixed convection; and (iii) New correlations on heat transfer of laminar free/forced mixed convection. They have wide coverage of Prandtl number and mixed convection parameter, and are suitable for all gases and important liquids including water for laminar free/forced mixed convection. The reported heat transfer correlations are so reliably because they are produced based on combination of theoretical analysis equations with the correlations formulated rigorously according to system of groups on accurate numerical solutions.

Keywords: Hydrodynamics, Heat transfer, Local-similarity transformation, Laminar mixed convection, Heat transfer correlations, New similarity method, Core similarity variables

1. Introduction

Free/forced mixed convective (for simplicity, hereinafter referred to as mixed convection）is a coupled phenomenon of free and forced convections, having even much challenging and wide practical application background. So far, the history of its study has over a half century. In the books of Gebhart et al. [1], Bejan [2] and Pop and Ingham [3] the detailed reviews were done for concerning the boundary layer equations on mixed convective flow. Meanwhile, numerous academic papers were published for investigation of mixed convection, and only some of them, such as refs. [4 -24], are listed here for saving space. In these investigations, the combined convection in a boundary layer flow was analysed. The studies, such as those of Acrivos [4], Sparrow et al. [5], Szewcyk [6], Chen [15], Merkin et al. [18], Ali [21], and Aydin el al. [22] investigated the effect of the parameter Gr_x/Re_x^2 on mixed convection. In most of these analyses, the stream function was induced based on the Falkner-Skan type transformation for obtaining the governing partial differential equations. Additionally, in some of the studies, the finite difference method such as in works of Oosthuizen [7], Chen [15], Hossain [17], Sami [20] and Anilkumar [24], the perturbation method such as in work of Yao [12] and Hossain [17], and the finite element method such as in work of Rana [23] were applied. The studies are also often found for free-forced mixed boundary layer flow in a porous medium, such as those presented by Cheng [8], Hsieh et al. [13] and Harris et al.[16], which broadened the range of application of mixed convection. Raju [10], Chen [11], Harris et al. [16], and Merkin et al. [18] studied the boundary-layer flow of mixed convection with the buoyancy forces either adding (positive flow) or opposing (negative flow) type to the bulk flow. Some of studies, such as those by Raju [10], Chen [11], and Aydin el al. [22] developed the expressions correlating Nusselt number in terms of the related parameters for further application of heat transfer.

For obtaining the correlations for heat transfer application of mixed convection, this present work will use three new approached. They are (i) local-similarity analysis and transformation based on our developed new similarity analysis method replacing the traditional Falkner-Skan type transformation; (ii) New governing local-similarity mathematical model, which is first applied in study of laminar free/forced mixed convection. It is more conveniently obtained and applied compared with that based on the Falkner-Skan type transformation for investigation of free/forced mixed convection; and (iii) New correlations on heat transfer of laminar free/forced mixed convection. They have wide coverage of Prandtl number and mixed convection parameter, and are suitable for all gases and important liquids including water for laminar free/forced mixed convection. We expect these developments will be in favor of meeting the challenges encountered for deep investigation of free/forced mixed convection.

2. Basic Conservation Equations

2.1. Governing Partial Differential Equations

In Fig.1 the physical model and co-ordinate system of boundary layer with two-dimensional mixed convection are shown schematically. A flat plate is vertically located in parallel fluid flow with its main stream velocity $w_{x,\infty}$. The plate surface temperature is t_w and the fluid bulk temperature is t_∞. Then, a coupled velocity boundary layer is produced near the plate. If the value of t_w is different from that of t_∞, a temperature boundary layer will occur near the plate. Then, Eqs. (1) to (3) can be taken as follows for governing partial differential equations of the laminar free/forced mixed convection boundary layer without consideration of variable physical properties and viscous thermal dissipation:

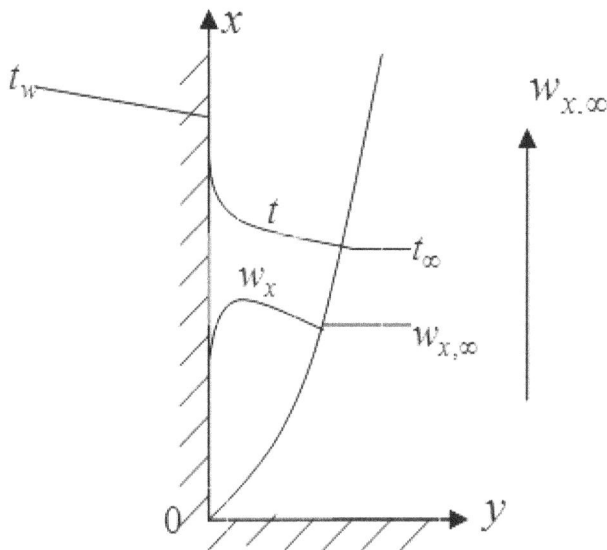

Fig. 1. *Physical model and coordinate system of boundary layer for laminar free/forced convection on a vertical flat plate*

$$\frac{\partial}{\partial x}(w_x) + \frac{\partial}{\partial y}(w_y) = 0 \tag{1}$$

$$w_x \frac{\partial w_x}{\partial x} + w_y \frac{\partial w_x}{\partial y} = v_f \frac{\partial^2 w_x}{\partial y^2} + g\beta(t - t_\infty) \tag{2}$$

$$w_x \frac{\partial t}{\partial x} + w_y \frac{\partial t}{\partial y} = \frac{v_f}{\Pr_f} \frac{\partial^2 t}{\partial y^2} \tag{3}$$

with the boundary condition equations

$$y = 0: \ w_x = 0, \ w_y = 0, \ t = t_w \tag{4}$$

$$y \to \infty: \ w_x = w_{x,\infty} \text{ (constant)}, \ t = t_\infty \tag{5}$$

where Eqs. (1) to (3) are continuity, momentum, and energy equations respectively, and the subscript f of the physical property variables denotes that the related reference temperature is the average temperature $t_f = \frac{t_w + t_\infty}{2}$.

3. Supposed Similarity Variables for Local-Similarity Transformation

For laminar mixed convection, full similarity transformation of the governing partial differential equations can not be achieved, and the local-similarity transformation should be done. The next work is further analysis for supposed similarity variables. We find that there are two key issues for mixed convection. They are: (i) The governing equations are attributed to free convection form; (ii) The boundary conditions belong to forced convection form. In this case, we select the core similarity variables with forced convection form for coincidence to the boundary conditions. Consulting our new similarity analysis method reported in refs. [25], the following equations with the core similarity variables are given for the similarity transformation of the velocity field:

$$w_x = w_{x,\infty} W_x(\eta, x) \tag{6}$$

$$w_y = w_{x,\infty} \left(\frac{1}{2} \mathrm{Re}_x\right)^{-1/2} W_y(\eta, x) \tag{7}$$

$$\theta(\eta, x) = \frac{t - t_\infty}{t_w - t_\infty} \tag{8}$$

Here, $W_x(\eta, x)$ and $W_y(\eta, x)$ are the core similarity variables for similarity transformation of the velocity field respectively in x and y directions, and $\theta(\eta, x)$ is supposed as the dimensionless temperature variable. While, the dimensionless coordinate variable η and the local Reynolds number $\mathrm{Re}_{x,f}$ are respectively described as

$$\mathrm{Re}_{x,f} = \frac{w_{x,\infty} x}{v_f} \tag{9}$$

$$\eta = \frac{y}{x}(\frac{1}{2}\text{Re}_{x,f})^{1/2} \qquad (10)$$

In addition, for the local-similarity analysis of the buoyancy term, the local Grashof number will be set as follow:

$$Gr_{x,f} = \frac{g\beta(t_w - t_\infty)}{v_f^2} \qquad (11)$$

It is indicated that the independent coordinate variable η and x related to the supposed similarity variables $W_x(\eta,x)$, $W_y(\eta,x)$, and $\theta(\eta,x)$ are taken to express the localn-similarity issue, and the subscript f denotes the reference temperature is average temperature $t_f = \frac{t_w + t_\infty}{2}$.

4. Local-Similarity Transformation of the Governing Equations

For local-similarity analysis on the supposed dimensionless coordinate variable η for the governing partial differential equations (1) to (3) of laminar free/forced mixed convection, the following transformation will be done:

4.1. Transformation of Eq.(1)

With Eqs.(6) we have

$$\frac{\partial w_x}{\partial x} = w_{x,\infty}(\frac{\partial W_x(\eta,x)}{\partial \eta} \frac{\partial \eta}{\partial x})$$

and with Eqs. (9) and (10) we have

$$\frac{\partial \eta}{\partial x} = -\frac{1}{2} y(\frac{1}{2}\frac{w_{x,\infty}}{v_f})^{1/2} x^{-3/2} = -\frac{1}{2}\eta x^{-1}$$

Then,

$$\frac{\partial w_x}{\partial x} = -\frac{1}{2} w_{x,\infty}\eta x^{-1} \frac{\partial W_x(\eta,x)}{\partial \eta}$$

With Eq. (7) we have

$$\frac{\partial w_y}{\partial y} = w_{x,\infty}(\frac{1}{2}\text{Re}_x)^{-1/2}(\frac{\partial W_y(\eta,x)}{\partial \eta} \frac{\partial \eta}{\partial y})$$

Where

$$\frac{\partial \eta}{\partial y} = x^{-1}(\frac{1}{2}\text{Re}_{x,\infty})^{1/2} \text{ , then,}$$

$$\frac{\partial w_y}{\partial y} = w_{x,\infty}(\frac{1}{2}\text{Re}_{x,\infty})^{-1/2} \frac{\partial W_y(\eta,x)}{\partial \eta} x^{-1}(\frac{1}{2}\text{Re}_{x,\infty})^{1/2}$$

$$= w_{x,\infty} \frac{\partial W_y(\eta,x)}{\partial \eta} x^{-1}$$

Therefore, Eq. (1) is transformed to

$$-\frac{1}{2}w_{x,\infty}\eta x^{-1}\frac{\partial W_x(\eta,x)}{\partial \eta} + w_{x,\infty}\frac{\partial W_y(\eta,x)}{\partial \eta}x^{-1} = 0$$

$$-\frac{1}{2}w_{x,\infty}\eta x^{-1}\frac{\partial W_x(\eta,x)}{\partial \eta} + w_{x,\infty}\frac{\partial W_y(\eta,x)}{\partial \eta}x^{-1} = 0$$

i. e.

$$-\eta\frac{\partial W_x(\eta,x)}{\partial \eta} + 2\frac{\partial W_y(\eta,x)}{\partial \eta} = 0 \qquad (12)$$

4.2. Transformation of Eq. (2)

With Eqs. (6) and (10) we have

$$\frac{\partial v_x}{\partial y} = w_{x,\infty}[\frac{\partial W_x(\eta,x)}{\partial \eta}\frac{\partial \eta}{\partial y}] = w_{x,\infty}\frac{\partial W_x(\eta,x)}{\partial \eta}x^{-1}(\frac{1}{2}\text{Re}_{x,\infty})^{1/2}$$

$$\frac{\partial^2 w_x}{\partial y^2} = w_{x,\infty}\frac{\partial^2 W_x(\eta,x)}{\partial \eta^2}x^{-1}(\frac{1}{2}\text{Re}_{x,\infty})^{1/2}\frac{\partial \eta}{\partial y}$$

$$= w_{x,\infty}\frac{\partial^2 W_x(\eta,x)}{\partial \eta^2}x^{-2}(\frac{1}{2}\text{Re}_{x,\infty})$$

Then, Eq. (2) is changed to

$$w_{x,\infty}W_x(\eta,x)[-\frac{1}{2}w_{x,\infty}\eta x^{-1}\frac{\partial W_x(\eta,x)}{\partial \eta}]$$

$$+w_{x,\infty}(\frac{1}{2}\text{Re}_x)^{-1/2}W_y(\eta,x)w_{x,\infty}\frac{\partial W_x(\eta,x)}{\partial \eta}x^{-1}(\frac{1}{2}\text{Re}_{x,\infty})^{1/2}$$

$$= v_f w_{x,\infty}\frac{\partial^2 W_x(\eta,x)}{\partial \eta^2}x^{-2}(\frac{1}{2}\text{Re}_{x,\infty}) + g\beta(t-t_\infty)$$

The above eq. should be

$$w_{x,\infty}W_x(\eta,x)[-\frac{1}{2}w_{x,\infty}\eta x^{-1}\frac{\partial W_x(\eta,x)}{\partial \eta}]$$

$$+w_{x,\infty}(\frac{1}{2}\text{Re}_{x,\infty})^{-1/2}W_y(\eta,x)w_{x,\infty}\frac{\partial W_x(\eta,x)}{\partial \eta}x^{-1}(\frac{1}{2}\text{Re}_{x,\infty})^{1/2}$$

$$= v_f w_{x,\infty}\frac{\partial^2 W_x(\eta,x)}{\partial \eta^2}x^{-2}(\frac{1}{2}\text{Re}_{x,\infty}) + g\beta(t-t_\infty)$$

i.e.

$$w_{x,\infty}W_x(\eta,x)[-\frac{1}{2}w_{x,\infty}\eta x^{-1}\frac{\partial W_x(\eta,x)}{\partial \eta}]$$

$$+w_{x,\infty}W_y(\eta,x)w_{x,\infty}\frac{\partial W_x(\eta,x)}{\partial \eta}x^{-1}$$

$$= v_f w_{x,\infty}\frac{\partial^2 W_x(\eta,x)}{\partial \eta^2}x^{-2}(\frac{1}{2}\text{Re}_{x,\infty}) + g\beta(t-t_\infty)$$

The above equation is divided by $\frac{1}{2}x^{-1}w_{x,\infty}^2$, and changed to

$$[W_x(\eta,x)[-\eta\frac{\partial W_x(\eta,x)}{\partial\eta}]+2W_y(\eta,x)\frac{\partial W_x(\eta,x)}{\partial\eta}]$$

$$=v_f\frac{\partial^2 W_x(\eta,x)}{\partial\eta^2}(\frac{1}{v_f})+2g\beta(t-t_\infty)\frac{x}{w_{x,\infty}^2}$$

The above equation is changed equivalently to

$$[W_x(\eta,x)(-\eta\frac{\partial W_x(\eta,x)}{\partial\eta})+2W_y(\eta,x)\frac{dW_x(\eta,x)}{\partial\eta}]=\frac{\partial^2 W_x(\eta,x)}{\partial\eta^2}+\frac{t-t_\infty}{t_w-t_\infty}2\frac{g\beta(t_w-t_\infty)x^3}{v_f^2}\mathrm{Re}_{x,f}^{-2}$$

i.e.

$$[W_x(\eta,x)(-\eta\frac{\partial W_x(\eta,x)}{\partial\eta})+2W_y(\eta,x)\frac{\partial W_x(\eta,x)}{\partial\eta}]$$

$$=\frac{\partial^2 W_x(\eta,x)}{\partial\eta^2}+2\theta(\eta,x)\cdot Gr_{x,f}\,\mathrm{Re}_{x,f}^{-2}$$

or

$$[-\eta W_x(\eta,x)+2W_y(\eta,x)]\frac{\partial W_x(\eta,x)}{\partial\eta}]=\frac{\partial^2 W_x(\eta,x)}{\partial\eta^2}+2\theta(\eta,x)\cdot Mc \quad (13)$$

where

$Mc=Gr_{x,f}\,\mathrm{Re}_{x,f}^{-2}$ is Richardson number or called mixed convection parameter. It demonstrates the effective rate of the free convection in the mixed convection. Theoretically, the value for the mixed convection parameter Mc can be in a large range $-\infty<Mc<+\infty$. The free/forced mixed convection is divided to two types, respectively for positive (adding) and negative (opposing) flows. The former flow is corresponding to $Mc>0$, and the latter flow is corresponding to $Mc<0$. Obviously, $Mc=0$ and $Mc\to\pm\infty$ are corresponding to net forced convection and free convection respectively.

4.3. Transformation of Eq. (3)

With Eq. (8) we have

$$\frac{\partial t}{\partial x}=(t_w-t_\infty)\frac{\partial\theta(\eta,x)}{\partial\eta}\frac{\partial\eta}{\partial x}$$

Therefore,

$$\frac{\partial t}{\partial x}=(t_w-t_\infty)\frac{\partial\theta(\eta,x)}{\partial\eta}(-\frac{1}{2}\eta x^{-1})$$

$$=(t_w-t_\infty)(-\frac{1}{2}\eta x^{-1})\frac{\partial\theta(\eta,x)}{\partial\eta}$$

With Eq. (8) we have

$$\frac{\partial t}{\partial y}=(t_w-t_\infty)\frac{\partial\theta(\eta,x)}{\partial\eta}\frac{\partial\eta}{\partial y}$$

Therefore,

$$\frac{\partial t}{\partial y}=(t_w-t_\infty)\frac{\partial\theta(\eta,x)}{\partial\eta}x^{-1}(\frac{1}{2}\mathrm{Re}_{x,\infty})^{1/2}$$

$$[W_x(\eta,x)(-\eta\frac{\partial W_x(\eta,x)}{\partial\eta})+2W_y(\eta,x)\frac{\partial W_x(\eta,x)}{\partial\eta}]$$

$$=\frac{\partial^2 W_x(\eta,x)}{\partial\eta^2}+2\frac{g\beta(t-t_\infty)x^3}{v_f^2}\frac{v_f^2}{w_{x,\infty}^2 x^2}$$

The above equation is further equivalently changed to

$$\frac{\partial^2 t}{\partial y^2}=x^{-2}(\frac{1}{2}\mathrm{Re}_{x,\infty})(t_w-t_\infty)\frac{\partial^2\theta^2(\eta,x)}{\partial\eta^2}$$

Then, Eq.(3) is changed to

$$w_{x,\infty}W_x(\eta,x)(t_w-t_\infty)[-\frac{1}{2}\eta x^{-1}\frac{\partial\theta(\eta,x)}{\partial\eta}]$$

$$+w_{x,\infty}(\frac{1}{2}\mathrm{Re}_x)^{-1/2}W_y(\eta,x)(t_w-t_\infty)\frac{\partial\theta(\eta,x)}{\partial\eta}x^{-1}(\frac{1}{2}\mathrm{Re}_{x,\infty})^{1/2}$$

$$=\frac{v_f}{\mathrm{Pr}_f}x^{-2}(\frac{1}{2}\mathrm{Re}_{x,\infty})(t_w-t_\infty)\frac{\partial^2\theta(\eta,x)}{\partial\eta^2}$$

i.e.

$$w_{x,\infty}W_x(\eta,x)(t_w-t_\infty)[-\eta x^{-1}\frac{\partial\theta(\eta,x)}{\partial\eta}]+2w_{x,\infty}W_y(\eta,x)(t_w-t_\infty)\frac{\partial\theta(\eta,x)}{\partial\eta}x^{-1}$$

$$=\frac{v_f}{\mathrm{Pr}_f}x^{-2}(\frac{xw_{x,\infty}}{v_f})(t_w-t_\infty)\frac{\partial^2\theta(\eta,x)}{\partial\eta^2}$$

The above equation is divided by $x^{-1}w_{x,\infty}(t_w-t_\infty)$, and transformed to

$$W_x(\eta,x)[-\eta\frac{\partial\theta(\eta,x)}{\partial\eta}]+2W_y(\eta,x)\frac{\partial\theta(\eta,x)}{\partial\eta}=\frac{1}{\mathrm{Pr}_f}\frac{\partial^2\theta(\eta,x)}{\partial\eta^2}$$

or

$$[-\eta W_x(\eta,x)+2W_y(\eta,x)]\frac{\partial\theta(\eta,x)}{\partial\eta}=\frac{1}{\mathrm{Pr}_f}\frac{\partial^2\theta(\eta,x)}{\partial\eta^2} \quad (14)$$

Then, with the local-summary transformation, Eqs. (1) to (3) are transformed to the following equivalent governing ordinary differential equations:

$$-\eta\frac{\partial W_x(\eta,x)}{\partial\eta}+2\frac{\partial W_y(\eta,x)}{\partial\eta}=0 \quad (12)$$

$$[-\eta W_x(\eta,x)+2W_y(\eta,x)]\frac{\partial W_x(\eta,x)}{\partial\eta}]=\frac{\partial^2 W_x(\eta,x)}{\partial\eta^2}+2\theta(\eta,x)\cdot Mc \quad (13)$$

$$[-\eta W_x(\eta,x)+2W_y(\eta,x)]\frac{\partial\theta(\eta,x)}{\partial\eta}=\frac{1}{\mathrm{Pr}_f}\frac{\partial^2\theta(\eta,x)}{\partial\eta^2} \quad (14)$$

with the following dimensionless equations on the boundary layer conditions:

$$\eta=0:\ W_x(\eta,x)=0,\ W_y(\eta,x)=0,\ \theta(\eta,x)=1 \quad (15)$$

$$\eta \to \infty : \ W_x(\eta, x) = 1, \ \ \theta(\eta, x) = 0 \qquad (16)$$

5. Rewriting the Second Independent Coordinate Variable

From the defined equation of the mixed convection parameter Mc, it is seen that the independent coordinate variable x is included in the mixed convection parameter Mc. From the transformed governing equations (12) to (14), it is found that the supposed second coordinate variable x exists in the transformed governing equations only in form of the mixed convection parameter Mc. Then, the mixed parameter Mc is reasonably taken as the second independent coordinate variable to replace the independent coordinate variable x. In this case, the transformed dimensionless governing Eqs.(12) to (14) are equivalent to the following ones respectively

$$-\eta \frac{\partial W_x(\eta, Mc)}{\partial \eta} + 2\frac{\partial W_y(\eta, Mc)}{\partial \eta} = 0 \qquad (12)^*$$

$$[-\eta W_x(\eta, Mc) + 2W_y(\eta, Mc)]\frac{\partial W_x(\eta, Mc)}{\partial \eta}]$$
$$= \frac{\partial^2 W_x(\eta, Mc)}{\partial \eta^2} + 2\theta(\eta, Mc) \cdot Mc \qquad (13)^*$$

$$[-\eta W_x(\eta, Mc) + 2W_y(\eta, Mc)]\frac{\partial \theta(\eta, Mc)}{\partial \eta} = \frac{1}{Pr_f}\frac{\partial^2 \theta(\eta, Mc)}{\partial \eta^2} \qquad (14)^*$$

with the following equivalent dimensionless equations on the boundary conditions

$$\eta = 0: \ W_x(\eta, Mc) = 0, \ W_y(\eta, Mc) = 0, \ \theta(\eta, Mc) = 1 \quad (15)^*$$

$$\eta \to \infty: \ W_x(\eta, Mc) = 1, \ \ \theta(\eta, Mc) = 0 \qquad (16)^*$$

and the corresponding equivalent similarity variables $W_x(\eta, x)$, $W_y(\eta, x)$ and $\theta(\eta, x)$ with the following relations:

$$w_x = w_{x,\infty} W_x(\eta, Mc) \qquad (9)^*$$

$$w_y = w_{x,\infty}(\frac{1}{2}Re_x)^{-1/2}W_y(\eta, Mc) \qquad (10)^*$$

$$\theta(\eta, Mc) = \frac{t - t_\infty}{t_w - t_w} \qquad (11)^*$$

6. Numerical Results

The governing dimensionless local-similarity equations $(12)^*$ to $(14)^*$ with their boundary condition equations $(15)^*$ and $(16)^*$ are solved by a shooting method with fifth-order Runge-Kutta integration. For solving the nonlinear problem, a variable mesh approach is applied to the numerical calculation programs. It can be seen that for ignoring the variable physical properties, the solutions of the dimensionless velocity components

$W_x(\eta, Mc)$ and $W_y(\eta, Mc)$ and the dimensionless temperature $\theta(\eta, Mc)$ are dependent on Prandtl number Pr_f and mixed convection parameter Mc. The accurate numerical results of the velocity and temperature fields are obtained in the wide ranges of Prandtl number Pr_f ($0.3 \le Pr_f \le 20$) and mixed parameter Mc ($0 \le Mc \le 10$). Some of the selected numerical results for the dimensionless velocity component $W_x(\eta, Mc)$ and temperature $\theta(\eta, Mc)$ fields are plotted in Figs. 2 to 7 with variation of the dimensionless coordinate variable η.

It is seen that with increasing the Prandtl number, the velocity will decrease monotonically, and the velocity and temperature boundary layer thickness will decrease monotonically, too. In addition, with increasing the mixed convection parameter Mc, the velocity will increase monotonically. However, the velocity and temperature boundary layer thickness will not obviously change with variation of the mixed convection parameter Mc.

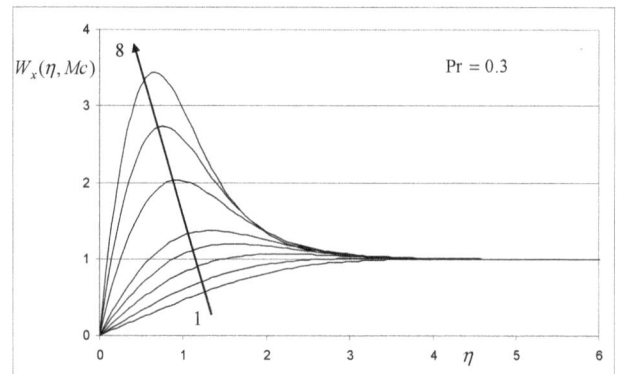

a

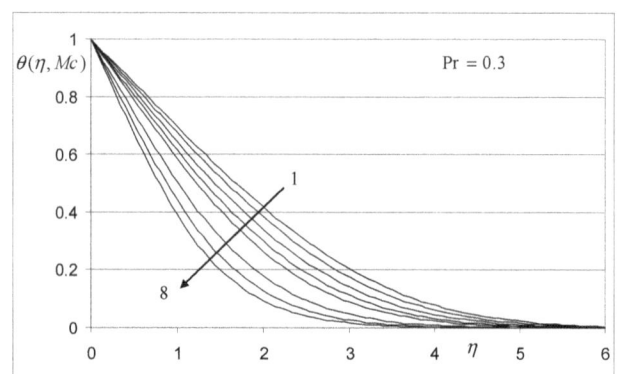

b

Note: Lines 1 to 8 denote Mc = 0, 0.1, 0.3, 0.6, 1, 3, 6 and 10 respectively

Fig. 2. *Numerical results of (a) velocity component $W_x(\eta, Mc)$ and (b) temperature $\theta (\eta, Mc)$ profiles for laminar fixed convection with fluid for Pr=0.3 on a vertical flat plate*

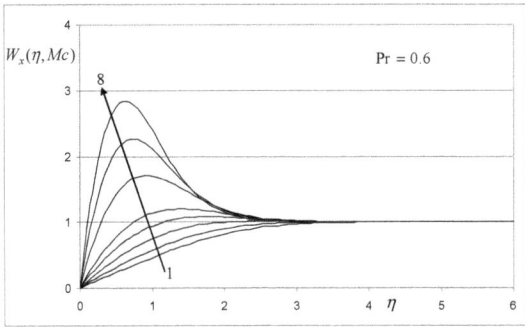

a

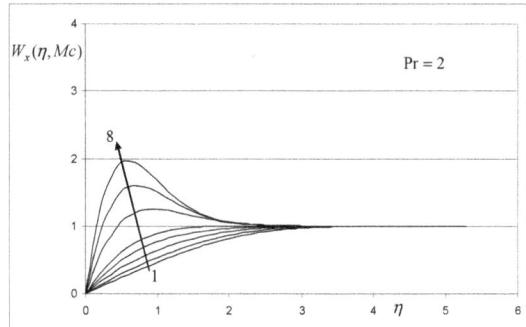

b

Note: Lines 1 to 8 denote Mc = 0, 0.1, 0.3, 0.6, 1, 3, 6 and 10 respectively

Fig. 3. *Numerical results of (a) velocity component $W_x(\eta, Mc)$ and (b) temperature $\theta (\eta, Mc)$ profiles for laminar fixed convection with fluid for Pr=0.6 on a vertical flat plate*

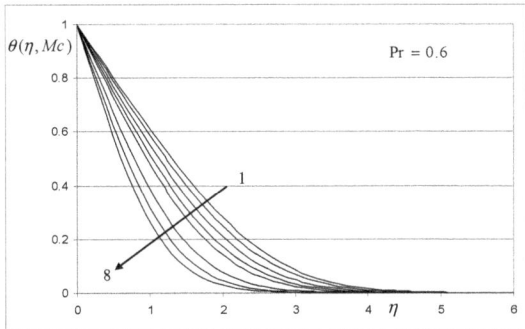

a

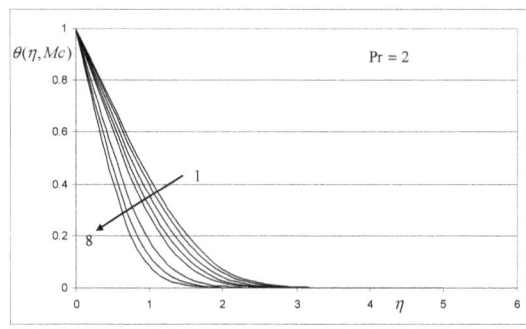

b

Note: Lines 1 to 8 denote Mc = 0, 0.1, 0.3, 0.6, 1, 3, 6 and 10 respectively

Fig. 4. *Numerical results of (a) velocity component $W_x(\eta, Mc)$ and (b) temperature $\theta (\eta, Mc)$ profiles for laminar fixed convection with fluid for Pr=1.0 on a vertical flat plate*

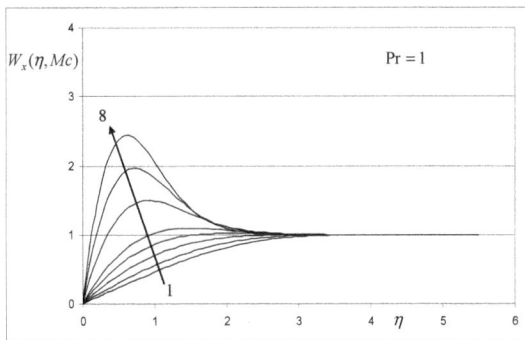

a

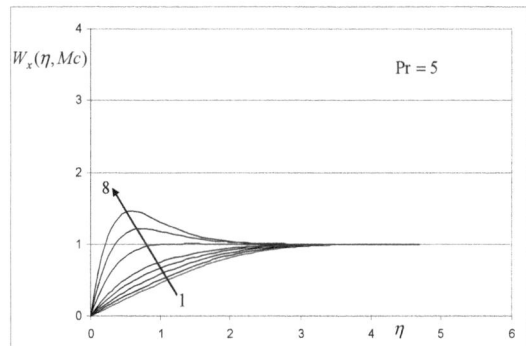

b

Note: Lines 1 to 8 denote Mc = 0, 0.1, 0.3, 0.6, 1, 3, 6 and 10 respectively

Fig. 5. *Numerical results of (a) velocity component $W_x(\eta, Mc)$ and (b) temperature $\theta (\eta, Mc)$ profiles for laminar fixed convection with fluid for Pr=2.0 on a vertical flat plate*

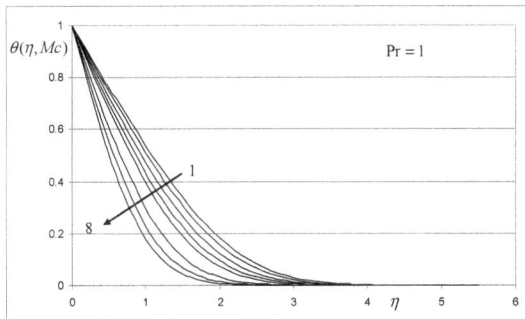

a

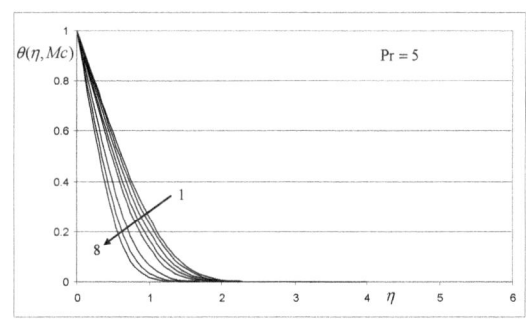

b

Note: Lines 1 to 8 denote Mc = 0, 0.1, 0.3, 0.6, 1, 3, 6 and 10 respectively

Fig. 6. *Numerical results of (a) velocity component $W_x(\eta, Mc)$ and (b) temperature $\theta (\eta, Mc)$ profiles for laminar fixed convection with fluid for Pr=5.0 on a vertical flat plate*

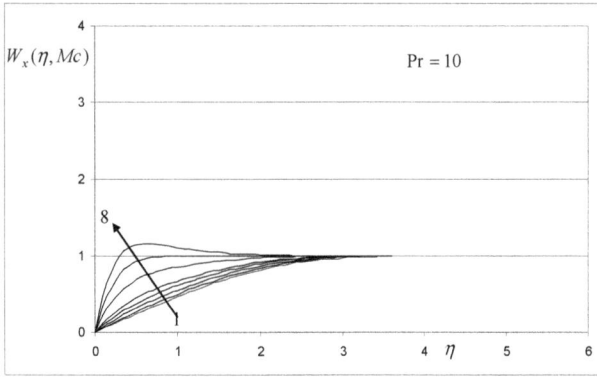

a

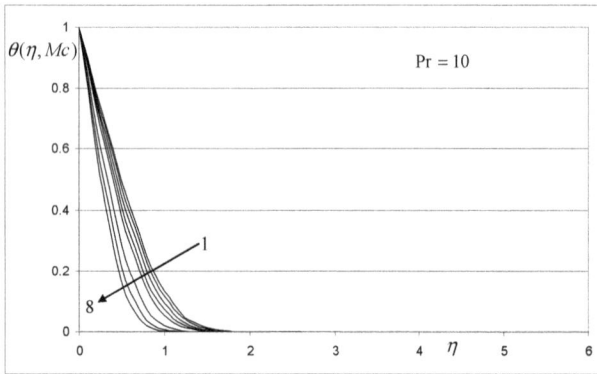

b

Note: Lines 1 to 8 denote Mc = 0, 0.1, 0.3, 0.6, 1, 3, 6 and 10 respectively

Fig. 7. *Numerical results of (a) velocity component $W_x(\eta, Mc)$ and (b) temperature θ (η, Mc) profiles for laminar fixed convection with fluid for Pr=10.0 on a vertical flat plate*

7. Skin-Friction Coefficient

So far, the skin friction coefficient has been analyzed by means of Falkner-Skan type transformation for free/forced mixed convection. In the present work, we will analyze it based on the present new similarity analysis procedure. For this analysis, the velocity gradient at the wall is important characteristic of the solution, and the local skin-friction coefficient $C_{x,f}$ is a dimensionless measure of the shear stress at the wall, i.e.

$$C_{x,f} = \frac{\tau_{w,x}}{\frac{1}{2}\rho_f w_{x,\infty}^2} = \frac{\mu_f (\frac{\partial w_x}{\partial y})_{y=0}}{\frac{1}{2}\rho_f w_{x,\infty}^2} \qquad (17)$$

According to the related derivation in section 3, we have

$$(\frac{\partial w_x}{\partial y})_{y=0} = x^{-1}(\frac{1}{2}Re_{x,f})^{1/2} w_{x,\infty} (\frac{\partial W_x(\eta, Mc)}{\partial \eta})_{\eta=0}$$

Therefore, the local skin-friction coefficient $C_{f,x}$ is expressed as

$$C_{x,f} = \frac{\mu_f x^{-1}(\frac{1}{2}Re_{x,f})^{1/2} w_{x,\infty} (\frac{\partial W_x(\eta, Mc)}{\partial \eta})_{\eta=0}}{\frac{1}{2}\rho_f w_{x,\infty}^2}$$

$$= \sqrt{2}\frac{v_f}{x w_{x,\infty}}(Re_{x,f})^{1/2}(\frac{\partial W_x(\eta, Mc)}{\partial \eta})_{\eta=0}$$

i.e.

$$C_{x,f} = \sqrt{2}(Re_{x,f})^{-1/2}(\frac{\partial W_x(\eta, Mc)}{\partial \eta})_{\eta=0} \qquad (18)$$

A system of the rigorous numerical solutions on the wall velocity gradient $(\frac{\partial W_x(\eta, Mc)}{\partial \eta})_{\eta=0}$ has been obtained numerically. The selected solutions of them are plotted in Fig. 8 with variation of Prandtl number Pr_f and mixed convection parameter Mc. Then, it is seen that the velocity gradient on the wall, $(\frac{\partial W_x(\eta, Mc)}{\partial \eta})_{\eta=0}$ increases with increasing the mixed convection parameter Mc, and decreases with increasing the Prandtl number Pr_f . In addition, with increasing the mixed convection parameter Mc, the velocity gradient on the wall, $(\frac{\partial W_x(\eta, Mc)}{\partial \eta})_{\eta=0}$ will decrease in an accelerative pace.

From Eq. (18) it is seen that the wall velocity gradient $(\frac{\partial W_x(\eta, Mc)}{\partial \eta})_{\eta=0}$ is the only one unknown variable for evaluation of the skin-friction coefficient.

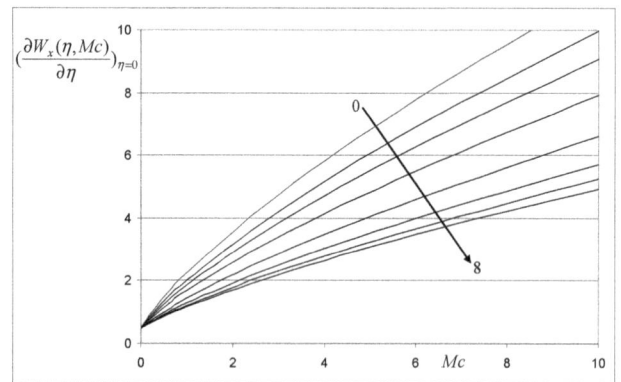

Fig. 8. *Numerical results of velocity gradient on the plate surface, $(\frac{\partial W_x(\eta, Mc)}{\partial \eta})_{\eta=0}$ varying with Prandtl number, Pr and mixed convection parameter Mc (Lines 1 to 8 denote Pr_f = 0.3, 0.6, 1, 2, 5, 10, 15 and 20 respectively).*

8. Heat Transfer

8.1. Theoretical Equations of heat transfer

The local heat transfer rate $q_{x,f}$ at position x per unit area from the surface of the plate without consideration of variable physical properties can be calculated by Fourier's law as $q_{x,f} = -\lambda_f (\frac{\partial t}{\partial y})_{y=0}$. The subscript f denotes the case that the boundary layer average temperature $t_f = \frac{t_w + t_\infty}{2}$ is taken as the reference temperature with ignoring the variable physical properties.

With Eqs. (8) and (10) we have

$$q_{x,f} = -\lambda_f (t_w - t_\infty)(\frac{\partial \theta(\eta, Mc)}{\partial \eta})_{\eta=0} \frac{\partial \eta}{\partial y}$$

where

$$\frac{\partial \eta}{\partial y} = x^{-1}(\frac{1}{2} Re_{x,\infty})^{1/2}$$

Then,

$$q_{x,f} = \lambda_f (t_w - t_\infty)(\frac{1}{2} Re_{x,f})^{1/2} x^{-1}(-\frac{\partial \theta(\eta, Mc)}{\partial \eta})_{\eta=0}$$

The local heat transfer coefficient $\alpha_{x,f}$, defined as $q_{x,f} = \alpha_{x,f}(t_w - t_\infty)$, will be given by

$$\alpha_{x,f} = \lambda_f (\frac{1}{2} Re_{x,f})^{1/2} x^{-1}(-\frac{\partial \theta(\eta, Mc)}{\partial \eta})_{\eta=0}$$

The local Nusselt number, defined by $Nu_{x,f} = \frac{\alpha_{x,f} \cdot x}{\lambda_f}$, will be

$$Nu_{x,f} = (\frac{1}{2} Re_{x,f})^{1/2} (-\frac{\partial \theta(\eta, Mc)}{\partial \eta})_{\eta=0} \qquad (19)$$

Here, Equation on local Nusselt number(19 is heat transfer theoretical equation.

8.2. Formulization of Dimensionless Wall Temperature Gradient

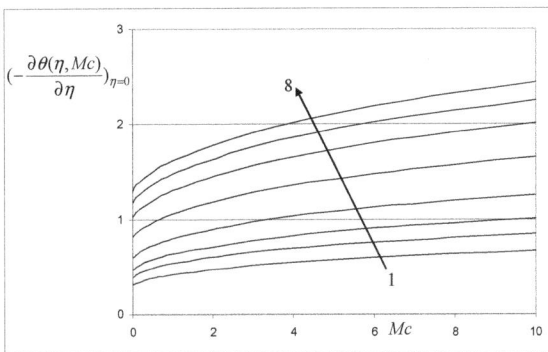

Fig. 9. Numerical results of wall temperature gradient on the plate surface, $(-\frac{\partial \theta(\eta, Mc)}{\partial \eta})_{\eta=0}$ *varying with Prandtl number Pr and mixed convection parameter Mc (Lines 1 to 8 denote Pr = 0.3, 0.6, 1, 2, 5, 10, 15 and 20 respectively)*

It is seen from the heat transfer theoretical equation (19) that the dimensionless temperature gradient $(-\frac{\partial \theta(\eta, Mc)}{\partial \eta})_{\eta=0}$ as the only one unknown variable dominates the prediction of mixed convection heat transfer. In this present work, a system of rigorous solutions on the dimensionless temperature gradient $(-\frac{\partial \theta(\eta, Mc)}{\partial \eta})_{\eta=0}$ have been obtained numerically, and their selected values are plotted in Fig. 9. It is seen that the wall temperature gradient $(-\frac{\partial \theta(\eta, Mc)}{\partial \eta})_{\eta=0}$ increases with increasing the mixed parameter Mc, and increases with increasing the Prandtl number Pr_f.

The system of rigorous numerical solutions of the wall temperature gradient $(-\frac{\partial \theta(\eta, Mc)}{\partial \eta})_{\eta=0}$ are formulated by means of a curve-fitting approach, and the obtained formulated correlations are shown as follows with the wide ranges of Prandtl number between 0.3 and 20 and the mixed convection parameter Mc between 0 and 10:

$$(-\frac{\partial \theta(\eta, Mc)}{\partial \eta})_{\eta=0} = a Pr_f^{\ b} \qquad (0.3 \le Pr \le 20) \quad (20)$$

where

$$a = \frac{(-0.0002Mc^4 + 0.0039Mc^3 - 0.0262Mc^2 + 0.6179Mc + 0.4699)}{(1 + Mc)^{0.75}}$$

$$(0 \le Mc \le 10) \qquad (21)$$

$$b = \frac{(0.00009Mc^4 - 0.0018Mc^3 + 0.0117Mc^2 + 0.2771Mc + 0.3433)}{1 + Mc}$$

$$(0 \le Mc \le 10) \qquad (22)$$

Here, Eqs.(19) to (22) are heat transfer correlations of laminar mixed convection on a vertical flat plate.

If Mc = 0, Eqs. (21) to (23) will be attributed to the forced convection issue as follows:

$$(-\frac{\partial \theta(\eta, Mc)}{\partial \eta})_{\eta=0} = 0.4699 Pr_f^{\ 0.3433} \qquad (20)^*$$

It is seen that Eq. (20)* is equivalent to Eq. (5.12) derived in [26] for laminar forced convection, and obviously is corresponding to well-known Pohlhausen equation [26]. Obviously, forced convection is only a special case of mixed convection.

9. Conclusions

In this work, the heat transfer correlations of laminar mixed convection on a vertical flat plate are obtained by combination of the theoretical heat transfer equations with the formulization correlations based on the system of accurate numerical solutions of dimensionless temperature gradient, which contains the following three novel approaches:

First, a new local-similarity analysis method is provided for extensive studies on laminar mixed convection to replace the traditional Falkner-Skan type transformation.

Second, a novel governing local-similarity model is derived based the above local-similarity analysis method for extensive study on laminar mixed convection. This model is easier to be derived and can be more conveniently applied than that based on the Falkner-Skan type transformation.

Third, the novel correlations of heat transfer are devoted for heat transfer application of laminar mixed convection. The correlations have so wide coverage berceuse with the ranges of Prandtl number from 0.3 to 20 and mixed convection parameter from 0 to 10 they are suitable for prediction of heat transfer of mixed convection of all gases and some important liquids including water. In additions, the heat transfer correlations are produced rigorously because they are based on combination of heat transfer theoretical equation with formulization correlations created according to a system of reliable numerical solutions of dimensionless temperature gradient.

The present work provides a foundation on theoretical methodology and governing mathematical model for our further investigation of challenge issue on mixed convection heat transfer with consideration of coupled effect of variable physical properties.

Nomenclature

b	exponent in equation
$C_{f,x}$	local skin-friction coefficient
g	Gravity acceleration, m/s^2
$Gr_{x,f}$	local Grashof number, $Gr_{x,f}$
Mc	Mixed convection parameter, $Mc = Gr_{x,f}\,\mathrm{Re}_{x,f}^{-2}$
$Nu_{x,f}$	local Nusselt number
Pr	Prandtl number
$\mathrm{Re}_{x,f}$	local Reynolds number
s	width, m
t	temperature, ^{o}C
t_w	wall temperature, ^{o}C
t_∞	bulk temperature, ^{o}C
t_f	mean temperature, ℃
x, y	coordinate variables, m
w_x, w_y	velocity components in x and y directions, respectively, m/s
w_x, w_y	dimensionless similarity variables on velocity components w_x, w_y, respectively

$w_{x,\infty}$	velocity of bulk flow, m/s
η	dimensionless coordinate variable
β	expansion coefficient, 1/K
ν	kinetic viscosity, m^2/s
θ	dimensionless temperature variable
$\tau_{w,x}$	local wall skin shear force, N/m^2
$(-\dfrac{\partial\theta(\eta,Mc)}{\partial\eta})_{\eta=0}$	dimensionless temperature gradient on the wall

Subscript

f	average value
w	wall surface
∞	far from the wall surface

References

[1] B. Gebhart and Y. Jaluria, Mahajan RL, Sammakia B. Buoyancy-induced flows and transfort. Hemisphere; 1988.

[2] A. Bejan, Convective heat transfer. Willey Inter Science; 1994.

[3] I. Pop and D.B. Ingham, Convective heat transfer: mathematical and computational odelling of viscous fluids and porous media. Elsevier UK, 2001.

[4] A. Acrivos, Combined laminar free and forced convection heat transfer in external flows, AIChE J. 4, 1958, 285-289

[5] E.M. Sparrow, R. Eichhorn, J.L Gregg, Combined forced and free convection in a boundary layer flow. Phys. of Fluids 2 (1959) 319-328

[6] A.A. Szewcyk, "Combined Forced and Free Convection Laminar Flow," Jour. Heat Transfer, Trans. ASME, Series C, Vol. 86, No. 4, November 1964, pp. 501-507.

[7] P.H. Oosthuizen and R. Hart, A numerical study of laminar combined convection flow over flat plates. J. Heat Transfer 95 (1973) 60 63

[8] P. Cheng, Combined free and forced boundary layer flows about inclined surfaces in a porous medium, Int. J. Neat Mass Transfer 20, 807-814 (1977).

[9] T.S. Chen and C.F. Yuh, A. Moutsoglou, Combined heat and mass transfer in mixed convection along vertical and inclined plates, International Journal of Heat and Mass transfer, Volume 23, Issue 4, 1980, Pages 527–537

[10] M.S. Raju, X.Q. Liu and C.K. Law, A formulation of combined forced and free convection past horizontal and vertical surfaces. Int. J. Heat Mass Transfer 27 (1984) 2215-2224

[11] T.S. Chen and B.F. Armaly, N. Ramachandran, Correlations for laminar mixed convection flows on vertical, inclined, and horizontal flat plates, ASME J. Heat Transfer 108 (1986) 835–840.

[12] L.S. Yao, Two-dimensional mixed convection along a flat plate, ASME J. Heat Transfer. 109 (1987) 440–445.

[13] J.C. Hsieh, T.S. Chen and B.F. Armaly, "Non-similarity solutions for mixed convection from vertical surfaces in porous media: variable surface temperature or heat flux", International Journal of Heat and Mass Transfer, Vol. 36, 1993, pp. 1485-93

[14] N. G. Kafoussias and E. W. Williams, The effect of temperature-dependent viscosity on free-forced convective laminar boundary layer flow past a vertical isothermal flat plate, Acta Mechanica Volume 110, Numbers 1-4, 123-137, 1995

[15] C.H. Chen, Laminar mixed convection adjacent to vertical, continuously stretching sheets, Heat and Mass Transfer, Volume 33, Issue 5/6, pp. 471-476 (1998).

[16] S.D. Harris, D.B. Ingham and I. Pop, Unsteady mixed convection boundary-layer flow on a vertical surface in a porous medium, Int. J. Heat Mass Transfer J. Heat and Mass Transfer, 42: 357 – 372, 1999

[17] M. A. Hossain and M. S. Munir, Mixed convection flow from a vertical flat plate with temperature dependent viscosity, Int. J. Therm. Sci. (2000) 39, 173–183

[18] J.H. Merkin and I. Pop, 2002. Mixed convection along a vertical surface: similarity solutions for uniform flow, Fluid Dyn. Res. 30, 233–250.

[19] H. Steinr'uck, About the physical relevance of similarity solutions of the boundary-layer flow equations describing mixed convection flow along a vertical plate, Fluid Dynamics Research 32 (2003) 1–13

[20] Sami A. Al-Sanea, Mixed convection heat transfer along a continuously moving heated vertical plate with suction or injection, International Journal of Heat and Mass Transfer 47 (2004) 1445–1465

[21] M.E. Ali, The effect of variable viscosity on mixed convection heat transfer along a vertical moving surface, Int. J. Therm. Sci. 45 (2006) 60–69

[22] Aydin O, Kaya A (2007) Mixed convection of a viscous dissipating fluid about a vertical flat plate. Appl Math Model 31:843–853

[23] P. Rana, R. Bhargava, Numerical study of heat transfer enhancement in mixed convection flow along a vertical plate with heat source/sink utilizing nanofluids, Commun Nonlinear Sci Numer Simulat 16 (2011) 4318–4334

[24] Devarapu Anilkumar, Nonsimilar solutions from a moving vertical, /Commun Nonlinear Sci Numer Simulat 16 (2011) 3147–3157

[25] D.Y. Shang, "Theory of heat transfer with forced convection film flows", Springer-Verlag, Berlin, Heldberg2011

[26] E. Pohlhausen, Der Warmeaustausch zwischen festen Korpern und Flussigkeiten mit kleiner Reiburg und kleiner Warmeleitung. Z. Angew. Math. Mech.1, 115-121, 1921.

Wind Speed Forecasting in China

Huiru Zhao, Sen Guo*

School of Economics and Management, North China Electric Power University, Changping District, Beijing, China

Email address:

guosen324@163.com (Sen Guo), guosen@ncepu.edu.cn (Sen Guo)

Abstract: China's wind power has developed rapidly in the past few years, the large-scale penetration of which will bring big influence on power systems. The wind speed forecasting research is quite important because it can alleviate the negative impacts. This paper reviews the current wind speed forecasting techniques in China. The literature (written in Chinese) sources and classification were firstly analyzed, and then the wind speed forecasting techniques in China were detailed reviewed from four aspects, which are statistical method, soft computing method, hybrid forecasting method and other forecasting methods. This paper can rich the current research in the field of wind speed forecasting.

Keywords: Wind Speed Forecasting, Forecasting Techniques, China

1. Introduction

In the past few years, the wind energy, as a kind of renewable energy, has experiences a rapid development. At the end of 2013, the installed capacity of wind power in the world has amounted to 318105 MW. Of which, the installed wind power capacity of China reached to 91412 MW, taking a 28.74% share of the whole world [1]. In 2010, China's cumulative installed wind power capacity amounted to 41827MW, which surpasses that of USA in term of installed capacity to rank first in the world and has kept the first place till now [2]. It is foreseeable that under the strong support of renewable energy supporting development policies from central government and local governments, the wind power in China will still develop fast in the next few year. It is predicted that the installed wind power capacity in China will increase to 0.2 billion kW at the end of 2020, and the wind power generation will account for 5% in the total power generation [3].

Wind power has the intermittent and fluctuant characteristics. Large-scale wind power penetration will bring new challenges to the safe and stable operation of power systems. When the penetration rate of wind power exceeds to a certain value, it will certainly affect the electric power quality and safe operation of power systems. Therefore, accurately forecasting the wind speed and wind power generation play an important role in planning and designing of wind farms, performing unit commitment, keeping the safe operation of power systems and improving the economic and social benefits.

China's wind power develops fast, which accesses to grid will take a huge impact on the power systems. To reduce the side effects, it is quite necessary to forecast the wind speed and wind power generation. In the past few years, Chinese researchers and scholars have performed some studies on forecasting related to wind power, and made some achievements. Currently, there are several papers reviewing the wind power forecasting issues, which focus on the peer-review journal articles written in English [4-6]. However, considering the rapid development of China's wind power and substantial articles written in Chinese which also contribute to the wind power forecasting issue, this paper performs the review on forecasting issue related to wind power in China, namely the current journal articles written in Chinese. The wind power generation will be largely affected by wind speed. When the wind speed is forecasted, the wind power generation can be calculated given the wind turbine type and wind farm location. Moreover, most of the journal articles written in Chinese are related to wind speed forecasting. Therefore, this paper focuses on the wind speed forecasting, which will fill the international review gap in the field of wind speed forecasting.

2. Literature Sources and Classification

The purpose of this paper is to review the wind speed forecasting in China. The reviewed literatures are selected from the high quality peer-reviewed journals written in

Chinese, which are indexed by EI (Engineering Index) and SCI (Science Citation Index) databases.

By indexing the wind speed forecasting in the EI and SCI databases subjected to the peer-reviewed articles written in Chinese, 50 article are finally obtained, which mainly come from Journals "Power System Technology ('电网技术' in Chinese)", "Acta Energiae Solaris Sinica ('太阳能学报' in Chinese)", "Power System Protection and Control ('电力系统保护与控制' in Chinese)", "Proceedings of the CSEE ('中国电机工程学报' in Chinese)", and so on. The detailed sourced journals are shown in Figure 1. It can be seen the Power System Technology journal has published the most articles related to wind speed forecasting in the past few years, which account for 24%. The journals Acta Energiae Solaris Sinica, Power System Protection and Control, Proceedings of the CSEE have also published some wind speed forecasting articles, which account for 18%, 18% and 16%, respectively.

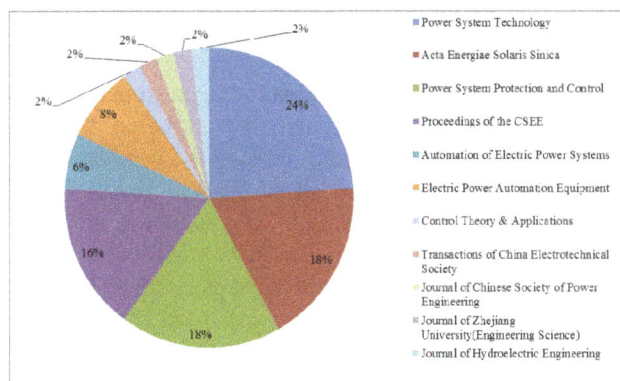

Figure 1. The Chinese high-quality journals publishing wind speed forecasting papers.

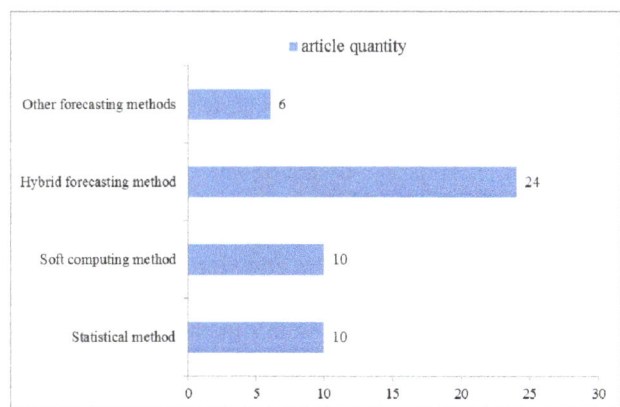

Figure 2. Article distribution sorted by different forecasting techniques.

According to the proposed forecasting techniques of these 50 literatures, four classifications can be obtained, which are statistical method, soft computing method, hybrid forecasting method and other forecasting methods. Different articles employing different wind speed forecasting techniques will be reviewed in detail in the following Sections 3-6, and the distribution is shown in Figure 2. It can be seen the articles employing the hybrid computing method are most, accounting for 48%; the following is statistical methods (20%) and soft computing method (20%). Therefore, we can

safely say that the hybrid forecasting technique for wind speed is dominated in China. There are also some researchers employ the statistical forecasting technique and soft computing forecasting technique to forecast wind speed. This also implies that the hybrid wind speed forecasting method which combines different methods may has a better forecasting performance in the field of wind speed forecasting.

3. Statistical Method

The statistical approach is based on training with measurement data and uses difference between the predicted and the actual wind speeds in immediate past to tune model parameters.

Ref. [7] considered the wind speed serial has the characteristics of sequence and autocorrelation, and then forecasted the wind speed based on the time serial analysis technique. Meanwhile, AIC (An Information Criterion) was adopted to verify the effectiveness of this wind speed forecasting model. Ref. [8] proposed a hybrid algorithm integrating time series analysis with Kalman filter to forecast wind speed, and the case study results show the forecasting accuracy of this proposed model can be improved and the time delay in the forecasting can also be well solved. Ref. [9] combined the Wavelet transform theory and time series analysis technique to forecast the wind speed, and the result shows the wind speed is firstly preceded by Wavelet transform theory can improve the forecasting accuracy. Ref. [10] presented a short-term wind speed simulation model based on Kaimal's wind speed power spectrum. The simulated wind speed series were compared to field measured wind speed time series. The results show good performance for simulation collectivity from the model, while it failed to reflect the specific wind speed fluctuation characteristic related to the simulated site. Hence, this paper proposed a method that employs the field measured wind speed to correct the simulated wind speed. After correcting, the simulated wind speed series not only possessed good performance for simulation collectivity, but reflected excellently the specific site characteristic wind speed fluctuation as well. The proposed method can enormously improve the performance of the model.

Ref. [11] used the wavelet decomposition method to decompose the wind speed into several different frequency bands, and then employed the different recursive least square (RLS) models to forecast each band, finally these forecasting results of high frequency bands and low frequency bands were combined to obtain the final forecasting results. The simulation experiment shows the average value of the mean absolute percentage error (MAPE) of wind speed forecasting is 12.25%. Ref. [12] used the method named as vectorization of univariate hourly wind speed time series for eliminating diurnal non-stationary, and vectorized hourly wind speed was expressed as a vector auto regression (VAR) model to forecast hourly wind speed. The results showed that the presented VAR model can yield satisfactory hourly wind speed forecast as long as 72 h ahead under normal weather conditions. Ref. [13] used the ARIMA (autoregressive integrated moving average) technique to forecast the wind speed, and the identified coefficient and AIC (Akaike information criterion) were employed to deal with to establish the ARIMA model. The forecasting result on the next one day's wind speed shows this method is effective. Ref. [14] employed the statistical clustering analysis technique to forecast the wind speed. By processing the sample data using maximal similarity criteria of statistical clustering approach, the prediction of wind speed based on ARIMA model was performed. The prediction accuracy of this proposed method is improved compared to the conventional ARIMA process. Based on study of long-term wind speed prediction using time series model, a new prediction scheme is proposed in Ref. [15]. The wind speed signal was firstly decomposed into trend signal and no-trend random signal, then the latter on was processed and compared respectively using sliding filter method and wavelet analysis method, and finally both the trend signal and previously processed signal were separately predicted and superposed so as to forecast the wind speed. Results show that both the sliding filter method and wavelet analysis method can realize long-term wind speed prediction with a high precision. A novel multi-step prediction for wind speed based on empirical mode decomposition (EMD) is presented in Ref. [16]. By means of the EMD technique, the original wind speed sequences were firstly decomposed into a series of functions with more stationary variation. Thus the interferences among the characteristic information embedded in the wind speed can be weakened. Then these functions were reconstructed into three components (high-middle-low frequency components) according to their run-lengths. After that, three multi-step prediction models were built on the basis of their respective variation rules. Finally, the prediction values corresponded to the three components were adaptively superposed to obtain the predicted wind speed. The forecasting results show that the proposed approach possesses higher accuracy and the prediction performance is satisfied when the wind speed sharply fluctuates.

4. Soft Computing Method

The soft computing method for wind speed forecasting refers to employing the machine learn technique or artificial intelligence technique to forecast the wind speed, such as neural network, support vector machine, and so on.

Ref. [17] proposed a wind speed forecasting for wind farm based on least squares support vector machine (LS-SVM) with the atmospheric pressure and temperature as the input. Meanwhile, the grid search was used to determine the parameters of the LS-SVM model. The obtained results of this proposed method are basically in accordance with the values of actual wind speed. Ref. [18] used the least squares support vector machine (LS-SVM) and based on actual wind speed data measured in a certain wind farm to forecast the hour-ahead wind speed of this wind farm. The prediction result shows the mean average percentage error of the predicted wind speed is

only 8.55%. Ref. [19] used the support vector machine (SVM) technique to forecast the short-term wind speed. The forecasting result shows the mean average percentage error of the predicted wind speed is only 10.07%. Ref. [20] proposed a new wind speed forecasting method which combines the GMDH neural network and fuzzy logic theory. This method fuzzy the neurons and introduced the feedback loop, which hybrid the Low-dimensional computing capacity of GMDH and the High-dimensional reasoning ability of fuzzy logic theory. The forecasting result shows this method can improve the forecasting accuracy. Ref. [21] proposed a support vector machine (SVM) method based on wavelet analysis and similar data. The training samples were built by extracting the similar data from large amounts of data. By using the wavelet decomposition, the original wind speed data was decomposed into trend signal of low frequency band and random signal of high frequency band. Finally, different SVM speed forecasting models were trained respectively and combined to obtain forecasting results. The simulation experiment shows this method can improve the forecasting accuracy.

Ref. [22] used the pattern recognition technique to select the sample, and employed the adaptive-network-based fuzzy inference system (ANFIS) model to forecast the wind speed. The forecasting result on Maui Island, Hawaii shows this method has a certain practicality. Ref. [23] used the Artificial Neural Network (ANN) model to forecast the wind speed. The parameters of ANN were determined by repeated training and testing on wind speed. The short-term wind speed forecasting result shows ANN method can achieve better result compared to that of ARMA model, but the treatment capability on mutation information is still limited. A wind power prediction system based on artificial neural network was established in Ref. [24]. The system relies on the numerical weather prediction, has friendly man-machine interface, and realizes seamless connection to the energy management system (EMS). The wind speed forecasting results indicate the prediction system is reliable and the root of mean square error (RMSE) is about 15%. Ref. [25] proposed a LSSVM (Least squares support vector machines)-based wind speed forecasting model. The historical wind speed, pressure and temperature were select as the input variables, and the grid search method was employed to determine the optimal parameters of LSSVM model. The forecasting result shows this method has a good forecasting performance. Ref. [26] presented a multivariate local predictor for short-term wind speed prediction of wind farm. It sifted multivariate time series by correlation principle to reconstruct multivariate phase space, and searched the neighborhood of the prediction state points to build the support vector regression models. The calculation results show this proposed method can improve the searching efficiency of local predictor that can find much more similar neighbor points and also can effectively improve the accuracy of short-term wind speed prediction.

5. Hybrid Forecasting Method

In general, the hybrid forecasting model refers to the combination of different approaches for wind speed forecasting, such as mixing statistical and soft computing approaches or combining SVM and PSO (Particle swarm optimization) method, etc.

Ref. [27] hybrid the time series technique and neural network to forecast the short-term wind speed. The time series model was used to select the input variables and multi-layer back propagation neural network and generalized regression neural network were used to conduct forecasting. The 10min, 20min and 30min wind speed forecasting results show this method possesses higher accuracy. Ref. [28] proposed a short-term wind speed forecasting method based on wavelet decomposition (WD) and least square support vector machine (LSSVM). The verification results of wind farms in Hong Kong region and the Hexi Corridor in Northwest China show that the proposed method can improve the accuracy of one-step ahead wind speed forecasting. Based on similarity curves, Ref. [29] proposed a back propagation (BP) neural network wind speed forecast method to improve the accuracy of wind speed forecast by using the seasonal cycle fluctuation of wind speed. Combining with time series analysis and grey forecasting method, a wind speed forecasting method was performed. Simulation results demonstrate that the proposed methodology can improve the accuracy of wind speed forecast. Ref. [30] combined the genetic algorithm and back propagation (BP) neural network to propose a new wind speed forecasting model. The wind speed, intensity, humidity, and barometric pressure were taken as the input of BP neural network model, and the genetic algorithm was used for determine the initial weights and thresholds. The 1h-ahead, 2h-ahead and 3h-ahead of wind speed were forecasted, and the results show this method is effective.

Ref. [31] proposed a new wind speed prediction scheme that uses rough set method. The key factors that affect the wind speed prediction were identified by rough set theory. Then the rough set neural network prediction model was built by adding the key factors as the additional inputs to the pure chaos neural network model. The forecasting result on a wind farm of Heilongjiang province show that the prediction accuracy of this proposed method is the best compared to that of chaos neural network model and persistence model. Ref. [32] proposed a hybrid wind speed forecasting model based on the wavelet decomposition, differential evolutionary and support vector machine. Compared to that of Cross-validation support vector machine method and BP neural network model, this proposed method has higher wind speed forecasting accuracy. Ref. [33] proposed a time series ANN (artificial neural network) method for wind speed forecasting. In the proposed method, the mathematical model was built by time series method to obtain the basic parameters of wind speed characteristics, then these parameters was used to choose input variables of ANN. Meanwhile, a rolling method to adjust weight factors was put

forward. The forecasting result shows this method can effectively improve the wind speed forecasting accuracy. Ref. [34] built the wind speed forecasting model based on least squares support vector machine theory, and tried to use Ant Colony Algorithm theory to optimization choice for parameters. Using the wind farm observed wind speed (sampling interval is 30 minutes) of the day before four days to forecast the 48ind wind speed of the fifth day through this proposed wind forecasting model, and the MAPE is only 9.53%. A wind speed forecasting model for wind farm based on wavelet decomposition, support vector machine and genetic algorithm is proposed in Ref. [35]. Through wavelet decomposition, the data were preprocessed and through genetic algorithm, the parameters were optimized. The simulation results show that the forecast wind speed is following the true value, what's more, the model can adapt to different wind data.

The spatial translation method was presented to make the data collection easier by reducing the number of correlative sites, which used RBF (Radial Basis Function) neural network to set the nonlinear relation of wind speed between correlative sites in Ref. [36]. A multi - interval forecast model based on the spatial correlation was built, which divided the forecast period into several intervals, analyzed the chronological correlation of wind speed between wind farm and correlative site and selected the best correlative site for each interval. The example calculation shows that the wind speed forecast accuracy is improved and the training time is reduced. A forecasting method based on LS-SVM considering the factors related is proposed to choose factors having significant impacts on the wind speed through correlation analysis as the reference quantity for information characteristics of the wind speed, to adopt grey relation analysis to pretreat the speed data, and search the historical speed with highly similar features to the forecasting day as the training samples of the LS-SVM model in Ref. [37]. The actual examples prove that the wind speed forecasting accuracy and reliability of this method are effectively improved. A short-term combination forecasting model of wind speed was developed based on the seasonal periodicity and time-continuity of wind energy in Ref. [38]. With the pattern recognition technique, two kinds of samples to seasonal periodicity and time-continuity were selected separately. The two kinds of samples were used respectively to train two back propagation (BP) neural network models to obtain the lateral and vertical wind speed forecasting values. By importing the two values to a BP neural network again, the wind speed value was finally predicted. Ref. [39] proposes a least squares support vector machine (LSSVM) model optimized by the particle swarm optimization (PSO).The phase space of the chaotic wind speed time series was reconstructed by calculating the embedding dimension and the delay time of the wind speed time series. The PSO was used to optimize the parameters of the LSSVM. Then the improved LSSVM model can be used to forecast the wind speed. The results show the improved LSSVM can meet the accuracy requirements, and has a better forecasting performance than SVM prediction model and the BP neural network prediction model. A wind speed forecasting method based on wavelet-neural network and is proposed n Ref. [40]. In the proposed method, the original waveform was decomposed in different scales by wavelet function and the decomposed periodic components were forecasted by time series, and the rest parts were forecasted by neural network, finally the signal series were reconstructed to obtain complete wind speed forecasting result. Adding differential evolution algorithm, the convergence speed of the proposed method was improved and the local minimum problem was also solved.

Ref. [41] used SVM and genetic algorithm to forecast the wind speed. The genetic algorithm was employed to optimize the penalty factor C and kernel parameter σ^2 of support vector machines. Compared with the general regression neural network(GRNN)method, this proposed forecasting method achieves better generalization ability and its average absolute value of relative error is only 8.32%. A hybrid method based on empirical mode decomposition (EMD) and time-series analysis was presented in Ref. [42]. The original wind speed sequences were firstly pretreated by EMD and decomposed into some intrinsic mode functions (IMF) and a residue. Then, according to their respective variation rule, each partition was modeled and forecasted using time-series analysis. Ultimately, all the forecasted values corresponded to these partitions were superposed to get the forecasted wind speed. The results indicate that the wind speed forecasting precision is improved. Ref. [43] used the wavelet package transform and support vector regression(SVR) to forecast the one hour to six hour-ahead wind speed forecasting of each ten minute. Firstly, in view of the non-stationary and nonlinear features of wind speed, the original wind speed sequence was decomposed into a series of sub-sequences; then these sub-sequences were respectively forecasted by SVR; finally, respective outputs were superposed to obtain final forecasted wind speed. The Adaboost algorithm was led in to improve back propagation (BP) neural network algorithm, and an Adaboost-based BP neural network method was proposed and applied to short-term wind speed forecasting in Ref. [44]. Results show using the proposed Adaboost-based BP neural network the accuracy of one or two hour-ahead wind speed forecasting was superior to respective forecasting accuracy by neural network and ARMA time series analysis, and the mean absolute percentage error of wind speed forecasting by the proposed algorithm was lower than 7.5% in high wind speed period (higher than 10 m/s).

A new approach for ultra-short-term wind speed forecasting is presented based on the phase space reconstruction technique and the local prediction method which can search the neighbors in the optimal neighborhood in the phase space and to build support vector regression (SVR) models in Ref. [45]. This approach finds the optimum neighborhood by considering of proportion of false neighbors, which guarantees the high similarity between neighbors and prediction state points, and the SVR model has a good capability of nonlinear fitness. A wind speed prediction model (WD-SVM) that uses data mining techniques of wavelet analysis and support vector machine, was developed in Ref. [46]. In this model, a given wind speed time series were decomposed by wavelet analysis into various layers that were predicted with support vector machines, and then by reconstructing the predicted values of each layer a prediction of wind speed was obtained. The average root-mean-square error of 10-minute average wind speed four hours in advance is 11.71%. Ref. [47] forecasted the wind speed for 1 h ahead at a resolution of every 10 min by means of mathematical morphology and support vector regression. The original wind speed sequences were decomposed into a series of subsequences with different frequencies and wave characters by adaptive multi-scale morphological algorithm. Then, the subsequences with the method of SVR were predicted respectively. Finally, the final predicted wind speed was calculated by the superposition of respective predictions. An interval type-2 fuzzy logic model was proposed to forecast short-term wind speed time series based on the singular value decompose and back-propagation (SVD-BP) hybrid iterative arithmetic in Ref. [48]. The BP arithmetic was used to tune input, antecedent and consequent parameters, the SVD method was used to choose reasonable rules. For testing the performance, both a type-1 fuzzy logic model and an interval type-2 fuzzy logic model using only BP arithmetic were designed as comparable benchmarks. The simulation results show that the interval type-2 fuzzy logic model using the SVD-BP hybrid iterative arithmetic has the best performance and practicability.

A wind power prediction method based on empirical mode decomposition (EMD) and support vector machine (SVM) is proposed in Ref. [49]. The wind speed data was firstly decomposed into a series of components with stationary by using EMD to reduce the influence between different feature information. Then, different models were built and different kernel functions and parameters were chosen to deal with each group of data by using SVM in order to improve the forecasting accuracy. Finally, short term wind power forecasting was made based on wind speed data through a practical wind power curve. A combined short-term wind speed forecasting model based on D-S evidence theory was proposed in Ref. [50]. The forecasting models of time series, BP neural network and support vector machine wer adopted to respectively forecast the wind speed. Based on the analysis of forecast errors, D-S evidence theory was applied to fuse these three models. The wind speed data for several days before were taken as the fusion samples to calculate the corresponding basic trust distribution functions, which were then fused. The results of fusion were taken as the weights of the wind speed forecasting model and the wind speed of the day to be forecasted was calculated. Simulative results show that, the proposed combined forecasting model has smaller forecasting error and better effect.

6. Other Forecasting Methods

Except the above wind power forecasting techniques, there are still some other forecasting methods developed by Chinese researchers to forecast the wind speed.

The short term wind speed prediction based on the Physical principles was performed in Ref. [51]. This approach employed the wind speed from the numerical weather prediction-NWP as an input data, a roughness change model and orographical change model were used to model the local effects of a wind farm. The predicted wind speed was compared with the measured wind speed under the typical wind conditions. The results show that the prediction value can basically meet the requirements of prediction precision. A combined model to predict wind speed based on improved fuzzy analytic hierarchy process (AHP) was presented in Ref. [52]. Taking predictive cycle of wind speed, the fluctuation of wind speed and the reliability of predictor to the predictive model as objectives, by means of fuzzy judgment matrix the optimal weight of the combined model was determined. The calculation results show that the prediction results by the proposed predictive model are more accurate than those from traditional single predictive model. The phase space reconstruction technology was used for short-term prediction of wind speed in Ref. [53]. Based on the fundamental relation of the delay time window Γ and m 、 τ, several sets of optimal combinations of m and τ were advanced, and the optimal combination was found. An effective method was presented to choose reference point, which determined near phase point by Euclidean distance and correlation degree among phase point. Some false neighboring points were kicked off to improve forecasting accuracy. On the selection of forecasting model, one-order local forecasting method and BP neural network model were used.

Based on grey predictor models, Ref. [54] presented one-step to four-step average ten-minute wind speed forecasting and give the residual error for steady wind. Wind speed predictions for unsteady wind and gust were also made. Then, taking the steady wind speed for instance, fitting parameters in various models that wind power changes with wind speed were obtained by modeling the relationship between real wind power and sequential wind speed. In order to enhance the forecasting precision of wind power, the function model fitting for wind power characteristic was established through comparing the precision of different models from the aspect of piecewise function and overall model. The forecasting result shows this method is effective. Ref. [55] proposed a chaos-based wind speed forecasting method. On the basis of applying phase space reconstruction to wind speed time series, the wind speed was forecasted by chaotic weighted zero-order local forecasting method; and then to remedy the insufficiency of above-mentioned method, namely the forecasting was carried out by searching near phase point under high embedded dimensions, a modified weighted zero-order local forecasting method, which determined near phase point by correlation degree among phase points, was proposed and a new approach to compute weighted coefficient was put forward to improve forecasting accuracy. Chaos theory and methods were used to solve the wind speed prediction problem in Ref. [56]. Firstly, the time delay and the embedding dimension were calculated by correlation integral approach for reconstructing phase space of wind speed time series. Then, wind speed chaotic prediction model of optimal neighborhood was proposed which gives overall consideration to the nearest neighbors' weights and generalized degrees of freedom, also an improved criterion for selecting optimal neighborhood. The practical calculation shows that the proposed model has superior predictive capability under the appropriate model parameters.

7. Conclusions

China's wind power has developed rapidly in the past few years, and ranked first in terms of cumulative installed capacity in the world since 2010. Due to the intermittent and fluctuant characteristics, the wind power penetration will bring big impact on power systems, and accurately forecasting wind power generation can alleviate the negative effects. To fill the wind speed forecasting review gap, this paper reviews the current wind speed forecasting techniques in China from four aspects, which are statistical method, soft computing method, hybrid forecasting method and other forecasting methods. The articles written in Chinese which has been published in high-quality peer-reviewed Chinese journals are analyzed in details. The hybrid wind speed forecasting method which combines different methods may have a better forecasting performance in the field of wind speed forecasting, which has caused widespread concerns of China's researchers.

Acknowledgements

The authors thank the editor and reviewers for their comments and suggestions.

References

[1] 2014 Wind power development report published by GWEC. http://www.gwec.net/wp-content/uploads/2014/04/%E5%85% A8%E7%90%83%E9%A3%8E%E7%94%B5%E7%BB%9F %E8%AE%A1%E6%95%B0%E6%8D%AE2013.pdf

[2] Zhao H, Guo S, Fu L. Review on the costs and benefits of renewable energy power subsidy in China [J]. Renewable and Sustainable Energy Reviews, 2014, 37: 538-549.

[3] http://www.cwpc.cn/cwpp/cn/services/cwpc-news-service/1-2 020/

[4] Tascikaraoglu A, Uzunoglu M. A review of combined approaches for prediction of short-term wind speed and power [J]. Renewable and Sustainable Energy Reviews, 2014, 34: 243-254.

[5]   Soman, Saurabh S., Hamidreza Zareipour, Om Malik. A review of wind power and wind speed forecasting methods with different time horizons. North American Power Symposium (NAPS), 2010. IEEE, 2010.

[6]   Foley A M, Leahy P G, Marvuglia A, et al. Current methods and advances in forecasting of wind power generation [J]. Renewable Energy, 2012, 37(1): 1-8.

[7]   DING Ming, ZHANG Lijun, WU Yichun. Wind speed forecast model for wind farms based on time series analysis [J]. Electric Power Automation Equipment, 2005, 25(8): 32-34.

[8]   PAN Di-fu, LIU Hui, LI Yan-fei. A wind speed forecasting optimization model for wind farms based on time series analysis and kalman filter algorithm [J]. Power System Technology, 2008, 32(7): 82-86.

[9]   Zhang Yanning, Kang Longyun, Zhou Shiqiong, et al. Wavelet analysis applied to wind speed prediction in predicate control system of wind turbine [J]. Acta Energiae Solaris Sinica, 2008, 29(5): 520-524.

[10] WANG Yao-nan, SUN Chun-shun, LI Xin-ran. Short-term wind speed simulation corrected with field measured wind speed [J]. Proceedings of the CSEE, 2008, 28(11): 94-100

[11] WANG Xiao-lan, LI Hui. Effective wind speed forecasting in annual prediction of output power for wind farm [J]. Proceedings of the CSEE, 2010, 30(8): 117-122.

[12] SUN Chun-shun, WANG Yao-nan, LI Xin-ran. A vector autoregression model of hourly wind speed and its applications in hourly wind speed forecasting [J]. Proceedings of the CSEE, 2008, 28(14): 112-117

[13] JIANG Jin-liang, LIN Guang-ming. Automatic station wind speed forecasting based on ARIMA model [J]. Control Theory & Applications, 2008, 25(2): 374-376.

[14] FANG Jiang-xiao, ZHOU Hui, HUANG Mei, T.S. Sidhu. Short-term wind power prediction based on statistical clustering analysis [J]. Power System Protection and Control, 2011, 39(11): 67-73.

[15] YANG Xi-yun, SUN Han-mo. Wind speed prediction in wind farms based on time series model [J]. Journal of Chinese Society of Power Engineering, 2011, 31(3): 203-208.

[16] Liu Xingjie, Mi Zengqiang, Yang Qixun, et al. A novel multi-step prediction for wind speed based on EMD [J]. Acta Energiae Solaris Sinica, 2010, 25(4): 165-170.

[17] DU Ying, LU Ji-ping, LI Qing, et al. Short-Term wind speed forecasting of wind farm based on least square-support vector machine [J]. Power System Technology, 2008, 32(15): 62-66.

[18] ZENG Jie, ZHANG Hua. A wind speed forecasting model based on least squares support vector machine [J]. Power System Technology, 2009, 33(18): 144-147.

[19] Zhang Hua, Zeng Jie. Wind speed forecasting model study based on support vector machine [J]. Acta Energiae Solaris Sinica, 2010, 31(7): 928-932.

[20] WU Dong-liang, WANG Yang, GUO Chuang-xin, et al. Short-term wind speed forecasting in wind farm based on improved GMDH network [J]. Power System Protection and Control, 2011, 39(2): 88-93.

[21] YANG Xiyun, SUN Baojun, ZHANG Xinfang, et al. Short-term wind speed forecasting based on support vector machine with similar data [J]. Proceedings of the CSEE, 2012, 32(4): 35-41.

[22] WU Xing-hua, ZHOU Hui, HUANG Mei. Wind speed and generated power forecasting based on pattern recognition in wind farm [J]. Power System Protection and Control, 2008, 36(1): 27-32.

[23] Huang Xiaohua, Li Deyuan, Lv Wenge, et al. Wind speed forecasting with artificial neural networks model [J]. Acta Energiae Solaris Sinica, 2011, 32(2): 193-197.

[24] FAN Gao-feng, WANG Wei-sheng, LIU Chun. Artificial neural network based wind power short term prediction system [J]. Power System Technology, 2008, 32(22): 72-76.

[25] DU Ying, LU Ji-ping, LI Qing, et al. Short-Term wind speed forecasting of wind farm based on least square-support vector machine [J]. Power System Technology, 2008, 32(15): 61-66.

[26] GUO Chuangxin, WANG Yang, SHEN Yong, et al. Multivariate local prediction method for short-term wind speed of wind farm [J]. Proceedings of the CSEE, 2012, 32(1): 24-31.

[27] CAI Kai, TAN Lun-nong, LI Chun-lin, et al. Short-Term wind speed forecasting combing time series and neural network method [J]. Power System Technology, 2008, 32(8): 82-85.

[28] WANG Xiao-lan, WANG Ming-wei. Short-Term wind speed forecasting based on wavelet decomposition and least square support vector machine [J]. Power System Technology, 2010, 34(1): 179-184.

[29] ZHANG Guoqiang, ZHANG Boming. Wind speed and wind turbine output forecast based on combination method [J]. Automation of Electric Power Systems, 2009 (18): 92-95.

[30] WANG De-ming, WANG Li, ZHANG Guang-ming. Short-term wind speed forecast model for wind farms based on genetic BP neural network [J]. Journal of Zhejiang University (Engineering Science), 2012, 46(5): 837-841.

[31] GAO Shuang, DONG Lei, GAO Yang, et al. Mid-long term wind speed prediction based on rough set theory [J]. Proceedings of the CSEE, 2012, 32(1): 32-37.

[32] PENG Chunhua, LIU Gang, SUN Huijuan. Wind speed forecasting based on wavelet decomposition and differential evolution-support vector machine for wind farms [J]. Electric Power Automation Equipment, 2012, 32(1): 9-13.

[33] YANG Xiu-yuan, XIAO Yang, CHEN Shu-yong. Wind speed and generated power forecasting in wind farm [J]. Proceedings of the CSEE, 2005, 25(11): 1-5.

[34] Zeng Jie, Zhang Hua. Wind speed forecasting model study based on least squares support vector machine and ant colony optimization [J]. Acta Energiae Solaris Sinica, 2011, 32(3): 296-300.

[35] Luo Wen, Wang Lina. Short-Term wind speed forecasting for wind farm [J]. Transactions of China Electrotechnical Society, 2011, 26(7): 68-74.

[36] LI Wenliang, WEI Zhinong, SUN Guoqiang, et al. Multi-interval wind speed forecast model based on improved spatial correlation and RBF neural network [J]. Electric Power Automation Equipment, 2009 (6): 89-92.

[37] LI Ran, CHEN Qian, XU Hong-rui. Wind speed forecasting method based on LS-SVM considering the related factors [J]. Power System Protection and Control, 2010 (21): 146-151.

[38] JIANG Xiaoliang, JIANG Chuanwen, PENG Minghong, et al. A short-term combination wind speed forecasting method considering seasonal periodicity and time-continuity [J]. Automation of Electric Power Systems, 2010 (15): 75-79.

[39] SUN Bin, YAO Hai-tao. The short-term wind speed forecast analysis based on the PSO-LSSVM predict model [J]. Power System Protection and Control, 2012, 40(5): 85-89.

[40] YANG Qi, ZHANG Jianha, Wang Xiangfeng, et al. Wind speed and wind power generation forecast based on Wavelet - Neural Networks model [J]. Power System Technology, 2009, 33(17): 44-48.

[41] YANG Hong, GU Shi-fu, CUI Ming-dong, et al. Forecast of short-term wind speed in wind farms based on GA optimized LS-SVM [J]. Power System Protection and Control, 2011, 39(11): 44-48.

[42] Liu Xingjie, Mi Zengqiang, Yang Qixun, et al. Wind speed forecasting based on EMD and time-series analysis [J]. Acta Energiae Solaris Sinica, 2010, 31(8): 1037-1041.

[43] CHEN Pan, CHEN Haoyong, YE Rong, et al. Wind speed forecasting based on combination of wavelet packet analysis with support vector regression [J]. Power System Technology, 2011, 35(5): 177-182.

[44] WU Junli, ZHANG Buhan, WANG Kui. Application of Adaboost-based BP neural network for short-term wind speed forecast [J]. Power System Technology, 2012, 36(9): 221-225.

[45] WANG Yang, ZHANG Jinjiang, WEN Bojian, et al. An optimal neighborhood in phase space based local prediction method for ultra-short-term wind speed forecasting [J]. Automation of Electric Power Systems, 2012, 35(24): 39-43.

[46] ZHANG Hua, YU Yongjing , FENG Zhijun, et al. Wind speed forecasting model based on wavelet decomposition and support vector machine [J]. Journal of Hydroelectric Engineering, 2012, 31(1): 208-212.

[47] CHEN Pan, CHEN Hao-yong, YE Rong. Wind speed forecasting based on multi-scale morphological analysis [J]. Power System Protection and Control, 2010 (21): 12-18.

[48] Zheng Gao, Xiao Jian, Wang Jing, et al. Forecasting study of short-term wind speed based on interval type-2 fuzzy logic method [J]. Acta Energiae Solaris Sinica, 2012, 32(12): 1792-1797.

[49] YE Lin, LIU Peng. Combined Model Based on EMD-SVM for Short-term Wind Power Prediction [J]. Proceedings of the CSEE, 2011, 31(31): 102-108.

[50] Liu Yanan, Wei Zhinong, Zhu Yan, et al. Short-term wind speed forecast based on D-S evidence theory [J]. Electric Power Automation Equipment, 2013, 33(8): 131-136.

[51] Feng Shuanglei, Wang Weisheng, Liu Chun, et al. Short term wind speed prediction based on physical principle [J]. Acta Energiae Solaris Sinica, 2011, 32(5): 611-616.

[52] HUANG Wen-jie, FU Li, XIAO Sheng. A predictive model of wind speed based on improved fuzzy analytical hierarchy process [J]. Power System Technology, 2010, 34(7): 164-168.

[53] Lü Tao, TANG Wei, SUO Li. Prediction of short-term wind speed in wind farm based on chaotic phase space reconstruction theory [J]. Power System Protection and Control, 2010 (21): 113-117.

[54] LI Jun-fang, ZHANG Bu-han, XIE Guang-long, et al. Grey predictor models for wind speed-wind power prediction [J]. Power System Protection and Control, 2010, 38(19): 151-159.

[55] LUO Hai-yang, LIU Tian-qi, LI Xing-yuan. Chaotic forecasting method of short-term wind speed in wind farm [J]. Power System Technology, 2009, 33(9): 67-71.

[56] Ding Tao, Xiao Hongfei. Wind speed chaotic prediction model based on optimal neighborhood [J]. Acta Energiae Solaris Sinica, 2011, 32(4): 560-564.

PERMISSIONS

All chapters in this book were first published in AJEE, by Science Publishing Group; hereby published with permission under the Creative Commons Attribution License or equivalent. Every chapter published in this book has been scrutinized by our experts. Their significance has been extensively debated. The topics covered herein carry significant findings which will fuel the growth of the discipline. They may even be implemented as practical applications or may be referred to as a beginning point for another development.

The contributors of this book come from diverse backgrounds, making this book a truly international effort. This book will bring forth new frontiers with its revolutionizing research information and detailed analysis of the nascent developments around the world.

We would like to thank all the contributing authors for lending their expertise to make the book truly unique. They have played a crucial role in the development of this book. Without their invaluable contributions this book wouldn't have been possible. They have made vital efforts to compile up to date information on the varied aspects of this subject to make this book a valuable addition to the collection of many professionals and students.

This book was conceptualized with the vision of imparting up-to-date information and advanced data in this field. To ensure the same, a matchless editorial board was set up. Every individual on the board went through rigorous rounds of assessment to prove their worth. After which they invested a large part of their time researching and compiling the most relevant data for our readers.

The editorial board has been involved in producing this book since its inception. They have spent rigorous hours researching and exploring the diverse topics which have resulted in the successful publishing of this book. They have passed on their knowledge of decades through this book. To expedite this challenging task, the publisher supported the team at every step. A small team of assistant editors was also appointed to further simplify the editing procedure and attain best results for the readers.

Apart from the editorial board, the designing team has also invested a significant amount of their time in understanding the subject and creating the most relevant covers. They scrutinized every image to scout for the most suitable representation of the subject and create an appropriate cover for the book.

The publishing team has been an ardent support to the editorial, designing and production team. Their endless efforts to recruit the best for this project, has resulted in the accomplishment of this book. They are a veteran in the field of academics and their pool of knowledge is as vast as their experience in printing. Their expertise and guidance has proved useful at every step. Their uncompromising quality standards have made this book an exceptional effort. Their encouragement from time to time has been an inspiration for everyone.

The publisher and the editorial board hope that this book will prove to be a valuable piece of knowledge for researchers, students, practitioners and scholars across the globe.

LIST OF CONTRIBUTORS

Alice Ponchio
Department of Philosophy, University of Padua, Padua, Italy

Alberto Mirandola
Department of Industrial Engineering, University of Padua, Padua, Italy

Manal A. Sorour
Food Engineering and Packaging Dept., Food Technology Research Institute, Agric. Research Center, Giza, Egypt

Naglaa H. M. Hassanen and Mona H. M. Ahmed
Special Food and Nutrition Dept., Food Technology Research Institute, Agric. Research Center, Giza, Egypt

Huseyin Murat Cekirge
Department of Mechanical Engineering, Prince Mohammad Bin Fahd University, Al Khobar, KSA

Ammar Elhassan
Department of Information Technology, Prince Mohammad Bin Fahd University, Al Khobar, KSA

Halidini Sarakikya
Department of Electrical Engineering, Arusha Technical College, Arusha, Tanzania

Jeremiah Kiplagat
Department of Energy Engineering, Kenyatta University, Nairobi, Kenya

Emmanuel I. Bello and Tunde I. Ogedengbe
Department of Mechanical Engineering, Federal University of Technology, Akure, Nigeria

Labunmi Lajide
Department of Chemistry, Federal University of Technology, Akure, Nigeria

Ilesanmi. A. Daniyan
Department of Mechanical & Mechatronics Engineering, Afe Babalola University, Ado-Ekiti, Nigeria

Temesgen Atnafu Yemata
Institute of technology, Department of Chemical Engineering, Addis Ababa, Ethiopia
Addis Ababa University, Chemical Engineering Department, Addis Ababa, Ethiopia

Ahmed Farouk Abdel Gawad and Hamza Ahmed Ghulman
Mechanical Engineering Department, College of Engineering and Islamic Architecture, Umm Al-Qura Univ., Makkah, Saudi Arabia

Francesca Marin
Department of Philosophy, Sociology, Education and Applied Psychology (FISPPA), University of Padua, Padua, Italy

Alberto Mirandola
Department of Industrial Engineering, University of Padua, Padua, Italy

A. A. Abdullah
Department of Mathematical Sciences, Umm Al-Qura University, Makkah, Saudi Arabia

F. S. Ibrahim
Department of Mathematics, University College, Umm Al-Qura University, Makkah, Saudi Arabia

A. F. Abdel Gawad and A. Batyyb
Mech. Eng. Dept., College of Engineering and Islamic Architecture, Umm Al-Qura University, Makkah, Saudi Arabia

Sajjad Arefdehgani
Department of Mechanical engineering, Tabriz Branch, Islamic Azad University, Tabriz, Iran

Omid Karimi Sadaghiyani
Department of Mechanical engineering, Urmia Branch, Urmia University, Urmia, Iran

Jalaluddin
Department of Mechanical Engineering, Hasanuddin University, Makassar, Indonesia

Akio Miyara
Department of Mechanical Engineering, Saga University, Saga-shi, Japan

Khalid A. Albis, Muhammad N. Radhwi and Ahmed F. Abdel Gawad
Mechanical Engineering Department, College of Engineering & Islamic Architecture, Umm Al-Qura University, Makkah, Saudi Arabia

Mohammed Abdulwahhab Abdulwahid
Marine Engineering Department, Andhra University, Visakhapatnam, India

Sadoun Fahad Dakhil
Fuel & Energy Department, Basrah Technical College, Basrah, Iraq
I. N. Niranjan Kumar
Marine Engineering Department, Andhra University, Visakhapatnam, India

S. R. Mostafa
Chemical Engineering Dept., Faculty of Eng., Cairo Univ., Giza. Egypt

M. A. Sorour
Food Eng. and Packaging Dept., Food Tech. Research Institute, Agric. Research center, Giza, Egypt

S. M. Bo Samri
Public Authority of Education Applied & Training Healthy Science, Food Processing Nutrition, Kuwait, Kuwait

Ahmet Haxhiaj, Nyrtene Deva and Mursel Rama
Faculty of Geosciences, University of Mitrovica "Isa Boletini", Mitrovica, Republic of Kosovo

Jamil Ahmed
Department of Computer science and Engineering, P.A College of Engineering, VTU, Mangalore, India

Hasibur Rahman Sardar
Department of Electronics & Communication Engineering, P.A College of Engineering, VTU, Mangalore, India

Abdul Razak Kaladgi
Department of Mechanical Engineering, P.A College of Engineering, VTU, Mangalore, India

Saiful Islam and A. T. M. Shahidul Huqe Muzemder
Department of Petroleum & Mining Engineering, Shahjalal University of Science and Technology, Sylhet, Bangladesh

Macben Makenzi
Mechatronic Department, Jomo Kenyatta University of Agriculture and Technology, Nairobi, Kenya

Nelson Timonah
Physics Department, Jomo Kenyatta University of Agriculture and technology, Nairobi, Kenya

Mutua Benedict
Faculty of Engineering and Technology, Egerton University, Nakuru, Kenya

Ismael Abisai
Ubbink East Africa Ltd, Naivasha, Kenya

Rusul Dawood Salim and Jassim Mahdi AL-Asadi
Physics Department, College of Education, University of Basrah, Basrah, Iraq

Aqeel Yousif Hashim
Technical institute of Basrah, Southern Technical University, Basrah, Iraq

Izuchukwu Francis Okafor and Cosmas Ngozichukwu Anyanwu
National Centre for Energy Research and Development, University of Nigeria Nsukka, Enugu State, Nigeria

Masih Allahbakhshi and Habib Sadeghi
Department of Civil Engineering, Mazandaran University of technology, Babol, Iran
Department of Chemical Engineering, Isfahan University, Isfahan, Iran

Solomon Hailu and Solomon Kahsay G. Mariam
Department of Biological and Chemical Engineering, Mekelle Institute of Technology, Mekelle University, Tigray, Ethiopia

Tesfay Berhe
Department of Biological and Chemical Engineering, Mekelle Institute of Technology, Mekelle University, Tigray, Ethiopia
Department of Chemical Engineering, Kombolcha Institute of Technology, Wollo University, Amhara, Ethiopia

De-Yi Shang
136 Ingersoll Cres., Ottawa, ON, Canada K2T 3W9

Liang-Cai Zhong
Department of Ferrous Metallurgy, Northeastern University, Shenyang 110004, China

Huiru Zhao and Sen Guo
School of Economics and Management, North China Electric Power University, Changping District, Beijing, China

Index